MARKING TIME

Romanticism and Evolution

EDITED BY JOEL FAFLAK

Marking Time

Romanticism and Evolution

UNIVERSITY OF TORONTO PRESS
Toronto Buffalo London

Toronto Buffalo London
www.utorontopress.com

ISBN 978-1-4426-4430-4 (cloth)

Library and Archives Canada Cataloguing in Publication

Marking time : Romanticism and evolution / edited by Joel Faflak.

Includes bibliographical references and index.
ISBN 978-1-4426-4430-4 (hardcover)

1. Romanticism. 2. Evolution (Biology) in literature. 3. Literature and science.
I. Faflak, Joel, 1959–, editor

PN603.M37 2017 809'.933609034 C2017-905010-9

This book has been published with the help of a grant from the Federation for the Humanities and Social Sciences, through the Awards to Scholarly Publications Program, using funds provided by the Social Sciences and Humanities Research Council of Canada.

University of Toronto Press acknowledges the financial assistance to its publishing program of the Canada Council for the Arts and the Ontario Arts Council, an agency of the Government of Ontario.

Canada Council for the Arts | Conseil des Arts du Canada

Funded by the Government of Canada | Financé par le gouvernement du Canada

Contents

Part Three: Goethe and the Contingencies of Life

Part Four: Evolutionary Idealisms

Illustrations

Acknowledgments

This volume has had a long gestation, materializing in rather ironic but very real terms the time of evolution itself. Its inspiration was "Romanticism and Evolution," an international conference held 12–14 May 2011 at Western University, London, Ontario, at that time still called The University of Western Ontario – an evolution, to be sure. The co-organizers were three stellar graduate students – Naqaa Abbas, Chris Bundock, and Joshua Lambier – whom I had the great pleasure of shepherding through the organization of that three-day event, the planning and execution of which they handled superbly. I remain indebted to their incredible collegiality, intelligence, vision, patience, and administrative savvy. I especially want to thank Josh Lambier, whose background research and prose for the original conference proposal were the inspiration for the volume's Introduction.

It has been wonderful working again with the University of Toronto Press, where I am especially indebted to three people. Editor Richard Ratzlaff remained loyal to and excited about this project from the start, as well as patient with its long trajectory when others might have called "extinct!" I am very happy to have the opportunity to work a second time with Barb Porter, who oversaw the final stages of production with no fuss, and a first time with copy-editor Kel Pero, who displayed extraordinary professionalism, acute editorial sense, and a great sense of humour. I also thank three anonymous Press reviewers, whose honest, fair, and generous criticism helped to refine even further an already stellar group of essays. Among the support and encouragement I am extremely lucky to receive from the academic community, I want to single out Allan Pero and Richard C. Sha, the latter of whom offered invaluable feedback on the Introduction. They are colleagues and friends of the first order.

The Trustees of Boston University kindly granted permission to reprint Maureen McLane's essay, originally published in *Studies in Romanticism*, vol. 52, no. 3, 2013, pp. 337–62, and at *Studies in Romanticism* I thank Chuck Rzekpa and

Deborah Swedberg for their generosity. Publication of *Marking Time* has been supported by a grant from the Aid to Scholarly Publishing Program, overseen by the Federation for the Humanities and Social Sciences, and by funds partly earmarked for the volume's editing and publication from a grant through the Canada Aid to Research Workshops and Conferences in Canada Program, administered through the Social Sciences and Humanities Research Council of Canada. I am grateful to live in a country where such support for Humanities scholarship remains intact.

Above all, I thank my contributors. Your patience with this project is a thing of biblical proportion. If the readiness is all, I was not always ready, and you had every right to question the volume's slow time. Academic publishing often follows a glacial pace, but really. I am grateful that you survived to observe where evolution took things, and I feel confident that the brilliance of your work will adapt to future times. I take full responsibility for any unexpected mutations in the final product.

This volume is dedicated to my late friend and mentor, Ross Woodman. Ostensibly, its topic overlaps very little with his Romantic scholarship. But Ross knew, with rather uncanny insight, how history and the unconscious run deep and slow. Of his many gifts to me, one was the capacity to understand that even in – *especially* in – psychic terms, "extinction" is relative. Or as Harper Pitt says at the end of *Angels in America,* "Nothing's lost forever. In this world, there is a kind of painful progress. Longing for what we've left behind, and dreaming ahead. At least I think that's so."

And finally, I thank Norm and our menagerie, who keep me dreaming ahead and who always lighten extinction's sometimes dark shadow. And that *is* so.

MARKING TIME

Introduction – Marking Time: Romanticism and Evolution

JOEL FAFLAK

I.

Although Victorian studies have long explored the seismic impact of *On the Origin of Species* (1859) on nineteenth-century culture, Charles Darwin's text did not have an immaculate conception. The product of post-Enlightenment thought, *Origin* exerted considerable influence on post-1859 debates about the order of life and being well into the twentieth and twenty-first centuries. We less readily imagine this influence as extending *backward* to Romantic thought and writing. Yet mapping Romantic evolution is more than merely seeking eighteenth- and early nineteenth-century "forerunners" to the Darwinian revolution. *Marking Time: Romanticism and Evolution* explores Romanticism's liminal position between the classical idea of an immutable, atemporal "great chain of being," outlined by Arthur O. Lovejoy's 1936 landmark study of the same name, and the rise in Romanticism of modern historiographies. This volume presents Romanticism as its own age of evolution by revisiting our notions of organicism, life, vitalism, natural history, and natural philosophy in relation to less-acknowledged notions of change and transformation in the cultural, literary, philosophical, and scientific discourses of the period. At the same time, our contributors track the remainders of Romanticism in the works of Charles Darwin, from his early reading of Wordsworth, Scott and Percy Shelley, to his study of Goethe, Schiller and Humboldt, to his lifelong interest in shared modes of subjective experience in the investigation of art and science. In general, the following essays challenge prevailing histories of evolution. Our contributors pay close attention to emergent, evolutionary themes of Romantic-era science, such as Romantic-Idealist conceptions of degeneration, morphology, species change, and organismic archetypes; the scientist's intimate observation of the various forms of mobility and the analogical relations of plants and animals; the discovery of an anthropological concept of deep time in the art, artefacts, and travel narratives culled from voyages of exploration; and the reimagination of the interdisciplinary relationships among related fields of inquiry, past and present, including political economy, sociology, optics, archaeology, and modern genetics.

To put the matter in terms that this volume is the first to explore, Romanticism was intensely engaged in comprehending how things *evolve*, in apprehending evolution as both means and end of human understanding itself, and in entertaining how evolution challenged the ends of the human. As Elizabeth Grosz argues, "the Darwinian revolution in thought disrupts and opens up life to other forms of development beyond, outside, and after the human," so that the "open-ended but relentless force to futurity undoes all stability and identity while also retaining a fidelity to historical forces" (3). Grosz's assessment thus also "opens up" Darwin's thought by taking us back to the future of its impact in and on a Romanticism that casts the shadow of its own futurity on evolutionary thought. Yet despite critical interest in Romantic science in recent decades, Romantic evolution has remained, as Hermione de Almeida suggests, "a very large – largely untouched – field for future study" (131). This neglect is partly explained by the fact that the term "evolution," whose biological, scientific, and cultural specificity we now take for granted, existed in Romantic thought and writing in inchoate, embryonic form. Moreover, only later in the nineteenth century did the identities of distinct disciplines – specifically, theology and biology – emerge as if by unconscious selection from the conflict of sensations generated by the publication of *On the Origin of Species,* helping to forge the future shape of evolutionary thought in turn. Whereas Charles Darwin's theories, and the figure of Darwin himself, galvanized Victorian study and debate about evolution, then, pre-Victorian evolution was a rather amorphous topic investigated across a range of fields and writers. Put another way, for Darwin evolution marked comprehensively the comprehensive development of things. In Romanticism, by contrast, the term functioned less globally and more locally, but was also thus a placeholder for emergent notions that had not yet found complete theoretical or scientific expression at a time when science was beginning to find general laws to link ontogeny to phylogeny.

It makes sense that the burgeoning of comparative data in the later eighteenth and early nineteenth centuries eventually made possible a Charles Darwin to systematize the huge swathes of evidence culled from around the globe by the Romantic sciences. Yet as we shall see in the following essays, the idea of system itself – at least any simply mechanical or static notion of system – became rather inadequate to explain the florid economy of information for which notions of evolution were called upon to account. One might argue that Romantic ideas about evolution were that much more powerfully mercurial because of their symbolic resonance, and thus signified a natural mobility and randomness for which

Darwin's later theories attempted to account and that they attempted thus to contain. The OED traces the biological definition of the term "evolution" to the mid-eighteenth century: "Emergence or release from an envelope or enclosing structure; (also) protrusion, evagination." Even within strictly biological terms, then, we already sense a sense of life unfolding in both continuous and abrupt, less-predictable ways. Romantic evolution signifies both revelation and revolution, a shift in existence that at once transformed and transmogrified forms of life. It was thus a metonym for both emergent and mutating life forms and the force of life that informed such refashionings – the forms of change as well as change itself. To mark evolution is also to express, delineate, rationalize, symbolize, idealize, personify the dynamism of time *in the process of articulating itself to itself,* to mark how time leaves its mark upon time through both the ontogeny of individuals and the phylogeny of species, groups, and history itself. To assess Romantic evolution is thus also to assess how it was determined by the forms of its representation.

In the absence of any central and centralizing notion of evolution in Romanticism, one lightning rod for a variety of emerging ideas about the formation and reformation of life was the eighteenth-century struggle between epigenetic and preformationist theories of generation. Epigenesis gained scientific traction when embryologist Caspar Friedrich Wolff, in his 1759 dissertation at the University of Halle, "Theoria Generationis," which built on both Aristotle's *De Generatione Animalium* and William Harvey's seventeenth-century theories of the development of animal life, refuted preformationism. Preformationism argued for the coherent organization of life from blueprint to fully realized organism as the stereotype (as it were) of its prototypical form. Harvey's theory was preformational in that it saw the ovum as this archetype of developed life, but epigenetic in its account of this life's *developing* forms. It then fell to Wolff's germ layer theory to take Harvey's still rather mechanistic account of epigenetic structure in a more vitalist direction that tracks animal life through its successive stages. This organic unfolding did not, at least in principle, challenge preformationism's coherent composition of life from template to accomplished construct. Rather, epigenesis explored alternate accounts, such as Blumenbach's *Bildungstrieb*, that posited life as more than mere mechanism, yet still regularized its orderly nature. Or as Robert Richards argues in *The Romantic Conception of Life* (2002), the unifying drive of *Naturphilosophie,* explored by Richards and others in this volume, was crucial to Darwin's thinking, and itself a key conceptualization of Romantic evolution. Yet as Denise Gigante reminds us in *Life: Organic Form and Romanticism* (2009), the turn to epigenesis elicits an *emergent* organicism that threatens order and system with the possibility of monstrosities and their monstrous development. Ultimately informing preformation's orderly constitution were, of course, God and the notion of intelligent design, so that one can also see how the simple yet radical shift to a mobile from static conception of organismic constitution challenged basic assumptions about not just the biological but religious, political, cultural, historical, and ultimately cosmic organization and orientation of the planet's life. As clearly as a theory like Wolff's challenged preformationist understandings of life, then, the sense of immutable development and structure offered by preformationism entrenched itself against the challenge laid down by Romantic evolution's implicit notions of indeterminacy, mutation, transience, mobility, etc.

The *critical* neglect of Romantic evolution can also be traced to a persistent suspicion within cultural disciplines regarding "biology" and its associations with the inhuman ends of natural determinism or even social Darwinism. For example, René Wellek and Austin Warren survey the various attempts to import a concept of evolution into literary history as a series of failures, especially when organic analogies impose the freight and finality of natural teleology on the complex and variable development of genre, historical periods, individual artists, or single works (16). More recently, Ernst Behler questions the matter of viewing the Romantic period through the lens of evolution and its implications for "historical determinism." For Behler, Romanticism is best understood "as the agent of a revolution" rather than "the product of an evolution" (63). Such criticisms imply a reaction against the dominance of organic form in approaches like New Criticism, wherein the temporal multiplicity of parts is subordinated to an ahistorical unifying design, an updated iteration of Coleridge's "unity in multeity." In Cleanth Brooks's figure of the well-wrought urn, for instance, a work of art evolves (in a somewhat different sense from scientific notions of the word) toward wholeness by resolving its own internal tensions and oppositions. Like Darwin's restriction of loose analogies, Brooks's model suggests a closed system whose internal mutations generate but are ultimately overcome by organicism's self-correcting economy, whereas Darwin's model indicates a more open-ended trajectory in which the circle of life can be rather less self-confirming. Or we can think of evolution in terms of political life, in which the interconnected yet often conflicting parts that constitute civil society are dialectically worked through and out within the living organism of the state, as in Hegel and post-Hegelian theories of right. Novalis even proposed a "poetic state" as a way to transform the atomistic machine-state into "a living, autonomous being" (45),[1] a harmonizing aesthetic transformation of the political that has become a characteristic feature of what Jerome McGann would call the "Romantic ideology." Such abstraction of the organic from the life sciences converts the unifying structural metaphor into what Paul de Man calls a "totalizing principle" (32), which reifies the work's form by arresting its dynamic play of forces.

We thus find in Romantic attitudes toward evolution an exploratory spirit that defines the period as what Richard Holmes has called an "Age of Wonder." At the same time, however, the challenge of evolution's more radical implications to entrenched notions of historical permanence incited at the very least ambivalence, at the very most vehement resistance toward the cause and effects of broad-based historical change. The following essays explore the various facets of Romantic evolution's multiple personalities through the various writers, texts, and cultural phenomena whose governing concepts materialize the *Gestalt* of Romantic evolution from among both reciprocal and competing claims. In turn, these claims made possible the evolution of Charles Darwin's theories: the move from local to universal that generated the formative idea of a more global taxonomy to account for otherwise heterogeneous scientific findings; the mapping of teleologies with no straightforward path, whose origins defy explanation and whose futures are radically uncertain; expansion of the scale of change to engage notions of prehistoric existence; unexpected mutations within otherwise stable or "organic" forms; species transmutation across time and space to produce startlingly different ecologies that altered life forms and behaviours; the spectre of the extinction of life and life forms as a challenge to notions of historical change and progress. In short, Romantic evolution moves toward a unity that encompasses multeity, yet is aware that exploring the multeity makes unity problematic, if not impossible.

It makes sense that Romanticism is one of Wellek and Warren's prime test cases, for its heterogeneous character, which Lovejoy laments in "On the Discrimination of Romanticisms" (1927), offers the perfect challenge when tracking the historical evolution of literature. Eager to avoid any "mental confusion" (260) about Romanticism as "a unitary quality which spreads like an infection or plague" (255), Wellek and Warren nonetheless point to a "pathological" Romantic organicism that the critical field, having long since gone through its own process of both natural and artificial selection, has embraced as part of a rather less-restricted and more general economy of Romantic being and thought. David Farrell Krell, for instance, has explored contagion, disease, and death as productive catalysts within the corpus of German idealist thought, which, thanks to Coleridge and others, made its own impact on the natural selectiveness of British Romantic thought. Patrick Brantlinger tracks how the evolution of evolutionary theory throughout the nineteenth century, particularly its notion of species demise, became a racially charged biological metaphor for the cultural survival of the "fittest" (i.e., less-primitive) populations. In her essay for this volume, Tilottama Rajan, borrowing from Schelling, speaks of the "asystasy" or "inner conflict" at the very core of life, which in turn generates, mutates, and degrades systems of thought (like those of Schelling or Hegel, but also Coleridge) produced to speak of life. There is, to paraphrase Georges Canguilhem, a chiasmus between the normal and the pathological, what Joan Steigerwald in this volume takes up as the "degeneration" of idealism as it attempts to articulate the very form of life itself, or what Maureen McLane explores as Malthus's discipline of life forms. Organicism is thus less an artificial imposition on living forms or a delineation of their "natural" or inevitable formation, but rather a way to trace this development's at-once determined/ultimate and contingent/arbitrary manifestations. Or to paraphrase Theresa Kelley's essay, it is within the very nature of life to "take chances." One needs to speak instead, as Alan Bewell or Andrew Piper note in their contributions, of the migrations of life via the forms through which it travels. Indeed, this mobility, as taken up by Gillian Beer, Matthew Rowlinson, and Gábor Zemplén, determines the interface between life and thought as a profoundly determinate morphological *in*determinacy.

Our contributors thus treat Romantic thought and writing as a dynamic laboratory whose protean nature uncannily reflects the very evolutionary modification, mutation, and metamorphosis it was struggling to fathom and, by comprehending, contain. Such investigation and experimentation exemplified what is now broadly call interdisciplinarity, although this designation suggests boundaries and biases that in Romanticism had not quite been reified in terms of current understanding (perhaps transdisciplinarity would be a better term). Indeed, contemporary perspectives often miss Romanticism's multi-perspectivalism. As we have seen, as often as not the heterogeneity of Romantic exploration tended toward finding unifying principles, or least the governing structure that multeity in unity offered. But even while Coleridge's posthumously published *Hints towards the Formation of a More Comprehensive Theory of Life* (1816) attempted, as its title suggests, to work toward a philosophical *Ungrund* and architectonic to explain the unfolding of life at all levels of both natural existence and human understanding, such idealism, as Rajan explores, is both sustained and unworked by its own possibilities.[2] Any evolution toward a governing synthesis remains, as Coleridge's title also suggests, a series of "hints" troubled by the ongoing "formation" toward a "comprehensiveness" always something evermore about to be. Or as Dahlia Porter has argued about the hybrid anatomy of Erasmus Darwin's *The Loves of the Plants*, Part Two of *The Botanic Garden* (1791), confronting the nature of nature generates hybrid forms that attempt to classify the fecund diversity of life, but succumb to its prolix vitality.[3]

Of course, not everyone has applauded the way in which Romantic literary writers took up science in their own time. Writing a fresh chapter in the well-worn "two cultures" debate between science and literature, and thus reinscribing the very boundary he seems to (want to) elide, evolutionary biologist Richard Dawkins in *Unweaving the Rainbow: Science, Delusion, and the Appetite for Wonder* (1998), for instance, revisits the works of canonical Romantic poets to test his hypothesis that the experience of poetry evokes a sense of wonder akin to that of scientific discovery. Dawkins regrets Romanticism's marked resistance to what we now call science, such as Keats's criticism of Newton in *Lamia* (1820) for "unweaving the rainbow" (39), Coleridge's "postmodern" interpretations of scientific discourse (40), or Blake's mystical account of Bacon, Newton, and Locke in *Jerusalem* (1804–20), which Dawkins calls "a waste of poetic talent" (16).[4] Darwin's grandfather, Erasmus, who in his Advertisement to *The Botanic Garden* wants "to inlist Imagination under the banner of Science" (1), is given rather short shrift for turning inspired scientific achievement into flat rhyming couplets that "do not enhance the science" (Dawkins 17). Or, as Dawkins warrants, how could a poet "worthy of the title Romantic" (ix) not be stirred by discoveries coming to light during his own time? For Richard Holmes, however, what Coleridge in his *Philosophical Lectures* of 1819 designated as a "second scientific revolution" (Holmes xvi)[5] makes science and literature less hostile to one another. This revolution's capacity for wonder galvanizes a dialogue between the two fields to produce the altogether "new vision" of what we now recognize as "Romantic science" (Holmes xv), a still burgeoning critical field since Jardine and Cunningham's *Romanticism and the Sciences* (1990) or Jan Golinski's *Science as Public Culture, 1760–1820* (1992). David Knight's introductory essay in the former argues that Romantic science's most distinctive feature was that it "lacked sharp and natural frontiers" (13), was not always concerned with taking sides, and thus was open to discursive miscegenations. Two examples of Romantic evolutionists, and key figures for many of this volume's contributors, stand out in this regard: Erasmus Darwin, who published scientific treatises in the form of poetry, and Johann Wolfgang von Goethe, the greatest German poet and dramatist of his age, who, like many of his contemporaries, was also a polymath and scientific innovator. Such work past and present speaks to the broadly cultural, scientific, and sociopolitical nature of Romantic science as a response to that world's living, potential, or posthumous forms. Romantic experimenters were, as it were, thinkers without borders.

Ironically, the translation of complex science to the public sphere, one of Dawkins's key motivations, is very much a Romantic invention (albeit often politically volatile, as Golinski reminds us). Between 1770 and 1840, sensational travel narratives appeared in print for the voyages of Louis-Antoine de Bougainville and James Cook; popular lecturers on science like Humphry Davy and his student Michael Faraday attracted large crowds at the new Royal and London Institutions; and scientific societies, journals, and museums emerged to foster the study of new disciplines like geology, biology, physiology, chemistry, and comparative anatomy. As Jon Klancher explores in *Transfiguring the Arts and Sciences: Knowledge and Cultural Institutions in the Romantic Age* (2013), such developments signal an evolution of the disciplines and their institutionalization as at once adaptable, entrenched, and unstable forms of knowledge and its dissemination across the public sphere. Romantic art intermingled ideas of Romantic science and philosophy to produce new aesthetic hybrids. Or, as Friedrich Schlegel wrote in his 1797 *Kritische Fragmente*, putting the matter in ideal terms: "Alle Kunst soll Wissenschaft, und alle

Wissenschaft soll Kunst werden; Poesie und Philosophie sollen vereinigt sein" ("All art should become science and all science art; poetry and philosophy should be made one"; cited in Wheeler 43), suggesting once again the Romantic impulse toward multiple perspectives and an equal desire to unify this multeity.

British and European Romantic writers, as Holmes points out, far from strictly objecting to the explosion of scientific discourses, were well-versed in its discoveries and controversies, and often deeply involved in its critical evaluation on behalf of the public imagination. In *The Politics of Evolution: Morphology, Medicine, and Reform in Radical England* (1989), Adrian Desmond, like Golinski, argues that evolution emerged as much from sociopolitical turbulence as it did from scientific investigation and exploration. For Desmond, progressive evolutionary theories promulgated across the Channel reanimated the ghosts of a radical French politics threatening entrenched, often aristocratic (and staunchly Anglican) cultural values. For instance, the Cuvier-Geoffroy debate of 1830, a kind of culmination of evolving theories of the previous decades, famously epitomized this challenge. Étienne Geoffroy Saint-Hilaire's analogical notion of species development as the modified and modifying morphology of a unified plan, developed from the theories of earlier naturalists and proto-evolutionists such as Comte de Buffon's and Jean-Baptiste Lamarck's evolutionary biology, was eventually trumped by Georges Cuvier's rather more fixed notions of animal structure. But following in the spirit of Toby Appel's *The Cuvier-Geoffroy Debate: French Biology in the Decades before Darwin* (1987), Desmond explores how the debate itself, and particularly the fact that Geoffroy's views reflected republican sympathies, exerted the kind of influence on British scientific views that (similar to the way in which Darwin's 1859 text galvanized otherwise conflicting prior theories of evolution) evoked the triumph of a new ruling class. Darwin's theory "'ratif[ied]' the competitive, individualist Malthusian ideology of the arriviste merchant class then acquiring power" (2–3), an earlier instance of a later social Darwinism. Put somewhat differently, Darwinism evoked a spirit of social reform that responded to the material conditions and thus rising power, in turn, of the working class. Such reform, attuned to the historically present rather than to distant tradition or a transcendental future, in turn threatened notions of biblical revelation and precedence; this challenge became the polestar of evil against entrenched religious, social, cultural, scientific, and political values.

Yet Romanticism's interest in how scientific ideas were effecting a broader transformation of the public sphere went beyond the sociopolitical. For in a world in which the increasingly global exchange of all things – resources, commodities, populations, ideas – was rearranging and metamorphosing the material conditions and even the very material constitution of global relations and existence itself, as

Noah Heringman demonstrates in his essay, one begins to see how "natural selection" inevitably and, well, rather naturally, emerged in the Romantic period. To rephrase our earlier point, the discovery and appraisal of local evolutions were increasingly gathering force as global phenomena, a scrutinizing and mapping of the world's vitality that would eventuate in Charles Darwin's more universal and universalizing theory. Put still another way, all the world, not just its "natural" parts, had become an experiment in evolution. Needless to say, critics on Romantic science – Bewell, Heringman, Kelley, or Piper in this volume; Tim Fulford, Noel Jackson, Mark Lussier, Richard C. Sha, and Alan Richardson elsewhere – are less concerned merely with tracing the prehistory of modern scientific disciplines.[6] Rather, they teach us more productively to see Romantic science as, once again, a shifting laboratory of thought shaped moment by moment by the very evolutionary descent it is, in the process of registering the turbulent impact of thought's own realization *of* this historical force, built to observe. Indeed, in essays by Beer, Rajan, Rowlinson, Steigerwald, or Zemplén we witness Romanticism's awakening to the complex ways in which environment and the shaping forces of thought are mutually constitutive. Romantic science did not merely record the world it was observing; it also recorded, thought, *felt* the lived experience of this observation as subject *to* evolution – the felt nature of which very much animated, inspired, and agitated the time's various religious enthusiasms and evolving spiritual vs. material, religion vs. science debates.

This "new vision" in Romantic thought and writing produced what might more specifically be called the life sciences, which took their intellectual and imaginative cue from the issue of evolution. As we have seen, Romanticism, unhappy with the Enlightenment's rational, atomistic, or static models of nature, began to see the natural world, and our relationship to it, as organic, dynamic, and evolutionary, and life itself as this relationship's primary focus. Romantic ideas of matter and mind are thus a productive mutation in the evolution of literary and scientific thought, shifting our understanding of Romantic revolution to the revolutionary nature of Romantic evolution through a resurgent interest in organicism, vitalism, natural history, natural philosophy, and the aesthetic, as seen in the work of Beer, Rajan, and Zemplén in this volume, and Michel Chaouli, Denise Gigante, Amanda Jo Goldstein, David Farrell Krell, and Timothy Lenoir elsewhere. Perhaps most powerful about their approach is a resistance to classifying pre-evolutionary models as either incommensurable paradigms or necessary catalysts. Rather, debating this issue is at the core of this volume's exploration of Romantic science's speculative, interdisciplinary approach to evolution before Charles Darwin. *Marking Time* thus explores evolution not only as a theme of scientific development but as a metaphor for the evolution of Romantic/post-Romantic thought as it deals with radical historical change through multiple forms: ecology, geology, *Bildung*, biography, autobiography, confession, to name only a few among the range of topics addressed in the following essays.

It was also Friedrich Schlegel who in *Dialogue on Poetry*, an 1800 contribution to the *Athenaeum*, wrote: "I seek and find the romantic among the older moderns, in Shakespeare, in Cervantes, in Italian poetry, in that age of chivalry, love and fable, from which the phenomenon and the word itself are derived" (cited in Ferber 7). Here Schlegel refutes the application of the term specifically to his own time, but rather indicates its broader evolution from an earlier notion of the *romauns* (cited in Ferber 7), a "phenomenon" whose force field reverberates across time.[7] Perhaps it has always been more productive, and at the same time more volatile and less tangible, to see Romanticism as a power, spirit, force, modality, temperament of human endeavour and expression evolving from the past, mutating and transforming older or even extinct forms into a designation that suggests less the time of the present than its dynamic and mercurial exchange with temporality. Not just since the historicist turn in Romantic studies have we been able to see in Romantic thought and writing both a profound attention to one's historical moment and a period lost in and to time. That is to say, Romanticism pays increasing attention both to the all-too-present nature of the quotidian and to the nearly irrevocable currents of an inscrutably deep time, as in James Hutton's 1788 *Theory of the Earth*, or deep astronomical time implied by Johann Lambert's 1750 *An Original Theory or New Hypothesis of the Universe*, Kant's 1755 *Allgemeine Naturgeschichte und Theorie des Himmels*, or William Herschel's later telescopic focus on the cosmos. These instances suggest that the future of human environments of all kinds, to take up Grosz's idea of evolution's radical becoming, is a devoutly wished-for prophecy, yet one without measure.

Romanticism's desire and effort to mark the cryptic passage of time – its "*dynamische Evolution*," to use Schelling's term from *Erster Entwurf eines Systems der Naturphilosophie* (1799) – offers a powerful lesson: the form of life was not always visible in this life's forms. Hutton's theory of the earth as a superorganismic form evolving through deep time or Lambert's, Kant's, or Herschel's theories of an evolving cosmos demonstrates that life forms were constituted by forms hidden within these forms, and hidden from life itself. Perhaps most radical – and one of *this* volume's lessons about Romanticism's confrontation with (its own) history – was the implication that life forms made visible and were constituted by their own extinction, either by developing from a process of natural selection that had transpired long before their present materialization, or by (extension) anticipating, recording, and thus instantiating in their ongoing formation the superannuity and eventually superfluity of their very being. This "*dynamische Evolution*" implied what became both for Romanticism and its heirs a troubling miscegenation of forms, environments, and ideas, the mobility of which lay beyond the pleasure principle of a life that functioned efficiently, consistently, teleologically. As de Almeida writes, "In the evolutionary thinking of romantic scientists and artists, the boundaries of the rational were ever fluid and elastic" (124) between and among life forms and the forms life takes, physically, socially, culturally, ideologically. And we know well enough how such mobilities and transgressions were of great sociopolitical import, concern, and threat both to the time itself and certainly to its heirs. The great fecundity of life opened up by exploring, examining, and classifying the globe from at least the eighteenth century forward created an expansionist anxiety on all fronts. Or, as Almeida continues, "It was enlightened to have the prevailing classification of nature challenged and expanded by reported encounters of greater diversity abroad. It was quite another (and revolutionary) matter to have an overwhelming volume of unassimilated – and inassimilable – oddities and unknowns *bury* all familiar and preconceived classification" (126).

In short, a new global theory – Charles Darwin's account of organismic change from the minutely local to transformations of and between species across broad swathes of time – was replacing the shards of an earlier taxonomical approach that no longer sufficed to account for the epic scale of natural modification, mutation, and adaptation expressed by Darwin and the forebears by whose Romantic confrontation with historical change he was inspired. With Romanticism's wondering about evolution came anxious responses to its spectres of mutability and change. Rebecca Stott's *Darwin's Ghosts: The Secret History of Evolution* (2012), her own epic account tracing Darwin's scientific precursors back to Aristotle, uses as one of her epigraphs a citation from Darwin's Notebook C, one of his field notebooks from the *Beagle* voyage: "Once grant that species [of] one genus may pass into each other ... & whole fabric totters & falls." Here disaster's twin spectres of monstrosity and obsolescence loom as evolution's threat to all levels of scientific and historical knowledge, from local confirmations to the most global verities of how life is constituted and structured – or rather, most radically for the time, how it constitutes and structures *itself* as its own guiding (and thus potentially misguiding) principle. We can trace disaster in vitalist debates of the late eighteenth and early-to-later nineteenth centuries, typified by the epigenesis/preformation controversy mentioned earlier, and somewhat later by theories of and debates about life galvanized by the early evolutionist and comparative anatomist John Hunter and developed by Coleridge, John Abernethy, and Joseph Green, among others.[8] Mary Shelley's *Frankenstein* (1818), partly a riff on the fallout from Alessandro Volta's, Humphry Davy's, or Giuseppe Galvani's scientific discoveries, is a literary response to such debates, but so are the evangelical, even millennial enthusiasms of various religious retrenchments sparked throughout the Romantic period, to which tensions between the material and the spiritual Romanticism's often-apocalyptic nature is at once dire response and desperate solution.

The abyss of deep time opened by Romantic evolution is the yawning chasm of a history so vast as to belie the very term "history." Romantic evolution thus also implies a vertiginous temporality. Rather than move inexorably backward to a remote past even tentatively attainable as an "origin" or incalculably forward to a "goal" always waiting to be born(e) to the present, this trajectory bears forward by being borne back ceaselessly to the past, to paraphrase the final lines of Fitzgerald's *The Great Gatsby*, adrift on a fathomless sea of life, like Schopenhauer's *principium individuationis*.[9] Without reducing evolution to biology alone, or rather by seeing this biology in rather less, well, "biological" terms, the multivalent discursive terrain of Romantic evolution – literary, scientific, aesthetic, philosophical, religious, political, sociological, economic, cultural – becomes the rather pivotal missing link in evolutionary theory, an entangled bank of natural facts whose increasingly unassimilable nature Romanticism was at once fascinated by and anxious about. If it has been the case that we often see the "relevance" of Romanticism ruin itself upon this particular shoal of its own incommensurable nature – how the time was mesmerized by the compulsively repetitive movement back to the future of its own thought – it is also clear that the times saw in this process the very movement of time and history themselves, which they set about at once merely to observe and at times obsessively to master.

II.

Marking Time opens with three essays that address the later Darwin's thought and writing, then subsequent essays move back to this future in Romantic evolution, although the opening essay by Gillian Beer, "Romanticism's Darwin," immediately renders this temporal structure extinct. Reading the epigenesis of Charles's work in that of his grandfather Erasmus, Beer is less concerned with offering a family romance of evolution's descent from Romanticism to the Victorian period than with tracking the romance of familial likenesses in which it would take something like time-lapse photography to capture the "intensity of [the later Darwin's] repeated and continuous observation" of his grandfather's writings and ideas as the grandson's "mental companion and interlocutor." Such obsessive attention at the ontogenetic level parallels the phylogenetic trauma we have suggested above: How to make the natural world and its histories conform to our classifications when nature's florid, unassimilable nature resists taxonomy, and how to legislate classification itself when its analogical nature turns against nature, are persistent questions. If for Erasmus Darwin analogy was evidence of common descent, it taught the later Darwin that difference is founded on kinship. This meant in turn a differentiation of strict from loose analogies that would accommodate the vertiginous nature of life observed, but in the grandson necessitate an even more arduous pursuit of theories to account for and contain such vitality. At the very least, this means a resistance to "turning analogy into homology." That "process and perfection lie uneasily together," and that "nothing is stationary" because "every organism is caught in the propulsion of reproduction," in turn generates "difference, divergence, variation down the generations," as well as "likeness." The same perfection we often mis-associate with Romantic idealism or Victorian progress becomes instead perfection's imperfect or temporary nature, what the later Darwin saw as its "doubtful goal."

That Erasmus chose, as Beer notes, "prose for ratiocination" but "poetry for speculation" exemplifies his own Romantic awakening to the mobility of evolutionary descent. For Alan Bewell, analogy becomes the radical mobility of and within species. Mobility, that is to say, is how species mark their evolutionary territory. As Bewell argues with reference to Darwin's *On the Movements and Habits of Climbing Plants* (1865), which explores how plants with tendrils aspire to a life of their own and thus to reach beyond the borders of their own genus, "Darwin was not just interested in understanding the different ways in which plants and animals had spread across the earth. He was also attentive to the broader evolutionary questions raised by that mobility." Gazing into the depths of biological time, Darwin reads speciation as spatialization. This transmutation of species across and over time produces an ecology of forms and behaviours that facilitates mobility in the first place, allowing them not to roost, but to re-settle "a world that is already occupied with competing species." Such "biological colonization" marks a species' "evolutionary success." Morphology is thus less a record of advancement than of a species' "capacity to range widely." Since no one possesses land or place absolutely, however, evolution also indicts imperialism. The evolutionary theorist is thus a decolonizer whose "artificial laboratory" of "colonial ecologies" allowed him to show how continents, not islands, produce the greatest diversity of life through the "global migration and competition of biota."

If Bewell uncovers the mobility of a Darwinian heuristic in the service (once again) of both explaining and containing nature's prolix and often alien nature,[10] Matthew Rowlinson puts the idea of evolution under erasure by examining what we might call its crisis of witnessing in which the eye misses precisely what it sets out to see: the hidden (variable and extinct) forms that constitute evolution itself. Reading between the lines of Darwin's famous "tree of life" drawing, Rowlinson explores how science transforms the "continuous process" that is natural selection "into a series of discontinuous events" that inevitably exclude "the extinction of the intermediate forms without which species would not exist." The eternal present of the act of witnessing becomes, by virtue of our inability to grasp it, wholly imaginary. Put otherwise: it is impossible to mark the point of variation and extinction because one's witnessing of these phenomena *is* the point. Like the Lacanian Real, we miss the reality before our face, marking evolution as an idea always lying in wait for our discovery. For Darwin, ideas are thus "unconscious representations of evolutionary remainders," and "the theory of descent is understood as a raising to consciousness of what had been unconscious." Between natural and artificial selection lies Darwin's third term, unconscious selection. This suggests how life forms appear to choose one another as if unconsciously – the unconscious agency of evolution's *Geist* that determines history – less through the "act of selection" itself than via "its effects over time." Blind "to its own agency," and without a theory of heredity, as in Gregor Mendel's later paper "Experiments on Plant Hybridization" (1866),[11] unconscious selection thus makes indistinguishable "the agency of human beings and that of nature." By adapting to how we see it, nature anticipates this seeing's form, a shaping of "human action" by history that we apprehend only via "the alienated and misrecognized form under which human agency in history appears in any imaginable present."

Our next section, "Romantic Temporalities," moves past evolution's Romantic remainders to explore how evolution haunts Romanticism, whose modes of historical understanding have so seminally shaped how we know and see since then. Noah Heringman revisits Bewell's spacing of time by exploring Cook's "discovery" of Tahiti, a founding moment of Romanticism, as J.C. Beaglehole argued. Accounts and representations of Tahiti galvanized a post-Enlightenment attention to a "deep ethnographic time ... where geology and human prehistory meet." The orientalizing, racializing impetus for this cultural (re-)mapping charted a sanative space between civilization and its barbaric other as a now-extinct life form. This objective, "temporal distance" compensates for an inability to confront the proximity of "cultural difference." But this carbon dating of cultural history also maps a narrative of progress that allows culture to rally its primitive ground toward a later edifying figure. Somewhat different from Rowlinson's crisis of evolutionary witnessing, the altering eye of deep ethnographic time alters all it sees by conjecturing a historical thickness and consistency that temporalizes difference (to borrow Nicholas Thomas's phrase). In turn, this gaze "not only restructures the horizontal space of geography; it also anticipates early geology's restructuring of vertical space in the pursuit of primitive rocks." Such early Romantic lines of flight, materializing the real of history through books and ideas that *did* travel, thus became the empirical – and imperial – validation underwriting such later accounts as De Quincey's orientalizing imaginaries.

Somewhat less agential is how life and history meet in Maureen McLane's essay on Malthus as "our contemporary" and the temporality of Malthusianism as a way of thinking history, and the history of populations, as part of a restricted rather than general economy. McLane borrows Marjorie Levinson's distinction, from set theory, between counting and matching. The point is not how we count populations, but how we find matches within and between species in order to legislate their evolution – which is to say, their cultural and sociopolitical development. This is about how to turn "I" into "we," the inclusiveness of which, like Nancy's society as opposed to community, absorbs singularities into the common whole. This is also to mark in Malthus's principle of population a desire for happiness that has little to do with desire. In Malthus, McLane writes, "thinking is precisely not fucking" because it thinks of populations as a form of biopolitics whose ideology shapes species as well as individual evolution. Malthus's essay is thus also a theory of moral sentiments that reads "desire" as "heteronormative" and "foresight" as "the great mental condom" disciplining the future development of human groups. In the end, Malthus "maps for us" not a principle, but "a political economy of population." He thus marks the problem of "time of and in thought, as well as the time of and in labour," ensuring, like nineteenth-century tracts on the evils of onanism, that the ground of life is never wasted with the seed of time.

Essays in the next section, "Goethe and the Contingencies of Life," evolve around perhaps the Romantic period's greatest polymath, whose writing on life (and life-writing) diversifies literary and scientific species through a corpus that unfolds by the unconscious selection of its own inner tensions and agencies. If morphological mobility and temporality are the themes of our first two sections, here our contributors examine morphological contingency – the chance *of* morphology as confronted by its godfather. Each essay takes up a radical break with the past that signifies differently from progress or teleology. Andrew Piper examines Romantic science's fascination with the vertebra as a vitally metaphorical archetypal structure that anatomizes and turns upon the constitutive fractures and inevitable incompletion of Linne's taxonomy. Piper then analyses this interminable conversion through the vertebra's overdetermined and variable forms: spires, shells, torsos, ruins, bones, the Coliseum, books. From spire to shell, for instance, marks a "turn from the aspirational to the rotational," from "divinatory identity ... towards a far more terrestrially oriented theory of life." Between revolution and evolution, "radical change" and "categorical continuity," Goethe's emergent theory of life writes itself as conversion, a merging of autobiographical and natural scientific that migrates from the personal to the political, natural, aesthetic, and bibliographic. This vertigo between troping and turning, unlike the certainty of Augustinian conversion, misprisions form as life's fundamental manoeuvre, like bones or ruins as relative figures for relativity itself: avatars of decline both in and out of time, an endurance literally and metaphorically "past the ontogenetic boundaries of an individual life." In the end, conversion is the condition of life *as* autobiography, a circularity without completion. Or in Piper's final words on the Roman Coliseum, "in the conjunction of the circular and the erroneous, the *irren* or erring that belongs to the vertiginous quality of knowing life ... we can see the most concrete example of the way evolution recapitulates revolution."

In Theresa Kelley's essay, this "*irren*" is the time of chance or accident in evolution's progress – the Lucretian fact that things *will* move forward, or rather *on*. Or, as Kelley states of the "impediments" of time propelling Wordsworth's *The Prelude,* the "need to mark the liabilities of future talk became something like the shadow structure of its narrative." Like Piper and McLane, Kelley mentions Aristotle on the accidents of substance as conversions that alter substance without transmuting its essence. Or, as Kelley notes via Ross Hamilton's work on the accidental in Romantic thought, chance contributes vitally to the making of (one's) life. Here "regular metamorphosis" meets its developmental other to articulate evolution, not as a teleology or biblical trajectory of sublime origins and apocalyptic climaxes, but as Hutton's genealogy of a deep time without beginnings or ends, an accretion of sometimes catastrophic shifts and careers. In tracking the path of evolution as narrative in Erasmus Darwin's *The Botanic Garden*, Kelley draws upon Beer's sense of an "analogical engine" less a "precision instrument than a persistent catching up of possibilities whose very strangeness speaks for a potentiality that risks being too much, thereby avoiding the same dull round of too little." Darwin's poem "conveys" a "duration that is marked by chance interruptions" rather than the regularities and efficiencies of Aristotelian plot. Yet duration is not mere endurance, for the chance of evolutionary history is "what make[s] futurity happen, whatever that futurity is," so that "romantic narrative makes a place for what is (yet) untimely, but also for accidents as material and figurative witnesses that no history or narrative can proceed in full knowledge of futurity."

Gábor Zemplén then takes up how the observation of natural forms determines morphology, and how the methodology and form of morphological research – the evolving form of Goethe's writings – produces morphological content in turn. Writing and narrative do not merely obey or reflect nature's movement, but are intrinsic to this movement's evolution. Like taxonomy, morphology strives toward autonomy. By shaping what it describes, however, morphology plays a language game of nature whose structure "leaves its traces in both the theory and the description" of the perceived, but is never caught up in the phenomenal because ultimately the structure always remains unseen. By adding the fourth dimension of time to its 3D version of Linnaeus, morphology offers the "ever-changing landscape of science" as "a way of life." Morphology thus becomes historiography, a way of tracking "the forces that shape the individual thinker and science in general." To observe process is to mark the "constantly changing phenomenal and peculiar syntax and semantics of morphology," what Joan Steigerwald in her later essay calls the "tension between productivity and constraint," the phenomenological and the reductionist, the system and (to borrow from Rajan again) asystasy.

As the story of life's metamorphosis, morphology offers archetypal structure as a matrix of possibility: "the persistent explanatory framework becomes a motive of the autobiography's narrative structure," just as Goethe's later corpus repeats earlier insights in a finer tone that harnesses rather than purifies imperfection as the catalyst of generation.

The final section turns its attention to the evolutionary remainders of Romantic idealism, particularly those of German thought and the *Bildungstrieb* of its idealist nature. Here, the tutelary spirit is Schelling as a key mutation in idealist philosophy's body of thought and thus in Romantic thought in general. The abyss of the past opened by Schelling's philosophy takes us back interminably to our evolutionary futures, and thus to how our current moment remains shaped by a natural, artificial, and especially unconscious selection among forms of thought whose becoming recedes before us. Robert Richards reads the natural history of evolution in terms of its authors' struggle to claim priority for discovering transmutation as evolution's governing principle, if not for discovering evolution itself. Richards, among the first to argue for Schelling as evolution's godfather, reads Schelling *avec* Goethe as mutually reinforcing and adapting evolution's "transformational hypothesis" by tracing this family romance back to Aristotle and forward, pre-Darwin, to Charles Bonnet, Johann Friedrich Blumenbach, Kant, Erasmus Darwin, and Richard Owen and, post-Darwin, to Ernst Haeckel or Kuno Fischer, among others. For Richards, the somewhat less-than-systematic unfolding of the flora of evolutionary thought represents a tangled bank of influences that would find a later "conceptual tidiness" in Darwin's *On the Origin of Species*. What was for Kant an "uncontrolled fantasy" was for others a speculative morphological possibility, what Schelling "proposed [as] a principle of *dynamische Evolution*." Like Goethe's conception of the archetype, Schelling's was that of a plenum standard incorporating all its differentia. But it was also an ideal realized through the temporal development of a huge variety of types responding to natural forces.

That in late eighteenth-century Britain and Europe nature began to appear as radical flux, such that Goethe could endorse transition itself occurring "over a very long span of time," profoundly affected and effected knowledge formations, and how knowledge formed itself as an evolving response to both external and internal forces. Tilottama Rajan accounts for the intellectual biology of these transmutations within the broader context of Schelling and Hegel's (mis)adaptation of and to one another. Rajan historicizes Richards's history of ideas somewhat differently by exploring evolution's thought, less as a teleology toward its own ultimate or absolute form than as the shaping, mis-shaping, and reshaping of disciplines, not as specific sites of knowledge, but as the forces of (dis)articulation within and between their shifting boundaries. Such cognitive mobilities and transferences of the period's mindset speak to a process of unconscious selection that marks "a profound shift towards understanding forms of thought and culture as themselves in a process of evolution." For Rajan, Romanticism's discovery of the history of nature as the nature of history indicates a disciplinary species change in which "human history" confronts, not its origin, but its "mirror stage and primal scene" – an abyss that is the very shaping spirit of negativity within history and its formations, deformations, and reformations, less forward movement than the diastolic and systolic variations of movement itself within history.

In this volume's final essay, Joan Steigerwald addresses this shaping spirit of natural history, and of the history of nature – shaping as degeneration – not as catalyst for a later Victorian moral indictment, but as the oscillation and involution of life in "both directions, inverting teleology and involving it in the material and contingent." For Steigerwald, Romanticism's various attempts to name the time of life – Treviranus's *Lebenskraft*, Wolff's *vis essentialis*, or Blumenbach's *Bildungstrieb* – are boundary concepts, matrices of tensions whose opposition makes conjunction possible. To name life was to experiment with life, to breed the "problem of how to grasp the complex phenomena of self-organization"; in turn, to separate life from its activity was thus to make the human a deformation or degeneration of nature. For Kant, teleological judgment limits this self-generating nature of reason. For Oken or Schelling, however, as Steigerwald argues, the "circularity and tension" of teleological judgment generated "productive insights." But whereas Oken's mathesis of life left nature as an unthought absolute in a "world already differentiated and becoming," for Schelling idealist conception and materialist processes were entangled and mutually constitutive, if ultimately unparsable. Like the physiogony of asystasy in Rajan's essay, the irreconcilable self-difference that is any system's force or principle, this "tension between productivity and constraint," of existence and ground, which informed every aspect of mind and nature, marks within Schelling's philosophy of nature a "*mittleren Begriffe*" that speaks to the contradictions of life in the midst of things. Schelling thus prefigures Darwin as a form of speculative physics, an organization of nature that is contradictory within itself, a conception of nature as (de)generated by a ground that remains unruly and dark, blind and unspeakable, figured only negatively.

With Steigerwald's account of the antagonism of degeneration and generation, we come, as it were, to the heart of evolution's darkness as the ungraspable core at the matrix of life itself. Or, to cite Grosz a final time, we confront an at-once radical, turbulent, and inescapably productive and reproductive sense of life's becoming as a life never quite our own, leaving us with what Percy Shelley in *The Triumph of Life* calls thoughts that "must remain untold" (21). Yet it is within the time of life in our lives *to* tell. Or, as Steigerwald aptly suggests, giving life to the history of the world, then, is the dialectic giving life to all our judgments of that natural history. Within this give and take of life, and the life of thought, we mark time according to what after Steigerwald we might call a (de)generative temporality, necessary for any creature to exist at all. The study of life's timeline in Romantic thought produced no ultimate archetype or absolute origin but rather an intense engagement on the entangled banks – which is to say, at the margins – of life where we catch its fleeting, incommensurable, and essential stuff of becoming through acts of profound, ephemeral, constitutive perception. Perhaps most powerfully, the following essays evince how any act of speculation can itself shape the evolution of things, and thus how mind and matter are inextricably linked in the unfolding of an evolutionary development both within and beyond the bounds of the "human" to which we are only the partial but nonetheless vital witnesses. To return in history is to revisit the modes of historical understanding that formed our own forms of knowing and seeing. And with each return, we affect and effect the forward movement of our own histories and history.

NOTES

1 Novalis writes: "A state imbued with spirit will be poetic of itself" (45). Klaus Peter cites from the same source, Novalis's *Vermischte Bermerkungen* (*Miscellaneous Observations*), the following: "Der poetische Staat – ist der wahrhafte vollkommene Staat" ("The poetic state is the true and perfect state") (*Novalis Schriften* 2:468; cited in Peter 201).

2 For a seminal study that anticipates much later work on Coleridge as a key figure within the discursive development of Romantic evolution, see Levere.

3 Porter addresses Darwin's attempt to use eighteenth-century Linnean taxonomy in his footnotes to restrain the eighteenth-century empiricist work of analogy (from Bacon) that threatens to transform into a rather less precise, more florid rhetorical figuration as metaphor or simile. See also Shteir.

4 For what we take as a rather pointed rebuttal to something like Dawkins's somewhat limited sense of the otherwise profoundly material metaphoricity of Romantic science, specifically as explored in and by Blake, see Goldstein, who explores the deep time of Romantic biology. For Goldstein, Blakean epigenesis is less the abandonment of preformationatist theories for those that explain the "autotelic power" and nature of life, and more rather an understanding of (self-)generation "as a work of acute historical and circumstantial dependency." Blake's poetry formulates the labour of (what we now know as) ontogeny, less as pure biology than as a temporal duration and development compelled and responded to by custom, habit, environment, and social circumstance – a historical contingency in which "the pressures of inheritance, need, contingency, and repetition interact to sculpt [the living form's] organs from conception until death." As much as biology determines existence, existence – a kind

of counter-determinism with agency – marks the time of deforming, reforming, and transforming biology. In Bourdieuvian fashion, Blake's epigenetic body materializes its social surround by way of re-shaping the social. *This* transmogrification Blake's poetic *corpus* "literally" enacts as the comingling, warring, and mutual enlivening of its figures as biology's social actants and actors, a compelling adjunct to contemporary scientific romances of epigenetic and genomic research.

5 Holmes writes, "The first person who referred to a 'second scientific revolution' was probably the poet Coleridge in his *Philosophical Lectures* of 1819" (xvi). In Lecture 12, Coleridge speaks of a post-Baconian explosion in scientific research of which his own age of discovery is the result:

> [T]he Reformation sounded the second trumpet and the authority of the Schools [of scientific theory as opposed to empirical observation and documentation] sunk ... under the intellectual courage and activity which this great revolution had inspired. Powers once awakened cannot rest in one object. All the sciences partook of the new influence and the world of the [experimental philosophy was soon mapped out for posterity] by the comprehensive and enterprising genius of Bacon. (*Lectures 1818–1819* 508–9)

6 See esp. Fulford, Lussier, and Sha.

7 As Michael Ferber reminds us, more recent efforts to define Romanticism are simply the most recent salvo in critical wars over a period that never had an identity to begin with.

8 See Mitchell 74–103; Almeida, *Romantic Medicine and John Keats* 98–110; Gigante 208–46; and Rajan's essay in this volume.

9 Speaking of the frailty of human identity when confronted by the Dionysian forces of life, Nietzsche, in *The Birth of Tragedy* (1879), cites Schopenhauer from *The World as Will and Representation* (1818): "'Just as the boatman sits in his small boat, trusting his frail craft in a stormy sea that is boundless in every direction, rising and falling with the howling, mountainous waves, so in the midst of a world full of suffering a misery the individual man calmly sits, supported by and trusting in the *principium individuationis*" (16–17). At the same time as Nietzsche criticizes Darwin and Darwinism, his vision of human existence owes a rather profound debt to Darwinian notions of survival, if not power. For an account of this debt, and of how Nietzsche gets Darwin both right and wrong, see John Richardson.

10 A tutelary spirit in Bewell's essay is Brantlinger's *Dark Vanishings*, which explores (pre-)Darwinism's racial context and implications.

11 A further irony suggested by Rowlinson's essay is that Mendel's pioneering insights on what was to become the science of genetics, including his notions of recessive vs. dominant genetic factors, went relatively unnoticed, of course, until the twentieth century.

WORKS CITED

Almeida, Hermione de. "Romanticism and the Triumph of Life Science: Prospects for Study." *Studies in Romanticism* 43.1 (Spring 2004): 119–34. Print.

Almeida, Hermione de. *Romantic Medicine and John Keats*. Oxford: Oxford UP, 1991. Print.

Behler, Ernst. "The German Romantic Revolution." Ed. James Pipkin. *English and German Romanticism: Cross-Currents and Controversies*. Reihe Siegen 44. Heidelberg: Carl Winter Univ.-verl., 1985. 61–77. Print.

Brantlinger, Patrick. *Dark Vanishings: Discourse on the Extinction of Primitive Races, 1800–1930*. Ithaca: Cornell UP, 2003. Print.

Brooks, Cleanth. *The Well-Wrought Urn: Studies in the Structure of Poetry*. New York: Harcourt, Brace and World, 1947. Print.

Canguilhem, Georges. *The Normal and the Pathological*. Trans. Carolyn R. Fawcett, with Robert S. Cohen. Introd. Michel Foucault. New York: Zone Books, 1989. Print.

Chaouli, Michel. *The Laboratory of Poetry: Chemistry and Poetics in the Work of Friedrich Schlegel*. Baltimore: Johns Hopkins UP, 2002. Print.

Coleridge, Samuel Taylor. *Lectures 1818–1819: On the History of Philosophy II*. Ed. J.R. de J. Jackson. Vol. 8. *The Collected Works of Samuel Taylor Coleridge*. Princeton: Princeton UP, 2000. Print.

Darwin, Erasmus. *The Botanic Garden: A Poem, in Two Parts*. London: Printed for J. Johnson, 1791. Print.

Dawkins, Richard. *Unweaving the Rainbow: Science, Delusion, and the Appetite for Wonder*. Boston: Houghton Mifflin, 1998. Print.

De Man, Paul. "Form and Intent in the American New Criticism." *Blindness and Insight: Essays in the Rhetoric of Contemporary Romanticism*. Minneapolis: U of Minnesota P, 1971; rev. ed. 1983. 20–35. Print.

Desmond, Adrian J. *The Politics of Evolution: Morphology, Medicine, and Reform in Radical London*. Chicago: U of Chicago P, 1989. Print.

Ferber, Michael. *A Very Brief Introduction to Romanticism*. Oxford: Oxford UP, 2010. Print.

Fulford, Timothy. *Romanticism and Science: Subcultures and Subversions*. New York: Routledge, 2002. Print.

Gigante, Denise. *Life: Organic Form and Romanticism*. New Haven: Yale UP, 2009. Print.

Goldstein, Amanda Jo. "William Blake and the Time of Ontogeny." *Systems of Life: Economics, Politics and the Biological Sciences, 1750–1859*. Ed. Warren Montag and Rick Barney. New York: Fordham UP, forthcoming. Print.

Grosz, Elizabeth. *Becoming Undone: Darwinian Reflections on Life, Politics, and Art*. Durham, NC: Duke UP, 2011. Print.

Holmes, Richard. *The Age of Wonder: How the Romantic Generation Discovered the Beauty and Terror of Science*. London: Harper P, 2008. Print.

Jackson, Noel. *Science and Sensation in Romantic Poetry*. Cambridge: Cambridge UP, 2008. Print.

Klancher, Jon. *Transfiguring the Arts and Sciences: Knowledge and Cultural Institutions in the Romantic Age*. Cambridge: Cambridge UP, 2013. Print.

Knight, David S. "Romanticism and the Sciences." *Romanticism and the Sciences*. Ed. Andrew Cunningham and Nicholas Jardine. Cambridge: Cambridge UP, 1990. 13–24. Print.

Kreiter, Carmen S. "Evolution and William Blake." *Studies in Romanticism* 4.2 (Winter 1965): 110–18. Print.

Krell, David Farrell. *Contagion: Sexuality, Disease, and Death in German Idealism and Romanticism*. Bloomington and Indianapolis: Indiana UP, 1998. Print.

Lenoir, Timothy. *Instituting Science: The Cultural Production of Scientific Disciplines*. Stanford: Stanford UP, 1997. Print.

Levere, Trevor H. *Poetry Realized in Nature: Samuel Taylor Coleridge and Early Nineteenth-Century Science*. Cambridge: Cambridge UP, 1981. Print.

Lovejoy, Arthur O. "On the Discrimination of Romanticisms." *Publications of the Modern Language Association* 39 (1924): 229–53. Print.

Lovejoy, Arthur O. *The Great Chain of Being: A Study of the History of An Idea*. New York: Harper and Row, 1936; rpt. 1960. Print.

Lussier, Mark. *Romantic Dynamics: The Poetics of Physicality*. New York: St Martin's P, 2000. Print.

McGann, Jerome J. *The Romantic Ideology: A Critical Investigation*. Chicago: U of Chicago P, 1983. Print.

Novalis (Friedrich von Hardenberg). *Novalis: Philosophical Writings*. Ed. and trans. Margaret Mahony Stoljar. Albany, NY: SUNY P, 1997. Print.

Novalis. *Novalis Schriften: Historische-Kritische Ausgabe*. 6 vols. Ed. Richard Samuel, Hans-Joachim Mähl and Gerhard Schulz et al. Stuttgart: Verlag W. Kohlhammer, 1960–2006. Print.

Peter, Klaus. "History and Moral Imperatives: The Contradictions of Political Romanticism." *The Literature of German Romanticism*. Ed. Dennis F. Mahoney. Woodbridge: Boydell and Brewer, 2004. 191–208. Print.

Porter, Dahlia. "Scientific Analogy and Literary Taxonomy in Darwin's *Loves of the Plants. European Romantic Review* 18.2 (April 2007): 213–21. Print.

Richards, Robert J. *The Romantic Conception of Life: Science and Philosophy in the Age of Goethe*. Chicago: U of Chicago P, 2002. Print.

Richardson, Alan. *The Neural Sublime: Cognitive Theories and Romantic Texts*. Baltimore: Johns Hopkins UP, 2010. Print.

Richardson, John. *Nietzsche's New Darwinism*. Oxford: Oxford UP, 2004. Print.

Sha, Richard C. *Science and Imagination in Romanticism*. Baltimore: Johns Hopkins UP, 2017. Print.

Shelley, Percy Bysshe. *Shelley's Poetry and Prose.* 2nd ed. Ed. Donald H. Reiman and Neil Fraistat. New York: Norton, 2002.

Shteir, Anne B. *Cultivating Women, Cultivating Science: Flora's Daughters and Botany in England, 1760–1860.* Baltimore: Johns Hopkins UP, 1996. Print.

Stott, Rebecca. *Darwin's Ghosts: The Secret History of Evolution.* New York: Random House, 2013. Print.

Wellek, Rene, and Austin Warren. *Theory of Literature.* 2nd ed. New York: Harcourt, Brace and Company, 1956. Print.

Wheeler, Kathleen M., ed. *German Aesthetic and Literary Criticism: The Romantic Ironists and Goethe.* Cambridge: Cambridge UP, 1984. Print.

PART ONE

Romanticism's Darwin

Chapter One

Plants, Analogy, and Perfection: Loose and Strict Analogies

GILLIAN BEER

They were in the first instance, bulbs or seeds, and later, living things endowed with sex, and pores, and susceptibilities which adapted themselves by all manner of ingenious devices to live and beget life, and could be fashioned squat or tapering, flame-coloured or pale, pure or spotted, by processes which might reveal the secrets of human existence.

– Virginia Woolf, *Night and Day*

In his Advertisement to *The Loves of the Plants*, Erasmus Darwin declared his purpose to lead the votaries of Imagination "from the looser analogies, which dress out the imagery of poetry, to the stricter ones, which form the ratiocination of philosophy" (preamble). And in the Preface to *Zoonomia* he argues that "rational analogy" is founded on the recognition that "the whole [natural world] is one family of one parent" (1). In evolutionary theory, analogy is felt to be both an essential tool and a doubtful guide. It goes deep into the structure of Charles Darwin's theory, but it causes argumentative difficulties: evolutionary ideas insist on variability and variety, on the value of slight differences between specific examples in the natural world. It is through such differences that evolution is enabled. This delight in the specific example rather than the standard type complicates thinking about analogy and, even more, perfection.

Plants

Much of Charles Darwin's later research work is concentrated on plant life and, indeed, botany is already prominent in the Notebooks of the late 1830s. He is fascinated by the prodigious inventiveness of plant behaviour and by the exquisite forms that seem to exceed usefulness. My argument will, among other things, draw on Charles Darwin's own annotated copies of Erasmus Darwin's *The Botanic*

Garden (second edition 1790) and his *Zoonomia* (1794–6), both copies now housed in the Cambridge University Library.[1] And it is worth noting at the outset that Erasmus Darwin's older brother Robert Waring Darwin published *Principia Botanica: or, a concise and easy introduction to the sexual botany of Linnaeus* in 1787, which reached a third edition in 1810. In 1838 Charles Darwin indexed that volume for his own uses, as he did the first volume of Erasmus Darwin's *Zoonomia* in March of the following year (Darwin, *Correspondence* 4:456). In the August of 1839, at the height of his creative forecasting of his new theory, Charles recorded having read the *Transactions of the Horticultural Society of London* right through, all the way from volume 1 to volume 7 part 3 page 433, and considered it probably worth rereading them all "as I do not think I was quite aware of the many points of importance" (*Correspondence* 4:457).

Why did plant life matter so much to Charles Darwin? A first, perhaps unexpected, answer may be that he started his research life thinking of himself as a geologist: moulds and petrifactions and sedimentary rocks all preserved evidence of plants from periods far more distant than those displaying the relics of animal life. Darwin needed a vastly expanded history of the world for his theories to work, and in plants he found evidence of life that has turned out to be from around 400 million years ago as opposed to animal life of which there is evidence from only 2.5 million years ago. Paleobotany was an important element in his argument (though not yet so named), all the more because these ancient fossil plants could be compared with their living inheritors as he collected materials on the *Beagle* during his five-year voyage to many different countries, climates, and conditions (Kohn et al. 643–5).

Later in his life, Darwin's settled presence at his home, Down House, in Kent allowed him to undertake botanical experiments that required a long time frame, often stretching over several years. This was in contrast to his earlier travel experience, during which he felt always the pressure of moving on, leaving behind, seeing plants in only that moment of their growth, not in their full cycle. At that time, in the 1830s, he was preserving and transporting specimens; later, he was watching plant life in all the intimacy of day-by-day observation. Fossils carried him across aeons of time; tendrils performed their movements in the immense, slow, observed present. In *The Movements and Habits of Climbing Plants* (first published in 1865 as an essay, "On the Movements and Habits of Climbing Plants," for the *Journal of the Linnean Society* and later revised as a book in 1875), the descriptions mingle precision and enjoyment with an intensity that phenomenologically allows the erotic to bloom, as in passages such as this: "The extremity of the tendril is almost straight and sharp. The whole terminal portion exhibits a singular habit, which in an animal would be called an instinct; for it continually searches for any little crevice or hole into which to insert itself" (95–6).

Darwin was always distrustful of arguments that settled for "instinct," which he saw as tending to arrest investigation rather than press onwards. Here he uses two incipient analogies: "in an animal would be called an instinct," while "continually searches for" invokes human will and experience, attributing them more broadly to other life forms, too. The passage continues:

> I had two young plants; and, after having observed this habit, I placed near them posts, which had been bored by beetles, or had become fissured by drying. The tendrils, by their own movements and by those of the internodes, slowly travelled over the surface of the wood, and when the apex came to a hole or fissure it inserted itself; in order to effect this the extremity for the length of half or a quarter of an inch, would often bend itself at right angles to the basal part. I have watched this process between twenty and thirty times. (*Movements* 95–6)

The observer begins to be part of the activity, keeping watch for immense lengths of time. Then:

> The same tendril would frequently withdraw from one hole and insert its point into a second hole. I have also seen a tendril keep its point, in one case for 20 hours, and in another for 36 hours, in a minute hole, and then withdraw it. While the point is thus temporarily inserted, the opposite tendril goes on revolving. (*Movements* 95–6)

Exploring and withdrawing, the tendril seems alive in a way disconcertingly close to that of the manner of the observing human, a closeness that is rendered through the analogous hints of sexual dalliance. Observer and observed are in a dance of accord, and Darwin succeeds in establishing in words something of what would much later be offered to us through time-lapse photography in which we can watch (without spending twenty hours) the sway, creep, flurry, and movement of the plant. But the intensity of his repeated and continuous observation cannot be matched by time-lapse photography; human and plant in this writing seem intertwined.

The motion and reach of plants has fascinated observers for centuries, and botanical illustrations often seek to suggest movement. Charles Darwin thought that he lacked proficiency in such work, which was an important skill for a natural historian at that time, and writing became his substitute technique for exploring the intricacy of plant life. So he followed his grandfather Erasmus in representing the forms and categories, but also the energies and erotic life of plants, in writing. Erasmus used both prose and poetry; Darwin used prose alone, but a prose that records the intimate concentration of the observer and sometimes, tantalizingly, the responsiveness of the plant observed.

The Interludes in *The Botanic Garden* are preoccupied with the technical rhetorical properties of poetry and its differences from prose, and include extensive discussions of simile and analogy. Poetic analogy does best when it is not "on all fours," Erasmus Darwin suggests, whereas in his prose he is seeking equivalences that are more than suggestive. He emphasized "storge," natural or familial affection, in his prose accounts of the plants' seeking mates, alongside the imagery of courtship that his poetry explored. In *Zoonomia* he writes:

> I ask, by what means are the anthers in many flowers, and stigmas in other flowers, directed to find their paramours? How do either of them know, that the other exists in their vicinity? Is this curious kind of storge produced by mechanic attraction, or by the sensation of love? The latter opinion is supported by the strongest analogy, because a reproduction of the species is the consequence; and then another organ of sense must be wanted to direct these vegetable amourettes to find each other, one probably analogous to our sense of smell, which in the animal world directs the new-born infant to its source of nourishment, and they may thus possess a faculty of perceiving as well as of producing odours. (1:106)

The sensorium of plants, Erasmus Darwin suggests, is closely analogous to that of human beings – and despite his "probably" and "may," the analogy is presented as a strict, though as yet unmeasured, parallel.

Both Darwins are part of a long tradition of imagining plant life and its sensory and emotional habitudes, but each of them brings a fresh sensibility and a particular theoretical inclination to his descriptions. And for Charles, plants had a decisive place in his theory. Even if one were to concentrate only on his published scientific works, it would be clear how crucial plant life was to Charles Darwin's thinking: in the *Origin,* plant examples are important in the chapters on struggle, variation, natural selection, geographical distribution and mutual affinities. From 1862, when he published *On the Various contrivances by which British and Foreign Orchids are fertilised by Insects, and on the Good Effects of Intercrossing*; through "On the Movements and Habits of Climbing Plants" (1865); *The Variation of Plants and Animals Under Domestication* (1868); *The Descent of Man and Selection in Relation to Sex* (1871); *Insectivorous Plants* (1875); *The Effects of Cross and Self-Fertilization in the Vegetable Kingdom* (1877)*; The Different Forms of Flowers on Plants of the Same Species* (1876); through 1877, spent, as he said, on the "circumnutating movements of plants and blooms"; to the 1880 *The Power of Movement in Plants,* one begins to wonder how he had time to think of other things. And in 1879 he published a biographical study of his grandfather, Erasmus Darwin.

But that apparent back-loading of botanical studies into the latter years of his life is also somewhat misleading, as is that late appearance of Erasmus, for if we turn to the private notebooks of the 1830s, it becomes clear that his grandfather Erasmus is often his mental companion and interlocutor, as also is his father, the doctor. Charles reflects in the M Notebook in 1838 on beauty of form, in shapes and perspective; in seaweed and trees, "the leaves of the foreground either owe their beauty to absolute form or to the repetition of similar forms as in angular leaves." As he muses on this, his thought turns to the particular pleasures of imagination:

> Pleasure of imagination, which correspond to those awakened during music. – connection with poetry, abundance, fertility, rustic life, virtuous happiness: recall scraps of poetry; – former thoughts, & experienced people recall pictures & therefore imagining pleasure of imitation come into play. – the train of thoughts vary no doubt in different people, an agriculturalist in whose mind supply of food was evasive and ill-defined thought would receive pleasure from thinking of the fertility. – I a geologist, have ill-defined notion of land covered with ocean, former animals, slow force cracking surface etc truly poetical (V. Wordsworth about sciences being sufficiently habitual to become poetical.) . – the botanist might so view plants & trees. – I am sure I remember my pleasure in Kensington Gardens has often been greatly excited by looking at trees as great compound animals united by wonderful & mysterious manner. (*Metaphysics* 13)

The name of Wordsworth comes to the surface in this passage as Charles recalls Wordsworth's *Preface* to the *Lyrical Ballads* with its emphasis on the solitariness of the scientist and the social expansion of the poet. And unnamed but present is Erasmus Darwin also. The idea of trees as "great compound animals" recalls, consciously or not, the passage in *Zoonomia* in which Erasmus remarks that "the individuals of the vegetable kingdom may be considered as inferior or less perfect animals; a tree is a congeries of many living buds, and in this respect resembles the branches of coralline, which are a congeries of a multitude of animals" (1:102). Paul Barrett, in his commentary, nicely adduces a later letter from Emma Darwin in August 1860 in which as she remarks, "At present he is treating Drosera just like a living creature, and I suppose he hopes to end in proving it to be an animal" (Darwin, *Metaphysics* 39). Darwin himself referred to "my beloved Drosera; it is a wonderful plant, or rather a most sagacious animal" – beloved and carnivorous (Francis Darwin 341). Climbing and carnivorous plants, with their seeming *intent,* particularly fascinated Charles Darwin, as they had his grandfather also.

I have written elsewhere about Darwin's reading of Wordsworth, and in particular *The Excursion,* as well as his enthusiasm for other Romantic poets – Coleridge, Keats, Shelley, and Byron – during the formative years up to and including the

Beagle voyage and the private notebooks, when he was formulating his evolutionary ideas in the late 1830s ("Darwin and Romanticism"). I shall not repeat this material here, but it is striking in relation to Darwin's fascination with plant consciousness, for example, to know that he read Shelley's poem "The Sensitive Plant" in the 1820s. This plant had also been an object of study for Erasmus Darwin, and was to be so again for Charles.

When thinking about the relationships between Romanticism and evolution (including Darwinian evolution), it is important to bear in mind the degree to which Linnaeus (*Systema Naturae* 1735) and Goethe (*The Metamorphosis of Plants* 1790) both lie behind the Romantic-to-Victorian preoccupation with plant metamorphosis and sex in plants as a taxonomic tool. Charles Darwin was obliged to struggle with both these systems, elements of which told against his own thinking: in particular, Goethe's emphasis on the leaf as including all other plant manifestations, and Linnaeus's "ranked hierarchy" which separated and ordered species.

Charles Darwin, in contrast, developed a lateral order of kinship that recognized how close-knit and even levelled were all organic forms of life and that emphasized their evolutionary descent. His correspondence in the 1840s particularly shows him persistently in debate with both earlier masters. For example, he writes to his close friend the botanist Joseph Hooker on 2 June 1847 about a "tuft of the quasi-hybrid Laburnum, with two kinds of flowers on one stalk ... Is not this very curious and opposed to the morphological idea that a flower is a condensed continuous spire of leaves," as Goethe had argued (*Correspondence* 4:44)? Variety is emphasized instead of condensation. The passage continues with Darwin pointing to likenesses across different orders of being (flowers, starfish, insects) in a way that spurs his analogical thinking: "Does it not look, as if flowers were normally bilateral; just in the same way as we now know that the radiating star-fish etc are bilateral? The case reminds me of those insects with exactly half having secondary male characteristics & the other half female" (*Correspondence* 4:44–5).

But Darwin is also shrewd about the degree to which it is possible to amalgamate descriptions of diverse orders of life. His own copy of *Zoonomia*, much sidelined and annotated, shows him in conversation and debate with his grandfather on such matters as conformity between plants and animals. So, for example, Erasmus writes on page 507:

> Other plants, which in this contest for light and air were too slender to rise by their own strength, learned by degrees to adhere to their neighbours, either by putting forth roots like the ivy, or by tendrils like the vine, or by spiral contortions like the honey-suckle.

Charles underlines the word "learned" and writes in the margin "What an assumption!!!" Here he is probably resisting the Lamarckian implications of "learned," as if the organism can within its own lifespan learn and hand on to its progeny what has been learned. Certainly he is here resisting turning analogy into homology. But other examples of his marginalia show him pressing further some of the implications of Erasmus's observations – for example, where Erasmus writes on p. 58 about the inclinations of the senses (smell, taste etc.), Darwin adds, "Hope is mental desire"; and where Erasmus discusses the recurrent closing up and opening of plants by night and day, Darwin adds a question: "Does habit imply having ideas?" He is particularly interested in examples of mimicry and camouflage in plants, to judge from his sidelining of such passages. In Notebook N he continues to brood on the question of whether plants have "any notion of cause and effect/they have habitual action which depends on such confidence/ when does such notion commence? – " (*Metaphysics* 72), and he defends his grandfather's observations on the reasoning power of wasps.

Above all, he concurs with his grandfather's early idea of an "original living filament" (*Zoonomia* 1:502) that has over generations transformed into an immense variety of organisms. Darwin sidelines that passage, with the remark: "Bell Bridgewater Treatise argues against this." This refers to Charles Bell's treatise on "The Hand," which sought to demonstrate the perfection of design through an example less frequent than that of the eye.

In the Introduction to *Zoonomia* Erasmus Darwin connects his evolutionary idea to the uses of analogy:

> The great CREATOR of all things has infinitely diversified the works of his hands, but has at the same time stamped a certain similitude on the features of nature, that demonstrates to us, that the whole is one family of one parent. On this similitude is founded all rational analogy; which, so long as it is concerned in comparing the essential properties of bodies, leads us to many and important discoveries; but when with licentious activity it links together objects, otherwise discordant, by some fanciful similitude; it may indeed collect ornaments for wit and poetry, but philosophy and truth recoil from its combinations. (1:viii)

In the next paragraph, on the same page, Erasmus argues the need for "strict analogy" in medical theory, mainly as an aid to memory and to ordering:

> The want of a theory, deduced from such strict analogy, to conduct the practice of medicine is lamented by its professors; for, as a great number of unconnected facts are difficult to be acquired, and to be reasoned from, the art of medicine is in many instances less efficacious under the direction of its wisest practitioners.

Perfection

Before I turn at large to the issue of analogy, let me dwell first on the idea of perfection and its sometimes-adversarial place in Charles Darwin's thinking, particularly in its relation to analogy. Looking at Ernst Haeckel's dazzling illustrations of biological forms, for instance, it is hard to deny perfection. The intricacy, spontaneity, and reach of the plants, their implicit motion, their colours and their strange, dishevelled forms – as well as the punctilious neatness of tendrils and internodes – are interlocked with the technical prowess of the artists from different periods who figure them forth. The pleasures of the eye convince us that nothing could be bettered, in image or object: that is certainly one form of perfection.

Another sense of perfection is the one that his mentor John Stevens Henslow taught Darwin. Henslow, originally a crystallographer, was professor of botany at Cambridge, and famously encouraged the young Darwin and arranged for him to have a place on the expedition of the *Beagle*. It was to him that Darwin sent all the plants he was collecting on the *Beagle* voyage (now in the Herbarium at Cambridge), and in a letter of 15–21 January 1833 Henslow advised him on how to select and preserve botanical specimens: "Avoid sending *scraps*. Make the specimens as perfect as you can, *root, flowers, & leaves,* and you can't do wrong" (*Correspondence* 1:293; emphasis in original).

Here, "as perfect as you can" means respecting the completeness of the plant rather than selecting some one aspect. And in one of his holograph marginal comments on Erasmus Darwin's *Zoonomia,* Charles Darwin is musing on the activity of learning music: "There appears to be a perfect gradation from movements of which [one is] partly unconscious to those which with effort can be recollected yet, but one sees instinct, habit" (sideline to *Zoonomia* 1:191). Here, "perfect" signifies uninterrupted and without deviation. We need to be alert to the variety of usages, but all these examples suggest coherence and fullness.

Yet process and perfection lie uneasily together: evolution has at its core the energy of change. A strong sense of "perfect" is "finished," "completed," as we see in the grammatical term the "perfect tense," used when the situation described has already been brought to a close. In evolution, nothing is stationary; every organism is caught in the propulsion of reproduction. And "reproduction" produces not simply likeness but difference, divergence, variation down the generations.

Every organism is also caught in the net of its environment, which is itself composed of all the needs, desires, conditions, and interactions of the organisms that compose it, among them the single organism from whose standpoint we began the sentence. That is the perception that underlies Charles Darwin's apparently paradoxical assertion that "Natural selection tends only to make each organic being as perfect as, or slightly more perfect than, the other inhabitants of the same country with which it has to struggle for existence" (*Origin* 201).

"Perfection" here is on a sliding scale and is also severely local, not an absolute state that can be maintained in the face of incomers or shifts in conditions. That is, natural selection offers no guarantee of survival to organisms perfected to their current environment. Any slight shift may disadvantage them. Perfection is thus a tentative, temporary condition, not a platonic absolute in which the perfect example conforms to an ideal type.

Indeed, Darwin greatly distrusted the idea of the standard plant, since, to him, no small divergence was without significance. Evolutionary change occurs through lowly, slight variations being carried forward through successive generations. Strikingly, in the period when he was writing, plants were being more and more commercialized and bred to a single standard: apples, for example, were being freighted farther from their home orchards and were being graded for sale according to their variety. Now, in our supermarkets, we see that process taken to an extreme, so that all the apples offered for any variety are of one rotundity, one colour, and one taste.

It was also the custom among botanical illustrators to merge differences into a representation of the "standard plant," a practice of which, for example, Joseph Hooker, Darwin's close friend and Director of the Botanical Gardens at Kew, heartily approved. Jim Endersby has given a fascinating account of Hooker's struggle with colonial botanists who wished to insist on the variation of their local plants – and plant names – from the imperial norm (137–69).

Charles Darwin's practice would sympathize with the colonies. He did not identify perfection with the norm. His prodigious powers of observation and of prolonged attention caught any divergence, any detail, any anomaly in the individual plant and valued it. He sought out fresh varieties and, he thought, even as he investigated, species. He clearly helped his children to value these powers, too. In the Notebook devoted to his children's actions and sayings, he records Lenny, on 5 June 1855, when he was four: "Lenny found for me before Dinner a new Grass, so he said 'I are an extraordinary grass-finder, & I must keep it particularly by my side all dinner-time'" (*Correspondence* 4:430).

The individual example was testimony to the powers of diversification and hyper-abundance in the natural world, and the individual organism was the medium through which were transmitted the differences that fuelled the future. It was necessary for Charles Darwin to propose a history that ranged back through time long before the human if he were satisfactorily to explain the profusion of different kinds in the current world, and *imperfection* became an important tool in this argument. It functioned both to represent the immense passing of time, the frail and temporary being of "being in the world" – and yet each being's manifest sensory realisation in the present moment, with all its value. He needed swathes of time, and he needed examples that demonstrated that evolution was the outcome of multiple deviations enacted over aeons. In one of the most famous passages of *On the Origin of Species* he muses on this particular difficulty in his theory, a difficulty that he forthwith transforms into homely explanation:

> Natural selection will not produce absolute perfection, nor do we always meet, as far as we can judge, with this high standard under nature. The correction for the aberration of the light is said, on high authority, not to be perfect even in that most perfect organ the eye. If our reason leads us to admire with enthusiasm a multitude of inimitable contrivances in nature, this reason tells us, though we may easily err on both sides, that some contrivances are less perfect. Can we consider the sting of the Wasp or of the bee as perfect, which, when used against attacking animals, cannot be withdrawn, owing to the backward serratures, and so inevitably causes the death of the insect by tearing out their viscera? (202)

Is it significant that his examples here are all taken from the animal kingdom, not from plants?

In such a passage, he is setting himself against the whole tradition of natural theology, which displayed the workings of God in the material world, and which habitually took the eye as the prime example of God's designing. Indeed, so habitual was that move that Charles Bell deliberately chose the hand rather than the eye for his Bridgewater Treatise demonstration of design: *The Hand, its Mechanism and Vital Endowments as Evincing Design* (1833). Already in his notebooks of the late 1830s, Darwin, for his own eyes only, attacked *The Adaptation of External Nature to the Moral and Intellectual Condition of Man* (1833), by Thomas Chalmers. In his essay on "Theology and Natural Selection," he rejects the necessity for an argument from design. Reading John Macculloch's *Proofs and illustrations of the attributes of God* (1837), Darwin resists his "long rigmarole about plants being created to arrest mud etc. at deltas" (*Metaphysics* 157). Instead, "All flow from some grand and simple laws." And: "I look at every adaptation, as the surviving one of ten thousand trials – each step being perfect ... to the then existing conditions" (160). Implicit is the affirmation that those conditions are not constant.

In Erasmus Darwin's account of the bodily changes of animals over many generations, "which may have been effected to accommodate them to new ways of procuring their food," he muses on "the existence of teats on the breasts of male animals," and from that speculates that "[p]erhaps all the productions of nature are in their progress to greater perfection?" (*Poetical Works* 9). Here, evolutionary change is understood as a movement *toward* perfection. That sanguine hope is only occasionally indulged by his grandson; for Charles Darwin, perfection itself becomes a doubtful goal.

So static perfection is inimical to the process of evolution and is identified, in contrast, with the idea of design and creation *ex nihilo*. It is striking that Charles's emphasis is often not just on *imperfection* but on the *temporary* nature of perfection, since it often appears as aptness within an environment: "perfect ... to the then existing conditions." And any environment is a shifting set of conditions.

Perfection, in his thinking, is momentarily realized as poise, but lost again in the onward tumble of procreation. Indeed, Darwin's understanding of the world is closer to Walt Whitman's assertion in section 3 of his "Song of Myself" (1855) in *Leaves of Grass*:

> There was never any more inception than there is now,
> Nor any more youth or age than there is now,
> And will never be any more perfection than there is now,
> Nor any more heaven or hell than there is now.
>
> Urge and urge and urge,
> Always the procreant urge of the world.
>
> Out of the dimness opposite equals advance, always substance and increase, always sex,
> Always a knit of identity, always distinction, always a breed of life. (28)

This is a different kind of perfection: the *always sufficient* nature of the world even as it strives onward. Perhaps here we begin to see the special value of analogy to the arguments of both Erasmus and Charles Darwin.

Analogy

Analogy uncovers connection but allows for change, "the procreant urge of the world." Analogy asserts kinship, but makes room for dis-analogy: "Always a knit of identity, always distinction, always a breed of life." It is neither homology nor hybridism. It is concerned with "opposite equals." In his private notebook on "Metaphysics, Materialism and Mind," Darwin quotes: "Ay Sir there is much in analogy we never find out" (*Metaphysics* 36). The form of the remark sounds like Dr. Johnson, but it indicates another of the aspects of analogy that appealed to Darwin: its super-abundance of material. There are always leftover elements that allow for change and new thinking. Precisely its *imperfection,* the limited coherence between terms, sets the mind racing.

Bishop Butler, in his great work *The Analogy of Religion, Natural and Revealed, to the Constitution and Course of Nature* (1736), set the terms for the propriety of analogy as evidence of design. Natural theology found evidence of God working in the natural world, and argued for the accord between such stable manifestation and the ordering will of God. Erasmus Darwin saw analogy as evidence, rather, of common descent:

> Shall we then say that the vegetable living filament was originally different from that of each tribe of animals above described? And that the productive living filament of each of those tribes was different originally from the other? Or, as the earth and ocean were probably peopled with vegetable productions long before the existence of animals ... shall we conjecture that one and the same kind of living filament is and has been the cause of all organic life? (*Zoonomia* 1:507)

Charles Darwin followed that general argument, but took it much farther. Like his grandfather, he sought a rational foundation for analogy, but he also recognized analogy as a tool for thought that might allow leaps of insight unjustified by full equivalence: in Notebook N he remarks, "Children understand before they can talk, so do many animals. – analogy probably false, may lead to something. – " (*Metaphysics* 72).

That wonderfully unexpected free leap at the end of the sentence, "probably false – may lead to something – ," is a typical gesture in these early writings. In *Origin,* he repeatedly invokes analogy in his morphological explanations – for example, of neuter insects: "we may safely conclude from the analogy of ordinary variations" (239); "analogy makes me greatly doubt" (254); "analogy makes me believe" (387). And in the conclusion, he writes analogically, even while cautiously demurring at analogical argument:

> Analogy would lead me one step further, namely, to the belief that all animals and plants have descended from some one prototype. But analogy may be a deceitful guide. Nevertheless, all living things have much in common, in their chemical composition, their germinal vesicles, their cellular structure, and their laws of growth and reproduction. (484)

So substitute "in common" for "analogy" and the argument holds, indeed validates analogy as natural law. The law-like nature of analogy is precious to Darwin, as is its occasional aberrations, and while he has learned from his grandfather to discriminate between loose and strict analogies, he finds uses for them both.

Whereas perfection in its stasis and completion stands in the way of evolutionary understanding, analogy offers a freer tool for thought, as well as a substantial explanation of the relation of the variety of the present natural world to the long past. In Charles Darwin's thought, difference is as important as kinship, but is *founded* on kinship. Kinship was his grandfather's driving concern, too.

But there is one other important way in which Erasmus Darwin heartened his grandson and trained him to observe against the grain of respectable inhibition. Both Erasmus and Charles make light of "The Loves of the Plants," Erasmus's epic poem with scholarly botanical notes, the verses devoted to the amorous life of vegetable experience:

> Queen of the marsh imperial DROSERA treads
> Rush-fringéd banks, and moss-embroidered beds;
> Redundant folds of glossy silk surround
> Her slender waist, and trail upon the ground;
> *Five* sister-nymphs collect with graceful ease,
> Or spread the floating purple to the breeze;
> And *five* fair youths with duteous love comply
> With each soft mandate of her moving eye. (30–1)

As Erasmus explains it, he is reversing Ovid's metamorphoses so that the transforming analogies show not people turning into plants, but plants into people: "I have undertaken ... to restore some of them to their original animality, after having remained prisoners so long in their respective vegetable mansions" (x). He supplies careful botanical descriptions in his footnotes to ground his fanciful descriptions. He is an adept of both tight and loose analogy, and happy to use them alongside each other, but only as long as they remain within their separate genres of prose and poetry. However hybrid, extreme, multivalent the plant activities Erasmus Darwin observes, he separates out the *genres* in their description: poetry for speculation, prose for ratiocination.

Charles Darwin's descriptions of climbing and carnivorous plants raise some of the same issues about plant consciousness as do Erasmus's. Thus Erasmus:

> [V]egetable life seems to possess an organ of sense to distinguish the variations of heat, another to distinguish the varying degrees of moisture, another of light, another of touch, and probably another analogous to our sense of smell. To these must be added the indubitable evidence of their passion of love, and I think we may truly conclude, that they are furnished with a common sensorium belonging to each bud and that they must occasionally repeat those perceptions either in their dreams or waking hours, and consequently possess ideas of so many of the properties of the external world, and of their own existence. (*Zoonomia* 1:112–13)

In his notebooks, Charles Darwin wonders, like his grandfather, "Do plants have idea of cause and effect?" and he muses on whether they may even "in some senses" have "free will" (*Metaphysics* 18). That is, analogy for Charles, both loose and strict, now stretches to include the psychology of the human and of plants. He watches the tendrils of climbing plants bend and turn and "after several hours seize fast hold of the twigs, like a bird when perched" (*Movements* 88). The intensity of Charles's observation is sensually charged as well as coolly scientific; indeed, sensuality is an aspect of his scientific imagination. In that, he was close to his grandfather.

The later part of his theoretical career saw Charles Darwin placing sex at the centre of explanation: in sexual selection, sight, song, touch, smell, inclination, and withdrawal were understood as essential to the continuance of life. Extravagance becomes crucial, theatrical extremes of behaviour necessary. Did he draw on his grandfather's imagination for this newly disinhibited explanation of evolutionary process? He was fascinated by the sheer inventiveness of forms in nature; of orchids Darwin writes:

> Hardly any fact has struck me so much as the endless diversities of structure, – the prodigality of resources, – for gaining the very same end, namely, the fertilisation of one flower by the pollen from another plant. (*Effects* 284)

Sexual selection brings the two Darwins close in imagination, as they investigate the loves of the plants each in their own way. Whereas in the 1830s Charles needed to set bounds between himself and his grandfather's thinking, in the 1870s he returns to his grandfather's work and draws sustenance from it. "Analogy may be a deceitful guide," as he warns himself from time to time, yet he would (for different reasons) side with Wordsworth, who blamed Peter Bell, to whom

> The primrose by the river's brim
> A yellow primrose was to him
> And it was nothing more.

Each flower was intensely present to both Erasmus and Charles, but it was never isolated. It was one aspect of the plant, and part of its continuance. And, beyond that, it was set always in relation to other organisms, shifting from ancestral forms, revealed in kinship, through analogy both strict and loose.

NOTE

1 The edition Darwin owned and annotated, and bequeathed to Francis Darwin, is the two-volume set listed in the Works Cited below. Charles Darwin's signature, dated 1826, is on the flyleaf. Held in Cambridge University Library. Darwin also owned

and annotated the two-volume *Zoonomia* published 1794–6 by J. Johnson (see Works Cited below). This copy was first owned by Charles Darwin's father, Robert Waring Darwin, and the volumes are now also held in Cambridge University Library, to which I am grateful for their assistance.

WORKS CITED

Beer, Gillian. "Darwin and Romanticism." *The Wordsworth Circle* 41 (2010): 3–9. Print.

Darwin, Charles. *Charles Darwin's The Life of Erasmus Darwin.* Ed. Desmond King-Hele. Cambridge: Cambridge UP, 2003. Print.

Darwin, Charles. *Correspondence of Charles Darwin.* 17 vols. Ed. Fred Burckhardt, Sydney Smith et al. Cambridge: Cambridge UP, 1985–. Print.

Darwin, Charles. *The Descent of Man and Selection in Relation to Sex.* London: John Murray, 1871. Print.

Darwin, Charles. *The Different Forms of Flowers on Plants of the Same Species.* London: John Murray, 1876. Print.

Darwin, Charles. *The Effects of Cross- and Self-Fertilization in the Vegetable Kingdom.* London: John Murray, 1877. Print.

Darwin, Charles. *Insectiverous Plants.* London: John Murray, 1875. Print.

Darwin, Charles. *Metaphysics, Materialism, and the Evolution of Mind: Early Writings of Charles Darwin.* Transcribed and notes Paul. H. Barrett. Comm. Howard E. Gruber. Chicago: U of Chicago P, 1980. Print.

Darwin, Charles. *The Movements and Habits of Climbing Plants.* London: John Murray, 1906. Print.

Darwin, Charles. "On the Movement and Habits of Climbing Plants," *Journal of the Linnean Society* 9 (1865); rev. ed. *The Movements and Habits of Climbing Plants.* London: John Murray, 1875; popular edition, London: John Murray, 1906. Print.

Darwin, Charles. *On the Origin of Species. A Facsimile of the First Edition.* Cambridge, Mass.: Harvard UP, 1964. Print.

Darwin, Charles. *On the Origin of Species by Means of Natural Selection, or the Preservation of Favoured Races in the Struggle for Life.* London: John Murray, 1959. Print.

Darwin, Charles. *On the Various Contrivances by which British and Foreign Orchids are Fertilised by Insects, and on the Good Effects of Intercrossing.* London: John Murray, 1862. Print.

Darwin, Charles. *The Power of Movement in Plants.* London: John Murray, 1881. Print.

Darwin, Charles. *The Variation of Plants and Animals Under Domestication.* London: John Murray, 1868. Print.

Darwin, Emma. *A Century of Family Letters, 1792–1896.* Ed. Henrietta Litchfield. London: John Murray, 1915. Print.

Darwin, Erasmus. *The Botanic Garden; A Poem in Two Parts: Part I Containing the Economy of Vegetation.* London: Printed for J. Johnson, St Paul's Churchyard, 1791. *Part II The Loves of the Plants, with Philosophical Notes, The Second Edition.* London: Printed by J. Nichols, For J. Johnson, St Paul's Churchyard, 1790. Print.

Darwin, Erasmus. *Zoonomia.* 2 vols. London: J. Johnson, 1794–6. Print.

Darwin, Erasmus. *The Poetical Works of Erasmus Darwin Containing the Botanical Garden and the Temple of Nature, with Philosophical Notes and Plates.* London: British Library Historical Collection, digitized, nd. Print.

Darwin, Francis. *Charles Darwin, His Life Told in an Autobiographical Chapter and in a Selected Series of his Published Letters.* New York: D. Appleton and Co., 1892. Print.

Darwin, Robert Waring. *Principia Botanica: or, a concise and easy introduction to the sexual botany of Linnaeus.* Newark: printed by Allin and Co. and sold by G.G.J. and J. Robinson, London, and all other booksellers, 1787. Print.

Endersby, Jim. *Imperial Nature: Joseph Hooker and the Practices of Victorian Science.* Chicago: U of Chicago P, 2008. Print.

Whitman, Walt. Leaves of Grass *and Selected Prose.* Ed. Ellman Crasnow. London: J.M. Dent, 1993. Print.

Woolf, Virginia. *Night and Day.* Ed. Julia Briggs. London: Penguin, 1992. Print.

Chapter Two

Darwin and the Mobility of Species

ALAN BEWELL

It has recently been argued that a new "mobilities paradigm" is emerging in the social sciences, focused upon how mobility is produced and structures contemporary societies (Sheller and Urry). We have learned that, for the most part, mobility is not something that comes naturally, but instead is the product of invention. Much of the dynamism and power of modern societies is attributable to the invention of new communication and transportation technologies and networks that extend our capacity to carry people, things, and information across greater distances with greater speed and in a greater number of ways. Indeed, industrialization, globalization, and modernity have been integrally bound up with the rise of new forms of mobility. The majority of theoretical work in this area, from Georg Simmel to John Urry, has been done in the social sciences: in sociology, anthropology, geography, science and technology studies, migration studies, and transport studies.[1] Since the natural world is commonly seen as being antithetical to the technologically driven world of modernity, and since it is normally seen as something that is rooted in place, it is understandable that nature is rarely mentioned in mobility studies. The organic world may be in constant movement, but unlike a road, bridge or railway, which are constructed to lead somewhere, we do not usually tend to see the movement as having a direction. Animals may travel, but they do not really have anywhere to go; they are just wandering. Even the great migrations of birds and animals, like the planets, seem fixed in their course. Also, as Marx remarked in the *Grundrisse*, "Nature builds no machines, no locomotives, railways, electric telegraphs, self-acting mules etc. These are products of human industry; natural material transformed into organs of the human will over nature, or of human participation in nature. They are organs of the human brain, created by the human hand; the power of knowledge, objectified" (703). Given a context in which human history defines itself in opposition to natural history, as that which is produced versus that which is given or predetermined, it is not surprising that when theorists of mobility are discussed, Charles Darwin's name goes unmentioned, and yet, more than any other writer of the nineteenth century, Darwin set nature in motion. In *On the Origin of Species* (1859), he presented an extraordinarily modern conception of the evolution of the natural world as a rich and complex story of the appearance of increasingly more sophisticated organisms that had mastered the challenges of movement.

Against Marx's confident assertion that "Nature builds no machines" should be set a passage that appears near the conclusion of *On the Origin of Species*, in which Darwin, recalling his earlier encounter with indigenous people on the voyage of the *Beagle*, urges his readers to see nature through modern eyes:

> When we no longer look at an organic being as a savage looks at a ship, as at something wholly beyond his comprehension; when we regard every production of nature as one which has had a history; when we contemplate every complex structure and instinct as the summing up of many contrivances, each useful to the possessor, nearly in the same way as when we look at any great mechanical invention as the summing up of the labour, the experience, the reason, and even the blunders of numerous workmen; when we thus view each organic being, how far more interesting, I speak from experience, will the study of natural history become. (485–6)

Historians of science display some discomfort regarding Darwin's frequent use of industrial and technological metaphors to describe the operations of nature. In an 18 June 1862 letter to Friedrich Engels, Marx himself famously observed: "It is remarkable how Darwin recognises among beasts and plants his English society with its division of labour, competition, opening up of new markets, 'inventions,' and the Malthusian 'struggle for existence'" (Marx and Engels 128). With the exception of Silvan S. Schweber, who speaks of Darwin's "'biologizing' the explanations political economy gave for the dynamics of the wealth of nations" (212), historians have largely taken these comparisons as being essentially analogies, not statements about how nature actually works, and have thus not taken them seriously enough.[2] That Darwin would liken an organism to an ocean-going vessel tells us much about his perspective on the natural world. Instead of seeing the study of nature as being antithetical to the study of technological inventions, Darwin is here suggesting that they are as much the product of a long history of inventions as a ship, a "summing up of many contrivances, each useful to the possessor" and "of the labour, the experience, the reason, and even the blunders of numerous workmen." Every organism, its form, its behaviour, even its instincts, is, for Darwin, the extraordinary result of a complex history of engineering, the product of the improvements and modifications in form and behaviour introduced by less-successful precursors. This passage is also as much about ways of seeing as it is about the nature of organic beings, for Darwin evokes the idea of the "savage mind" as a figure of the way of seeing nature that he is seeking to displace. To understand nature, one must see it through eyes that can appreciate this history of inventions, modern eyes that see a living being as the "summation" or telos of a long history of predecessors and "contrivances." It is, perhaps, not beside the point that in speaking of the technological capacities of the Fuegians, Darwin complained that "their skill in some respects may be compared to the instinct of animals; for it is not improved by experience: the canoe, their most ingenious work, poor as it is, has remained the same, as we know from Drake, for the last two hundred and fifty years" (*Journal of Researches* 216).

In 1865, in the *Proceedings of the Linnean Society*, Darwin published a meticulously researched paper on plant physiology entitled *On the Movements and Habits of Climbing Plants.* Written at the same time as he was working on insectivorous plants, the paper, later published as a monograph in 1875, recalls the vitalistic philosophy of his grandfather, Erasmus, who believed that "vegetables are in reality an inferior order of animals," especially in regard to their capacity for sensation and feeling (*Phytologia* 1). Darwin's wife Emma commented to Mary Elizabeth Lyell on 29 July 1860 that "he is treating Drosera [the insect-loving sundew] just like a living creature, and I suppose he hopes to end in proving it to be an animal" (*Correspondence,* Letter 2880). Darwin was certainly interested in the organic continuity between plants and animals, and in this regard his interests were allied with those of his grandfather. Yet his experiments were aimed less at reinforcing the analogy between plants and animals than at studying how plants moved and how different forms of mobility had evolved in plants. The research project had its origin in Darwin's discovery that the tendrils and stems of climbing plants constantly moved. Further study allowed him to hypothesize that climbing plants used this movement to explore the space around them. By modifying parts that served other purposes in other plants, they had developed advanced sensory mechanisms that allowed them to react to and to grasp whatever they touched. Thus, Darwin was able to question the conventional Aristotelian notion "that plants are distinguished from animals by not having the power of movement," and, instead, he adopted the less-dichotomous, evolutionary view that "plants acquire and display this power only when it is of some advantage to them; this being of comparatively rare occurrence" (*Climbing Plants* 117–18). When movement proved to be an advantage to plants, they developed the biological means to do so.

Darwin was not claiming that plants move quickly. With the exception of the Venus Flytrap, which captures its unsuspecting insect prey by snapping shut its vegetable jaws in a split second, to see plants move, you must watch them in slow motion. For Darwin, who was at this point largely stationary and who had established his scientific reputation by studying barnacles, seemingly the most stationary of animals, much of the pleasure of this research lay in the ingenuity with which he was able to trace incremental movements that to others were imperceptible.[3] The evolutionary advantage of climbers, he believed, lay in their having repurposed their parts in order to allow them to move. Darwin writes: "The most interesting point in the natural history of climbing plants is the various kinds of movement which they display in manifest relation to their wants. The most different organs – stems, branches, flower-peduncles, petioles, mid-ribs of the leaf and leaflets, and apparently aërial roots – all possess this power" (115). While other plants expend their energy on strengthening their roots, trunks, and stems, climbing plants had developed highly efficient and advanced mechanisms that allowed them to clamber quickly over other plants in the competition for sunlight. Basing his classification on the different mechanisms that each group had evolved in order to move, Darwin distinguished four kinds of climbers. Least interesting to him were the two groups that moved via roots or hooks, because these techniques were more mechanical, and restricted these plants to growing in dense or tangled vegetation. More intriguing were the twining plants that used their stems to seek out and wrap themselves around supports, the most exceptional of these being the tendril-bearers, which had developed specialized leaves – tendrils – that allowed them to sense whatever lay in their reach and to clasp whatever they touched. Darwin arranged these plants on an evolutionary continuum based upon their respective degrees of mobility, concluding that leaf-climbers had developed from twiners, and that the more advanced tendril-bearers had once been leaf-climbers. Climbing had not come naturally to these plants. Instead, by modifying parts originally developed for other purposes, they had developed specialized mechanisms of touch and grasping that allowed them to explore the world around them and to take advantage of the trunks and stems of other species in their quest for light. Returning to this topic more than a decade later, in *The Power of Movement in Plants* (1880), Darwin would elaborate further on this idea, asserting, through the idea of "circumnutation," that all plants have, to a greater or lesser degree, the capacity to move: "the habit of moving at certain periods is inherited both by plants and animals" (572).

On the Movements and Habits of Climbing Plants indicates the degree to which the mobility of organisms was an ongoing concern of Darwin in his later years. Yet his commitment to understanding mobility as a key aspect of evolutionary theory can be seen as an expression of a much broader commitment to a mobile conception of nature. Most seventeenth-century naturalists, in accepting the biblical account of the manner in which plants and animals had populated the earth once Noah's Ark had touched ground, assumed that plants and animals had easily travelled to their present locations from a single place on earth, Ararat; naturalists during Darwin's time, however, were far more inclined to downplay the capacity of biota to migrate, arguing instead that they had been created in situ, with natures and forms perfectly adapted to the places that they inhabited.[4] Thus, Louis Agassiz argued that "there is only one way to account for the distribution of animals as we find them, namely, to suppose that they are *autochthonoi*, that is to say, that they originated like plants, on the soil where they are found ... To each species has been assigned a limit which it has no disposition to overpass so long as it remains in a wild state" (179, 177). The sheer diversity of species, many of them highly endemic to isolated localities, the affinities among species within generas in various parts of the world, the seemingly insurmountable physical barriers separating many biological populations, the puzzling instances in which the same species could be found in very different locations separated by vast distances, or the equally curious situations in which identical climates were populated by radically different floras and faunas, all these things led many of Darwin's contemporaries, particularly Charles Lyell, Agassiz, and Alphonse de Candolle, to conclude that instead of there having been a single centre of creation, there had been many, from whence plants and animals had migrated, to the best of their abilities, to nearby locations. Nature was rooted in place, and the capacity of biota to migrate to new places was limited by geographical and physical boundaries. Faced with the contradictory and confusing complexity of issues raised by the worldwide distribution of biota, most naturalists continued to believe in the rootedness of the natural world and instead resorted to ideas of fluctuating landmasses and hypothetical land bridges (now submerged) in order to explain how biota had come to occupy their present locations.

Darwin occupied a minority position among his contemporaries in his strong commitment to the idea that the natural world was inherently mobile. During the 1830s and 1840s, the capacity of plants and animals to travel widely was an important aspect of his work, but it remained relatively untheorized. Janet Browne notes, for instance, that as early as the *Sketch of 1842*, Darwin was claiming that "species were capable of extensive migration, that organisms could freely move to occupy areas made available through topographical change" (*Secular Ark* 196). By 19 March 1845, he was a confirmed migrationist, writing that "we cannot pretend, with our present knowledge, to put any limit to the possible & even probable migration of plants" (*Correspondence,* Letter 842), and he maintained this position in *On the Origin of Species,* writing, for example, that given "the vast geographical and climatal changes which will have supervened since ancient times, almost any amount of migration is possible" (351). Still, as Browne's *Secular Ark* makes abundantly clear, there is a great deal of difference between maintaining an *abstract* belief in the capacity of plants and animals to travel and a *scientific understanding* of how they have done so. A distinctive aspect of Darwin's later evolutionary thought is that rather than assuming that mobility comes naturally to organic beings, he made it an important subject of evolutionary inquiry. In attending to the myriad ways in which organisms had found ways to travel across the globe, Darwin was at the forefront in developing modern biogeography and providing the first account of the mobility of plants and animals in concrete evolutionary terms.[5] As I hope to suggest, however, Darwin was not just interested in understanding the different ways in which plants and animals had spread across the earth. He was now also attentive to the broader evolutionary questions raised by that mobility. Why had some species developed highly sophisticated modes of travel, while others had not? Why were some species so successful in migrating and settling in new places, while others remained restricted to highly localized habitats? What role had mobility played in the history of the organic world? Was the history of nature, like human history or even the history of shipbuilding, a story of the evolution of increasingly more sophisticated forms of mobility? And why, if mobility was an evolutionary advantage, had some species relinquished higher forms of mobility for more rudimentary forms of movement? Noting, for instance, that almost half of the beetles on the island of Madeira (two hundred out of five hundred species) had lost the ability to fly, Darwin concluded that on this small island, regularly affected by strong winds that could easily blow weak-flying beetles out to sea, many of its coleoptera had learned to survive by strategically grounding themselves.[6] The flightless birds of the islands of Mauritius, Bourbon, Rodriguez, and New Zealand similarly demonstrated that organisms were capable of trading their wings for a new set of legs when circumstances changed.

Particularly after 1844, when, through discussions with Joseph Hooker, he came to see biogeography – or Geographical Distribution, as it was then called – as a "key-stone of the laws of creation," Darwin began to explore in greater detail whether there might not be a fundamental relationship between mobility and speciation itself.[7] Instead of seeing movement as simply an *indirect* cause of evolutionary change, Darwin began to explore whether mobility might not itself be a fundamental motor of transformation. This dimension of Darwin's work has been obscured, first, by a tendency among historians of science to understand evolutionary speciation as something that occurs *in time* and *in place*, but not *across space*. The deep relationship between movement and speciation in Darwin's thought is thus minimized. There is no question that much of the power of Darwin's theory lay in its capacity to see into the depths of biological time, explaining the complex relationship existing between past and present species and the historical existence of species in time. As Darwin commented in *On the Origin of Species*, "on this same view of descent with modification, all the great facts in Morphology become intelligible" (456). Equally important to Darwin, however, was the idea that every species, no matter how widespread its current range, could be traced back to a single geographical and temporal origin, "each species having been produced in one area alone, and having subsequently migrated from that area as far as its powers of migration and subsistence under past and present conditions permitted" (353). In the manuscript "Natural Selection," which immediately preceded the drafting of *On the Origin of Species*, Darwin writes that if it could be "absolutely proved that the same species has ever appeared, independently of migration on two separate points of the earth's surface: if this were proved or rendered highly probable, the whole of this volume would be useless & we should be compelled to admit the truth of the common view of absolute actual creation" (*Natural Selection* 566). This emphasis upon the idea that every species (and ultimately every genera) had a local origin placed great importance upon migration and settlement as the means by which the earth had been populated by organic beings. Thus, alongside Darwin's commitment to understanding species in terms of their descent from local origins, there lay a complementary story, which would preoccupy him in later years, of the relationship between speciation and mobility, a story of movements and migrations, departures and arrivals, settlement and displacement, as species whose evolutionary advantage lay in their capacity to move came into contact with new environments and less-mobile species. This story may have been eclipsed by the rich biogeographical work of Ernst Haeckel's *History of Creation* (1876) and Alfred Wallace's *Geographical Distribution of Animals* (1876), but it captured the imagination of an imperial age, as it suggested that the earth had been settled and resettled over vast periods of time as the descendants of ancestral races, born in local circumstances, had spread outward to occupy new territories, often changing as they did so. By stressing that the earth had been inherited by those species that had proven themselves best able to move, Darwin made mobility a central component of evolutionary thought.

Mobility and Speciation

Although the term "mobility" eludes easy definition, whether one is speaking of movement, transportation, or communication, it is nevertheless concerned with the technologies we have invented that allow us to bridge distances of various kinds. That Darwin understood evolution as a science that sought to understand speciation across distance is quite clear from a July 1847 manuscript in which he diagrammed speciation in terms of a series of dots and spaces, writing that "The affinities of organisms are represented by distance – species being called dots . by their being placed thus:

... .

(*Darwin Online*, CUL-DAR205.5.120)

It is, perhaps, understandable that in thinking about speciation, our attention is primarily focused on the "dots" that we use to mark individual species, and we tend to think of the origin of a species as being identifiable with a specific "point" in space and time. But that is not how Darwin understood speciation. Because speciation produced *both* the "dots" or species *and* the gaps between them, he did not see his work as being akin simply to "the grouping of the stars in constellations" (*Origin* 411) or to the drawing of lines between dots; instead, he sought to explain what had happened *between* these points, examining the "laws of relations between organisms separated by time & space" (*Darwin Online,* DAR 205.9:252). For Darwin, speciation was fundamentally a series of steps (in form, in time, and in space), not something that happened suddenly, but because most of these steps were now lost in time, the process by which species had diverged from their predecessors appeared as blank spaces in the fossil record. Employing the complementary concepts of *descent* and *mobility,* Darwin sought to explain what had happened in those gaps. Speciation was a process of creative gapping, as the progressive divergence or branching of species descending from a common parent increased their differentiation in form even as it allowed them to spread across space and to increase their foothold in time.

Before discussing Darwin's most detailed graphic representation of speciation in the "Tree of Life" diagram that he discusses in the "Natural Selection" chapter of *On the Origin of Species*, it will be useful, as a basis for comparison, to discuss another representation of the relationship between transmutation and mobility, that of M.C. Escher in his woodcut print *Metamorphosis II* [Figure 2.1] (Locher). Although Escher's print is not a biological or an evolutionary study of transmutation, it is akin to Darwin's "Tree of Life" inasmuch as it is both an exploration of the relationships that forms (in this case, graphic) share with each other and a study of the steps by which one form metamorphoses or transmutes into another. In a manner similar to the questions that preoccupied Darwin, who in April 1856 described the study of plant and animal distribution as a "most splendid sport ... a grand game of chess with the world for a Board" (*Correspondence*, Letter 1856), Escher seeks to show how one form can become another in space. *Metamorphosis II* is structured as both a movement of forms and a reversible narrative: if one reads it from left to right, the repeated word "metamorphose" becomes black-and-white squares, and then turns into lizards that become hexagons, a honeycomb, bees, birds, and so on. Unlike evolution, in which the relationship between form and movement is also structured by time, forms move in both directions in Escher's print, and whether one reads the print from left to right or vice versa, it is essentially a circular movement ending in "metamorphose."

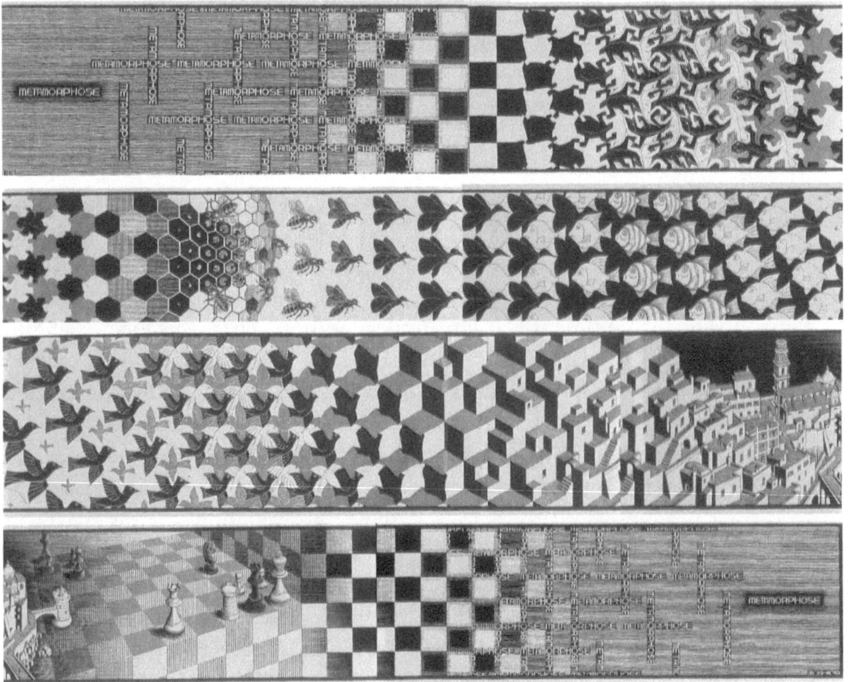

Figure 2.1 M.C. Escher, *Metamorphosis II* (1939). Reproduced by permission of The M.C. Escher Company.

Although Escher's print does not represent a history of how one form is transmuted into another, it is like evolution inasmuch as each shape or form in the print is not just an entity or "point" existing on its own, but instead is also a necessary step *from* and *toward* another form, as the squares progressively transform into honeycombs, and these eventually become bees, etc. Each variation in a form thus mediates another; each serves as a necessary step or stage in the appearance of another, because later forms incorporate elements introduced by earlier ones. Darwin saw organic forms in a similar way, for he also understood speciation as a series of *steps* whereby one variety or species mediates the appearance of another:

> Hence I look at individual differences, though of small interest to the systematist, as of high importance for us, as being the first step towards such slight varieties as

> are barely thought worth recording in works on natural history. And I look at varieties which are in any degree more distinct and permanent, as steps leading to more strongly marked and more permanent varieties; and at these latter, as leading to subspecies, and to species. (*Origin* 51–2)

Adopting Darwin's language, one might say that each form within a given sequence of forms in Escher's print, for instance, each variety of bee, bird, or fish in the "bee-to-bird" sequence or the "fish sequence," is an incipient form of a species that is no sooner realized than it is already in the process of becoming something else.

Putting aside for a moment Darwin's understanding of how this form-changing movement occurred in nature, I am interested here in the way in which Escher's print also demonstrates how changes in form are also the means by which these variations travel across the plane of the woodcut. In the same manner in which "still images" become "moving pictures" when together they form a sequence, the forms composing the bee-to-bird sequence or the fish sequence are not static, but instead their progressive differentiation – their transmutation – is the means by which they move across the print plane; they swim and fly through variation. In this sense, form is truly a step, establishing a sequence and a direction of movement. Changes of form and movement through space are thus allied. Though requiring substantial qualifications, it was this kind of movement, at least in the abstract, that I believe Darwin saw at work across the globe when he studied the life of organic forms in time.

What complicates this process and reinforces the analogy between Escher's formal experiments and Darwin's evolutionary thought is that the realization or differentiation of any given sequence of forms does not happen freely or independently of the other figures that are adjacent to it. Because Escher is working with a material medium, that is, a tiled or tessellated plane in which *all* the tiles fill the plane with no overlaps or empty spaces, any change in the boundary line defining any form automatically has an impact upon the forms that are adjacent to it. Consequently, the increasing differentiation of any given sequence of tiles is usually made in competition and at the expense of their neighbours. The transmutation and movement in a sequence of forms affects the form and movement of the other forms that border upon it. Without seeking to confuse biology and aesthetics, one might speak of the print as representing an ecology of forms, because each variation in a form struggles with the neighbouring forms within the tiled space – birds against fish, for instance – in an effort to differentiate itself fully. That is why we not only see forms coming into being, but others disappearing (i.e., becoming extinct) as they move from positive space to negative space in the tile plane. The fish located below the birds, for

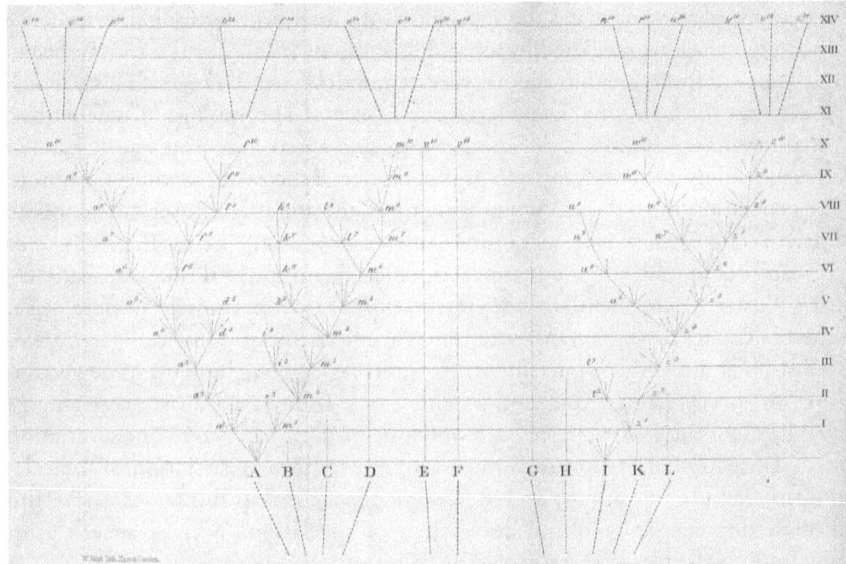

Figure 2.2 From the first edition of *On the Origin of Species*. Reproduced by permission of John van Wyhe, ed., *The Complete Work of Charles Darwin Online* (http://darwin-online.org.uk/), 2002–.

instance, disappear into the thin air that surrounds the bees. In developing this analogy I am not trying to suggest that Escher was seeking to illustrate evolutionary theory; he certainly is not making the claim that lizards descend from squares, or birds from bees. Instead, Escher was exploring the relationship between form and design, the formal constraints at play within the *closed system* of a tessellated surface, and what might be called the mathematical dimensions of the transmutation of form. The limitations of the tiled plane – like the limitations of a habitat – led him to explore the manner in which changes in form are bound up with movement (forms need space to be realized) and the ways in which the capacity of a series of forms to take shape or establish a direction – to move across a plane – affects all the forms of negative and positive space that surrounds it.

Darwin's analysis of the "Tree of Life" provides a dramatic culmination to his discussion of the Principle of Divergence in the chapter on "Natural Selection" (Figure 2.2). This diagram powerfully illustrates how the characteristic branching and clustering structure of a species taxonomy emerges in time. Like Escher's depiction of the metamorphosis of forms, Darwin's diagram also portrays speciation as a process by which the variation in an individual form is mediated by previous forms and, in turn, mediates the appearance of subsequent forms: in the sequence a^1 through a^{10}, a^2 is mediated by a^1, and it is also necessary for the appearance of a^3. Biological forms are thus the "very short and slow steps" (*Origin* 471) by which species pass through time. Every form can be seen as a beginning and a departure, a step toward the appearance of another form and a step away from another. What links variations is the transmission or inheritance of successful modifications from one generation to the next. At the same time, there are significant differences in Darwin's and Escher's representations of metamorphosis. Whereas Escher's woodcut explores graphically the manner in which forms move through space, Darwin's illustration is essentially a representation of their movement in time. Also, whereas Escher presents transmutation as a simple linear sequence, Darwin sees evolution as a divergent movement, the splitting or branching off of new variations or species from a common parent. Thus, from variation a^1 there emerge five different lines of variation, and only one of these leads to the new variation at a^2. Darwin's illustration thus provides a powerful representation of the relationship between taxonomy and the *descent* of species, but it is far less successful in depicting the geographical dimensions of speciation. Where Escher's print foregrounds the fact that the evolution of forms takes place within an ecology of competing forms produced by the fact that there is no open space, no room to move, in the tiled plane of the print without affecting another form, Darwin's diagram, filled as it is with what seems to be free and empty space, says nothing about how this branching plays out in space. And yet it is quite clear from the discussion of the principle of divergence, which this diagram was intended to illustrate, that Darwin believed that the "Tree of Life" depicted not only the branching of living forms back to the very beginnings of life but also the manner in which organisms, by transmutation, have populated the available space on the earth: "the great Tree of Life ... fills with its dead and broken branches the crust of the earth, and covers the surface with its every branching and beautiful ramifications" (130).

Like the figures in Escher's metamorphosis experiment, the living forms that Darwin portrays in the "Tree of Life" occupy space at the expense of other forms, and their ability to do so is fundamentally bound up with their success in having developed biological forms and modes of behaviour that, in Darwin's view, better fit them to take possession of places that would otherwise be occupied by others. The forms of different organisms may evolve in time, but they are fought for in space. The geographical dimensions of evolutionary speciation are explicit in Darwin's account of the manner in which the "principle of divergence" causes "differences, at first barely appreciable, steadily to increase, and the breeds to diverge in character both from each other and from their common parent" (112). Like artificial breeders, who tend toward extremes by selecting breeds that exhibit specific desired qualities while neglecting those varieties with intermediate characteristics, nature selects those aspects of an organism's form or behaviour that allow it to seize and occupy new space: "[T]he more diversified the descendants from any one species become in structure, constitution, and habits, by so much will they be better enabled to seize on many and widely diversified places in the polity of nature, and so be enabled to increase in numbers" (112). Darwin repeatedly stresses that diversification and variation are the means by which the descendants of a species gain access to new territory; "they become more diversified in structure, and are thus enabled to encroach on places occupied by other beings" (116); "the more diversified in structure the descendants from any one species can be rendered, the more places they will be enabled to seize on, and the more their modified progeny will be increased" (119). In a letter to Asa Gray written on 5 September 1857, Darwin explicitly links the branching of the "Tree of Life" to species variation and the seizure of territory that it makes possible:

> [T]he varying offspring of each species will try (only few will succeed) to seize on as many and as diverse places in the economy of nature as possible. Each new variety or species when formed will generally take the place of, and so exterminate its less well-fitted parent. This I believe to be the origin of the classification or arrangement of all organic beings at all times. These always *seem* to branch and sub-branch like a tree from a common trunk; the flourishing twigs destroying the less vigorous – the dead and lost branches rudely representing extinct genera and families. (*Correspondence*, Letter 2136)

Transmutation is thus the means by which a variety or species "take[s] the place of" others, even the place of "its less well-fitted parent." Inherently, Darwin's principle of divergence emphasizes the relationship between speciation and mobility, for it is through changes in their forms and behaviour that organisms achieve mobility, that is, that they can settle in a new place, in a world that is already occupied with competing species.

When in 1854 Darwin began to develop in detail the idea of a principle of divergence, he had plenty of evidence that variation was an essential aspect of life, because all individuals within domesticated and wild populations displayed an enormous range of differences. His major difficulty, therefore, was not in arguing for variation, but instead in explaining, as Janet Browne observes, how modifications in form could be accumulated in a manner that gave variation a direction ("Darwin's Botanical Arithmetic"). "How, then," he asked, "does the lesser difference between varieties become augmented into the greater difference between species?" (*Origin* 111). Darwin's answer, drawing upon the Malthusian idea that the fundamental imperative of every species is to maximize its numbers, was that successful variation was tested in the struggle of species for land. Since the population of any species is primarily kept in check by the availability of limited physical resources, the ultimate test of a variation in form lay in whether it made it possible for a species' descendants to increase their numbers either by occupying and better exploiting more ecological niches (or "stations") within a single area, or by extending their range through migration into new territories. The principle of divergence, in other words, was a theory of biological colonization, in which the evolutionary success of a species was determined by its ability to settle and colonize new regions. Only those changes in form – those steps – that increased a species' capacity to enter and exploit new stations or gain access to new geographic areas could be said to be profitable to a species in the long run. The degree to which new organic forms were able to depart from their ancestors and their ancestral places of origin in order to gain a greater access to new resources confirms the value of these form-driven seizures. The principle of divergence can thus be said to express a new kind of power that inheres in the capacity of a species to vary, and this power gained traction and direction through the territorial imperative to colonize new land.

Darwin was unwilling to understand the production of diversity in living forms as being entirely driven by the inherent biological capacity to vary. Instead, he argued that the diversity of organisms was also significantly a product of their struggle to seize upon and adapt themselves to the diversity of habitats that surrounded them. Thus Darwin combined a commitment to mobility with an equally important emphasis upon the power of places to shape organisms. He believed that modifications of form would enable new varieties of organisms to accommodate themselves to "vacated or 'not perfectly occupied'" niches in their environments (Browne, "Darwin's Botanical Arithmetic" 75). In the *Natural Selection* manuscript, Darwin stresses that "an unoccupied or not perfectly occupied place is an all important element in the action of natural selection" (252), and he remarks in *On the Origin of Species* that "places in the polity of nature ... can be better occupied" (108). This emphasis upon the idea that species are not perfectly adapted to the places they occupy allowed Darwin, as Dov Ospovat suggests, to understand evolution as the history of the constant improvement of species through their development of an increasingly specialized relationship to their environments. Darwin argued that a region can support a greater number of the descendants of a species "when greatly modified in different ways, in habits constitution & structure, so as to fill as many places, as possible, in the polity of nature, than when not at all or only slightly modified" (*Natural Selection* 228). In his most striking metaphor of this process, Darwin drew upon an image that first appeared in the 1838 Notebook, likening the earth to "a yielding surface, with ten thousand sharp wedges packed close together and driven inwards by incessant blows, sometimes one wedge being struck, and then another with greater force" (*Origin* 67). Here form is understood as a wedge that allows a species to drive itself into a space that is already occupied by other organic beings, often those species most closely related to it. In this far more competitive model of mobility, species change in order to move, and they move by virtue of their forms. But movement does not free them from the earth; it drives them more deeply into it. Successful variation makes greater mobility possible and allows species to make new space for themselves in a world occupied by others. That is why, when Darwin thinks about evolutionary progress or improvement, he rarely understands it in terms of advances in the morphology of a species; instead, he equates it with the capacity to range widely. Mobility is thus the biological gold standard of successful variation, the hard-won product of biological transmutations that have been tested, that is, selected, by the competition of new biological forms with others for limited space. Darwin's attempt to rethink nature in terms of the history of these forms' divergent struggle to expand the space that they inhabited set his theory against more traditional ideas about the manner in which species were rooted in nature by their perfect adaptation to place. At the same time, he was unwilling to relinquish the idea of adaptation to place because this ongoing struggle for place gave rise to the diversity of life, as biological forms fitted themselves to the new environments that they were also partly creating.[8]

There is also another reason why the "imperfect adaptation" was so important to Darwin. If species were perfectly adapted, there would be no reason for them to move or to improve, and certainly there would be no possibility for newcomers to displace less-successful varieties. Instead of seeing the natural world as being composed of a harmony of organisms that are all perfectly adapted to the places in which they are found, each remaining in its place, to cite Andrew Murray, by an "*inertia*, or instinctive regard for personal ease, which leads each creature to remain where it is while it is comfortable," Darwin's nature is composed of organic beings that are imperfect and restless, creatures that move and change as they constantly seek to improve their position in relation to others (Murray 10).[9] The many examples of alien species displacing indigenous biota led Darwin to conclude that organic beings were, at best, imperfectly adapted to the ecological niches in which they were found; their hold on these places, therefore, was not guaranteed by their inherent suitability to the places in which they were found, but was instead tenuous, reflecting the fact that a more successful competitor had not yet arrived on the scene. Thus, Darwin would take the position "that species in a state of nature are limited in their ranges by the competition of other organic beings quite as much as, or more than, by adaptation to particular climates ... the inhabitants of each country [are adapted] only in relation to the degree of perfection of their associates" (*Origin* 140, 472). Where others saw the fittedness of creatures to an ordered creation, structured by physical boundaries that kept creatures in their place, Darwin saw the traces of mobility registered in creatures whose forms indicated that they had originally come from somewhere else: a pampas woodpecker in a land where there were no trees; a terrestrial thrush, the water-ouzel, which dives and uses its wings to swim under water; upland geese with webbed feet living on dry land; and plants with hooked seeds found on islands where there were no fur-bearing animals to which they might attach themselves.

One of Darwin's biggest challenges in developing evolutionary theory was to explain how reproductive isolation, which he had recognized early on as being linked to speciation, functioned in a world where plants and animals were capable of extensive movement. In the *Essay of 1844*, Darwin stressed the need for "isolation as perfect as possible of such selected varieties; that is, the preventing their crossing with other forms" (183). Otherwise, the "races of most animals and plants, when unconfined in the same country, would tend to blend together" (71). In attempting to explain how isolation might occur in nature, Darwin looked to situations in which a small number of a species might become separated from the larger population, either through a small population colonizing a new region or in situations in which they might be cut off from the larger group – for instance, through the rise or fall of a land mass. Once this separation had occurred, Darwin thought that the descendants of a species would change as they adapted to their new circumstances or as the physical environment surrounding them changed. As Dov Ospovat has suggested, evolution at this time was primarily "a theory of organic response to environmental change" (210). This emphasis on isolation led Darwin to see islands, the very symbol of insularity, with their abundance of endemic species and their relative scarcity in the number and diversity of species and genera, as the fundamental cradles of speciation. It was on islands, Darwin believed, that incipient species, separated from their biological kindred, could respond to changing environmental factors without facing competition from other species. Ironically, the logic of isolation also led him to combine a model of colonization with an anti-migrationist stance, arguing that for speciation to occur it was necessary that there be physical barriers to check "the immigration of better adapted organisms" (*Origin* 104) that might exterminate an incipient species before it had adapted itself to an environment. Darwin was very uncomfortable with the idea that natural selection would operate best in a context in which the free movement and competition of species was held in check. So much of the intellectual thrust of *On the Origin of Species* was to retain this model while at the same time qualifying it in significant ways. Thus, he writes that no "great physical change, as of climate, or any unusual degree of isolation to check immigration, is actually necessary to produce new and unoccupied places for natural selection to fill up by modifying and improving some of the varying inhabitants" (82). Where a writer like Moritz Wagner sought to simplify Darwin's model by claiming that "the migration of organisms and their formation of colonies is ... the necessary condition of natural selection" (vii), and that "organisms that never leave their ancient area of distribution will undergo just as little change as certain other organisms to whom nature has granted a far too extensive means of transportal" (42–3), Darwin in *On the Origin of Species* was seeking to articulate a theory of speciation based, as David Kohn suggests, on "the biotic interactions of assemblages of organisms in small and uniform areas ... his attention goes to the ecology of crowding" (255).

Studying Weeds

Darwin's commitment to the principle of divergence led him to think about nature in a very different way. Instead of emphasizing the extent to which the natural world was kept in place by physical boundaries, Darwin sought to understand how speciation might occur in a world in which species and environments were constantly changing and moving, and where the fundamental factors shaping change were not just physical ones, such as changes in climate or the elevation or subsidence of land masses, but also the ever-evolving relationships among organic beings struggling with each other to extend their ranges. Like Charles Lyell, Darwin was an *actualist*, which required that he seek to interpret the past by observing the natural laws governing the organic world in the present. Consequently, rather than interpreting what was happening among the biota within colonial environments as being exceptional and anomalous, Darwin sought to align evolutionary history, as much as possible, with those forces that were producing the most radical changes in nature during his time, the most powerful of these being the mobility and competition among biota and human beings produced by European expansion and settlement of the globe. Colonial ecologies provided Darwin with an artificial laboratory wherein he could study the global migration and competition of biota. As he increasingly realized, what ultimately stood in the way of species settling elsewhere in the world was not physical barriers, but competing organisms.

Most of the changes that Darwin witnessed were due to human activities, such as travel, settlement, and biotic transfers. His challenge was to demonstrate that plants and animals could do the same under their own power, by devices of their own evolutionary making. He consequently did not take the mobility of biota for granted, but instead studied it with a view to making it an important aspect of evolutionary theory. From 1855 through 1856, he engaged in an ingenious series of experiments aimed at proving that plants and animals had many means at their disposal for travelling great distances.[10] Examining whether seeds might not be carried by ocean currents, Darwin soaked seeds in saltwater tanks for different lengths of time, and then checked their viability. "To my surprise," he writes in *On the Origin of Species*, "I found that out of 87 kinds, 64 germinated after an immersion of 28 days, and a few survived an immersion of 137 days" (358). In order to learn whether seeds might not be carried in the stomachs of birds, he grew seeds that he had found in the excrement of small birds. Forcing seeds into the bellies of dead fish that were then fed to ospreys, storks, and pelicans, he tested their germination rates once he had recovered them. Having discovered that three tablespoons of pond mud had enough seeds to grow 537 plants, Darwin postulated that seeds might also be carried by the dirty feet of migratory birds. He even weighed the dirt he found on the feet of some partridges, writing that he succeeded in removing "twenty-two grains of dry argillaceous earth from one foot of a partridge" (*Origin* 362). He also considered other forms of occasional transport: in tree roots, in flotsam, and on icebergs. Perhaps most gruesomely, he investigated whether freshwater snails could attach themselves to the feet of sleeping ducks by dangling a pair of duck's feet in an aquarium and then by counting how many immature snails clung to them. The experiments were tentative and inconclusive, but they did make it possible to counter other biologists, such as J.D. Hooker, who had been reluctant to accept the idea of long-distance plant or animal migration and instead explained plant and animal movement by having recourse to hypothetical land bridges and extended continents. In a letter to Hooker, Darwin insisted "against all the world that no man knows anything about [the] power of transoceanic power of migration" (*Correspondence*, Letter 2635).

At the same time as Darwin was examining the technologies of transportation developed by organic beings, he was also engaged in studying weeds, in the hope of answering one of the most important questions for a theorist of mobility – "why one species ranges widely and is very numerous, and why another allied species has a narrow range and is rare" (*Origin* 6). If all species began locally, why had some species been so successful in expanding their ranges, changing their forms and expanding into genera of allied species, while others had not? For Darwin, common weeds were evolutionary success stories, whose sheer numbers and extended ranges marked them as being among the most successful organisms on earth. By studying them, he hoped to discover the evolutionary relationship among mobility, dominance, and variation. Darwin's theory postulated that "all the grand leading facts of geographical distribution are explicable on the theory of migration (generally of the more dominant forms of life), together with subsequent modification and the multiplication of new forms" (408). In these studies, Darwin was seeking to understand the relationship between dominance and mobility. He set up experiments, notably a daily study of a two-foot by three-foot piece of uncultivated ground at Down, in order to learn how weeds seized unoccupied territory.

Of greater importance was his statistical discovery through "botanical arithmetic" that wide-ranging genera also produced the greatest number of new species and varieties.[11] Increases in variation and increases in mobility were strongly correlated. It seemed that, in nature, to those who had much, more was given.

Darwin used weeds to prove that the most common and widespread species and generas also produced the most variation: "Hence it is the most flourishing, or, as they may be called, the dominant species, – those which range widely over the world, are the most diffused in their own country, and are the most numerous in individuals, – which oftenest produce well-marked varieties, or, as I consider them, incipient species" (*Origin* 54). This is why Darwin increasingly saw continents, rather than islands, as the areas that produced the most powerful and dominant of life forms. Despite the relative poverty in the numbers of species to be found on isolated islands, like the Galápagos, they produced an enormous number of endemic species, found only on these islands. But these species had not proven themselves capable of extending their ranges beyond these islands. In contrast, Darwin saw those species that had originated on continents, particularly Eurasia, that great "manufactory of species" (470), as being intrinsically more capable of variation (if only through their sheer numbers) and more powerful than their insular competitors. These were species that had evolved in contexts where organisms were much more capable of movement, and thus subject to, and the products of, greater interspecies competition. These were also the species that had shown themselves capable not only of travelling to new regions, but also settling there. As Darwin reminded his readers, "we should never forget that to range widely implies not only the power of crossing barriers, but the more important power of being victorious in distant lands in the struggle for life with foreign associates" (405). It was through these means, Darwin concluded, that the North had produced those species and genera whose destiny it was to seize the world. These were "the more dominant forms, generated in the larger areas and more efficient workshops of the north" (380), and it was in this sense that Darwin would claim that "the productions of Great Britain may be said to be higher than those of New Zealand" (337); they were "advanced through natural selection and competition to a higher stage of perfection and dominating power, than the southern forms" (379). The beauty and the complex ways in which species adapt to their localities continued to fascinate Darwin, as is clear in his idea of the "entangled bank" (489–90). Yet in thinking about the direction of evolution, he increasingly believed that the future was reserved not for the unique, the rare and the fragile, but for those biota that were the natural equivalents of European settlers – those common, everyday weeds that had learned how to adapt themselves to new circumstances and had the power to settle in new places. Darwin commented: "Widely-ranging species, abounding in individuals, which have already triumphed over many competitors in their own widely-extended homes will have the best chance of seizing on new places, when they spread into new countries" (350).

In the late nineteenth century, Darwin and Wallace's work produced an idea of the history of life as the story of successive waves of Northern invaders colonizing and destroying the less-successful biota of the South. It was a biogeography suited to a people who saw themselves as a modern, specialized, competitive, and dominant race of Northerners fated to repopulate the globe with themselves and their companion plants and animals. Colonialism facilitated the world-wide migration of biota, and set immigrants against indigenous species. Darwin's nature is structured by this mobility of populations, and this struggle between native species and newcomers intent upon "beat[ing] & tak[ing] the place of the native occupants" (*Natural Selection* 235). In the ensuing struggle, "aboriginal" species seemed to be less adapted to holding their ground against the powerful onslaught of biotic intruders. The very success of these immigrants would provide Darwin with what he considered to be irrefutable evidence that colonial natures as a whole were inferior to European natures in their ability to compete for resources. In reflecting on colonial natures, he recognized the degree to which entire ecologies were being transformed, as plants and animals that had existed in their place for ages were being rapidly displaced by new ones coming from afar. Here the consequences of biotic migration could be seen being played out in fast motion. In this "great and complex battle of life" (*Origin* 80), a struggle of expansion and resistance, indigenous natures had no more claim to the spaces that they occupied than the invaders, for they, too, at some distant time in the past, had first forced their passage into these places, often by destroying their kindred and ancestors along the way. All species, at one time, had been settlers. In the modern world, where species seemed to have found the capacity to migrate across the globe, species that had been successful in the struggle against their competitors in highly isolated environments now had to struggle against a new set of highly mobile and highly competitive invaders.

Not everything, not everyone can move, however, so it is just as important to consider the people and things that are denied mobility in a society as much those that can travel. In *On the Origin of Species*, organisms that lack the ability to move through space and settle new territories maintain a precarious hold in time. They are the organisms that are categorized as being "the rare" or "the endemic," which means that they stand much closer to that other major Darwinian category, "the extinct." Darwinian evolution needs extinction, because it is the extinction of organisms that do not have the power to move that provides space in the polity of nature for those that do. Nowhere is Darwin's commitment to a modern nature, shaped by mobility, transformation, and change, more apparent than in how he speaks of species that lack the power to move. They are portrayed as being "living fossils" (*Origin* 107), stranded in time and space. At one point in *On the Origin of Species*, Darwin recalled a passage from Lyell and thought of the manner in which the Glacial period had made it possible for huge numbers of species to flow toward the equator from the north and the south. With the glacial retreat, many arctic species were left stranded, like "living drift on our mountain-summits." Here disjunct species are seen as a kind of "living drift," something passively laid down by glacial flows. Thinking about these creatures stranded in space and caught in a past time, Darwin immediately thought of savages: "The various beings thus left stranded may be compared with savage races of man, driven up and surviving in the mountain-fastnesses of almost every land, which serve as a record, full of interest to us, of the former inhabitants of the surrounding lowlands" (*Origin* 382). In Darwin's world, the only real imperative is to change so that you can move. Whether you are a human being or an alpine plant, the future is reserved for those who travel.

NOTES

1 See, for instance, Appadurai; Canzler, Kaufmann, and Kesselring; Clifford; Cresswell, *On the Move,* "Production"; Cresswell and Merriman; Cresswell and Uteng; Certeau; Giddens; Kaufman; Latour; Ohnmacht, Maksim, and Bergman; Schivelbusch; Sheller; Urry.

2 See also Young.

3 See Beer in this volume.

4 See Browne, *Secular Ark*. Browne notes that "most seventeenth-century commentators supposed that plants and animals had a marked capacity for migration; that they could establish themselves successfully in a new country and, in the case of some particularly exotic species, could even hybridize or degenerate into related forms" (10).

5 For a broader study of the understanding of movement in the early development of biogeography leading up to Darwin, see Browne, *Secular Ark*. See also Ospovat and Bowler.

6 See *Natural Selection* 291–3.

7 Charles Darwin to J.D. Hooker, 10 February 1845 (*Correspondence*, Letter 826).

8 Although one would not want to overstate the degree to which Darwin would have agreed with Richard Lewontin's view that an organism actively constructs the environment in which it is found, it is certainly the case that Darwin increasingly understood environments as being built upon the relationships of competing and co-operating organisms.

9 Murray later argues that "[t]he animal never voluntarily seeks the change which produces the development of a new species" (214).

10 For a more in-depth discussion of these activities, see Browne, *Charles Darwin: Voyaging* 516–21.

11 See Browne, "Darwin's Botanical Arithmetic."

WORKS CITED

Agassiz, Louis, and Augustus A. Gould. *Principles of Zoology: Part I, Comparative Physiology.* Boston: Gould, Kendall, and Lincoln, 1848. Print.

Appadurai, Arjun. *Modernity at Large.* Minneapolis: U of Minnesota P, 1996. Print.

Bowler, Peter J. "Geographical Distribution in the *Origin of Species.*" *The Cambridge Companion to the "Origin of Species."* Cambridge: Cambridge UP, 2009. 153–72. Print.

Browne, Janet. "Darwin's Botanical Arithmetic and the 'Principle of Divergence,' 1854–58." *Journal of the History of Biology* 13: 53–89. Print.

Browne, Janet. *The Secular Ark: Studies in the History of Biogeography.* New Haven: Yale UP, 1983. Print.

Browne, Janet. *Charles Darwin: Voyaging.* New York: Knopf, 1995. Print.

Canzler, Weert, Vincent Kaufmann, and Sven Kesselring, eds. *Tracing Mobilities: Towards a Cosmopolitan Perspective.* Aldershot: Ashgate, 2008. Print.

Certeau, Michael de. *The Practice of Everyday Life.* Trans. Stephen Rendall. Berkeley: U of California P, 1984. Print.

Clifford, James. "Traveling Cultures." *Routes: Travel and Translation in the Late Twentieth Century.* Cambridge, Mass.: Harvard UP, 1997. 17–46. Print.

Cresswell, Tim. *On the Move: Mobility in the Modern Western World.* New York: Routledge, 2006. Print.

Cresswell, Tim. "The Production of Mobilities." *New Formations* 43 (2001): 11–25. Print.

Cresswell, Tim, and Peter Merriman, eds. *Geographies of Mobilities: Practices, Spaces, Subjects.* Burlington, VT: Ashgate, 2010. Print.

Cresswell, Tim, and Tanu Priya Uteng, eds. *Gendered Mobilities.* Burlington, VT: Ashgate, 2008. Print.

Darwin, Charles. *Charles Darwin's Natural Selection; Being the Second Part of his Big Species Book Written from 1856 to 1858.* Ed. R.C. Stauffer. Cambridge: Cambridge UP, 1975. *Darwin Online.* Web. 3 September 2012.

Darwin, Charles. *The Darwin Correspondence Project.* Web. 2 September 2012.

Darwin, Charles. *Essay of 1844. The Foundations of "The Origin of Species." Two essays written in 1842 and 1844.* Ed. Francis Darwin. London: Cambridge UP, 1909. Print.

Darwin, Charles. *Journal of Researches into the Natural History and Geology of the Countries Visited During the Voyage of H.M.S. Beagle Round the World.* 2nd ed. London: John Murray, 1845. Print.

Darwin, Charles. *On the Movements and Habits of Climbing Plants.* London: Longman, Green, Longman, Roberts, and Green, and Williams and Norgate, 1865. Print.

Darwin, Charles. *On the Origin of Species: A facsimile of the 1st ed.* Introd. Ernst Mayr. Cambridge, Mass.: Harvard UP, 1964. Print.

Darwin, Charles. *The Power of Movement in Plants.* London: John Murray, 1880. Print.

Darwin, Erasmus. *Phytologia: or the Philosophy of Agriculture and Gardening.* London: J. Johnson, 1800. Print.

Giddens, Anthony. *Consequences of Modernity.* Stanford: Stanford UP, 1990. Print.

Kaufmann, Vincent. *Re-Thinking Mobility.* Aldershot: Ashgate, 2002. Print.

Kohn, David. "Darwin's *Principle of Divergence* as Internal Dialogue." *The Darwinian Heritage*, ed. David Kohn. Princeton: Princeton UP, 1985. 245–7. Print.

Latour, Bruno. *Science in Action.* Cambridge, Mass.: Harvard UP, 1987. Print.

Lewontin, R.C. *Biology as Ideology: The Doctrine of DNA.* Concord: Anansi, 1991. Print.

Locher, J.L. *The Magic of M.C. Escher.* New York: Harry N. Abrams, 2000. Print.

Marx, Karl. *Grundrisse: Foundations of the Critique of Political Economy (Rough Draft).* Translated with a Foreword by Martin Nicolaus. London: Penguin Books in association with *New Left Review*, 1973. Print.

Marx, Karl, and Friedrich Engels. *Selected Correspondence.* 2nd ed. Moscow: Progress Publishers, 1965. Print.

Murray, Andrew. *The Geographical Distribution of Mammals.* London: Day & Son, 1866. Print.

Ohnmacht, T., H. Maksim, and M.M. Bergman. *Mobilities and Inequality.* Aldershot: Ashgate, 2009. Print.

Ospovat, Dov. *The Development of Darwin's Theory: Natural History, Natural Theology and Natural Selection, 1838–1859.* Cambridge: Cambridge UP, 1981. Print.

Schivelbusch, Wolfgang. *The Railway Journey: The Industrialization of Time and Space in the Nineteenth Century.* Berkeley: U of California P, 1986. Print.

Sheller, Mimi. *Consuming the Caribbean.* New York: Routledge, 2003. Print.

Sheller, Mimi, and John Urry. "The new mobilities paradigm." *Environment and Planning A* 38 (2006): 207–26. Print.

Schweber, Silvan S. "Darwin and the Political Economists: Divergence of Character." *Journal of the History of Biology* 13 (1980): 195–289. Print.

Sulloway, Frank J. "Geographic Isolation in Darwin's Thinking: The Vicissitudes of a Crucial Idea." *Studies in History of Biology 3.* Baltimore: Johns Hopkins UP, 1979. 23–44. Print.

Urry, John. *Mobilities.* Cambridge: Polity, 2007. Print.

Young, Robert. *Darwin's Metaphor: Nature's Place in Victorian Culture.* Cambridge: Cambridge UP, 1985. Print.

Chapter Three

Darwin's ~~Ideas~~

MATTHEW ROWLINSON

I.

During 1838, the momentous year in which Darwin first formulated what was to become the theory of natural selection and also resolved to marry his cousin Emma Wedgwood, his reading included a significant engagement with the life and work of Sir Walter Scott. This engagement may have been prompted by the recent appearance of J.G. Lockhart's biography of Scott, which included substantial extracts from its subject's journal; Darwin read the first three volumes in 1838 and the remainder, except for volume 5, early in 1839 (Darwin, *Notebooks* 322, notebook C 269–70; see also his letter to Emma Wedgwood, dated 6–7 January 1839). The notebooks for 1838 include several references to the biography and also at least five to Scott's fiction – one to *Guy Mannering* (1815) and the other four to *The Antiquary* (1816). As well as observations and theoretical speculations on a variety of subjects, the notebooks refer extensively to Darwin's reading; Scott's are the only novels he mentions during 1838.

With one exception, Darwin's mentions of *The Antiquary* all refer to a single character, Elspeth Mucklebackit, whose role in the novel as a bearer of involuntary memory preoccupies Darwin in the reflections on consciousness, instinct, and heredity that he pursues throughout 1838. Here is a passage from Notebook M, in which Darwin refers to Elspeth:

> Now if a memory <<of a tune & words>> can thus lie dormant, during a whole life time, quite unconsciously of it, surely memory from one generation to another, as instincts are, is not so very wonderful. – ... Miss Cogan's memory of the tune, might be compared to birds singing, or some instinctive <or> sounds. – Miss. C memory cannot be called memory, because, she did not remembered [*sic*], it was a habitual action of thought-secreting organs, brought into play by morbid action. – Old

> Elspeth <<in Antiquary>> power of repeating poetry in her dotage is fact of same sort. (*Notebooks* 521; notebook M 7–8)

Five or six months later, in Notebook N, Darwin recalls the same character in a related context:

> Old people – (Antiquary Vol II p.77) remembering things of youth, when new ideas will not enter. is something analogous. to instinct, to the permanence of old heredetary ideas. – being lower faculty than the acquirement of new ideas. – (*Notebooks* 575; notebook N 46)

The romance-plot of *The Antiquary* involves a protagonist with a forgotten past; though she is a relatively minor character in the novel, it is Elspeth Mucklebackit who reveals the secret of his identity. She also reveals that his parents were not half-siblings, as they had been led to believe, and thus erases the taint of incest that had been the original reason for his bringing-up in secret and under an alias.[1]

Besides the narrative of the protagonist's birth, Elspeth Mucklebackit is also responsible for bringing to light memories and artefacts from the more distant past. As Darwin notes, Scott represents her as senile and frequently unaware of events taking place around her, but also as having such a powerful memory for old ballads and tales that she is described as speaking "like a prent buke" (Scott 4:250). Her connection to the past is symbolized by the yarn that it is her "habitual and mechanical occupation" to spin (4:250). As she spins, she sings, to be overheard in the last moments of her life by the antiquary of the novel's title, who transcribes her song in the service of his genealogical and historical research. It is the antiquary who eventually uses what he learns from Elspeth to work out the identity of the novel's hero and restore him to his birthright.

Elspeth appears in Scott's novel, then, as part of a sustained allegorical representation of the relation between Scotland's commercial, Protestant present – embodied in the antiquary, Jonathan Oldbuck – and its Catholic and feudal history. The novel makes an explicit figural link between the condition of Elspeth's mind and that of the material remains of this history:

> [A]uld Elspeth's like some of the ancient ruined strengths and castles that ane sees among the hills. There are mony parts of her mind that appear ... laid waste and decayed, but then there's parts that look the steever [firmer], and the stronger, and the grander, because they are rising just like to fragments amang the ruins o' the rest. (Scott 4:266)

Scott's entire conception of Elspeth's psyche is thus shaped by the internal allegory in which she features. In this passage, the unimpaired parts of her mind are figured as sublime remainders; she brings the past into the present, but in the mode of automatic repetition, whose lack of reference to its context Scott figures by making her memory not an organic faculty but a text – "a prent buke."

In Scott's novel, the preservation of historical remainders is not an organic process but a social one; Darwin's interest in his representation of Elspeth comes or appears to come without any interest in the historical meaning Scott uses it to convey. But we know that a central effect of Darwin's entire work was to trouble distinctions between the biological and the social, and in the final section of this paper I will argue that his career-long preoccupation with the topic of unconscious acts and ideas not only has affinities with Romantic historicism, but is itself a displaced form of historical argument. For much of the paper, however, our concern will be with epistemology rather than history. In the first section, we will see how his representations of automatic behaviour and unconscious memory work to stabilize the relation between the subject and the object of knowledge in the notebooks. In the second section, we will turn to the representation of the eye and the topic of vision in *On the Origin of Species* (1859) as a way into epistemological issues in that text. In the final section, however, we return to history to see how the figures of unconsciousness and blindness we have traced help explain Darwin's claims about the unconscious knowledge of natural selection to be found in the work of his scientific precursors, and ultimately to propose the existence of a historical unconscious in *On the Origin of Species* itself.

In both of the passages I quoted above from the notebooks, Darwin links the individual mind's ability to retain memories "unconsciously" with what he believes is the possibility that memories can be transmitted between individuals of different generations by inheritance. In the earlier of the two passages, Darwin supports his idea that memories can be inherited with an analogy to inherited instincts; in the latter, the two ideas are conflated so that instinct itself becomes "an old heredetary [*sic*] idea." I have discussed in a companion essay to this how Darwin held throughout his career that instincts could be formed by a process in which functional actions became habits, and then as habits were transmitted through inheritance ("Foreign Bodies: Or, How Did Darwin Invent the Symptom?"). Nothing in the theory of natural selection required this position, and, indeed, Darwin's account of instincts in *On the Origin of Species* opens by arguing that instincts are less commonly formed in this way than by accidental variation:

> As modifications of corporal structures arise from, and are increased by, use or habit, and are diminished or lost by disuse, so I do not doubt it has been with instincts. But

> I believe that the effects of habit are of quite subordinate importance to the effects of the natural selection of what may be called accidental variations of instincts. (157)

In subsequent editions of *On the Origin of Species*,[2] however, and in his later works, Darwin's emphasis on accidental variation as the origin of instincts diminished, and he became progressively more interested in understanding them as originating in voluntary behaviours. In *The Descent of Man* (1871), he describes such instincts as *degraded*: "Some intelligent actions, after being performed during several generations, become converted into instincts and are inherited, as when birds on oceanic islands learn to avoid man. These actions may then be said to be degraded in character, for they are no longer performed through reason or from experience" (88). In *The Expression of the Emotions in Man and Animals* (1872), the transformation of a functional behaviour into a habit provides the first of the three principles by which the work accounts for human and animal expressions: "Certain complex actions are of direct or indirect service under certain states of mind, or serve to relieve or gratify certain sensations, desires, etc.; and when the same state of mind is induced, however feebly, there is a tendency through the force of habit and association for the same movements to be performed, though they may not be of the least use" (34). Such habitual actions constitute the *expression* of a state of mind associated with them. Darwin goes on to treat them, as he does all habits, as heritable.

The growth during Darwin's career of his interest in behaviours that with the passage of time have lost their function derives from his increasingly explicit rejection of theology. Believers in the independent creation of each species, like William Paley or Georges Cuvier, described a world in which every trait of every organism is adapted by the creator to the organism's conditions of existence. Darwin's argument for descent with modification relies for evidence on traits that are maladapted or useless and thus suggest descent from earlier forms in which they served a purpose. In *On the Origin of Species* Darwin thus compares rudimentary organs, which have not been much affected by natural selection, to "letters in a word, still retained in the spelling, but become useless in the pronunciation, but which still serve as a clue in seeking for its derivation" (335). This is the rationale that leads Darwin to the study of involuntary expressions of emotion, which, unlike contemporary psychologists, he viewed as having no adaptive value. He views them, therefore, above all as evidence of biological kinship. The erection of the hair as a response to fright in human beings has no function, but gives evidence of our kinship to other animals – such as cats – in which the puffing-up of the fur under threat serves to terrify potential predators or adversaries. The blush as an expression of shame serves similarly, not to demonstrate human kinship with other animals, but as evidence of the shared humanity of different races of human beings.

Instincts and other kinds of involuntary behaviour, when they have been de-functioned, thus play the same role in Darwin's argument as residual and rudimentary organic structures, attesting to the descent and affinities of the organisms they affect. We have seen, however, that from Darwin's earliest notebook entries on evolution to the 1870s, in his writing on involuntary behaviours he consistently views them as the remainders of "intelligent actions," as he puts it in *The Descent of Man* (88), or as "hereditary ideas," to use the term of the notebooks. When he takes this view of involuntary behaviours, Darwin seems to engage in a form of projection, by which a trait that enables the *naturalist* to understand an organism's past is treated as an embodied memory of the past belonging to the *organism itself.* Hence Darwin's view in the notebooks that the heritability of instincts is evidence for a theory of inherited memory, and his personification of the useless trait that has outlasted its function as an old person, like Elspeth Mucklebackit, who has lived long enough to become an anachronism.

My claim that Darwin engages in projection when he represents involuntary behaviour as a form of memory is, however, complicated by that representation's internal contradictions. He introduces one of his references to Elspeth Mucklebackit by discussing the case of a woman in her "second childhood" whose recall of the songs of her youth he likens to the instinctive song of birds: "Miss C. memory cannot be called memory, because she did not remembered [*sic*], it was a habitual action of thought-secreting organ, brought into play by morbid action" (*Notebooks* 524, 521; notebook M 21, 8). Miss Cogan's memory that "cannot be called memory," like Elspeth's capacity to repeat old ballads, is thus an automatic behaviour – an "habitual action" of the mind – that *takes the place* of what had once been an effect of conscious intention. If we begin with the suggestion that Darwin's idea of instincts and automatic behaviours as a form of memory arises from a projection of his own scientific knowledge onto the object of that knowledge, then we need to add that what is projected onto the object is more precisely its own difference from the subject. That is to say, the difference between the scientist who knows the past and the organism that unknowingly provides the scientist's evidence is transformed into a contradiction within the latter – a "memory [that] cannot be called memory."

In the passages we have been discussing, an epistemological difference is figured as a temporal or even historical one. The need for this figuration arises because in Darwin's theory of descent, the subject and the object are, in fact, the same. In the absence of a transcendental object *outside* nature whose purposes or ideas natural history can understand itself as uncovering, the field comes to be characterized by self-reflexivity. As the distance between the subject and object of knowledge collapses, the difference between knowing and not knowing is obscured. Hence

Darwin's preoccupation in the notebooks with characters formed by a history of which, with the passage of time, they have become unconscious.

In the next section of this essay, we will see how in *On the Origin of Species* Darwin describes natural selection as a principle that, as it forms species, does so in ways that conceal its own operation. The result is that, in Darwin's dialectical account, even his predecessors' failure to recognize natural selection becomes evidence for his case. To close this section, however, I will consider some further passages on instinct and involuntary action from the notebooks in which, as in those we have already discussed, besides serving as evidence for the theory of descent with modification, they also trigger Darwin's self-reflexive preoccupation with his own position as theorist. In a passage I have already cited, he characterizes the ability of old age to remember "things of youth, when new ideas will not enter," as analogous "to instinct, to the presence of heredetary ideas," and then observes that this is a "lower faculty than the acquirement of new ideas" (*Notebooks* 575; notebook N 46). Throughout his work, Darwin understands the development of intelligence as entailing the gradual loss of instincts and their replacement by learned and voluntary behaviours. This he views, with some qualification, as a form of progress: "We must believe, that it requires a far higher & far more complicated organization to learn Greek, than to have it handed down as an instinct" (576; notebook N 48). Though both intelligence and instinct are modes of adaptation, they differ in that one is a form of memory – of "heredetary ideas" – while the other comes into being as an erasure of the past. Paradoxically, the very faculty by which Darwin discovers human descent – the faculty that acquires new ideas – comes into existence only where evidence for that descent has vanished:

> Man having some instincts of revenge <<& anger>>, which experience shows it must for his happiness to check ... nor is it odd he should have had them. – with lesser intellect they might be necessary and no doubt preservative, & are now, like all other structures slowly vanishing – the mind of man is no more perfect, than instincts of animals to all & changing contingencies, or bodies of either. (549–50; notebook M 122–3)

Here, in the form of conflict between human intellect and residual instincts, we find another version of the antithesis between consciousness and involuntary action with which we began. The antithesis between the subject who knows the past and the object-body in which the past is materialized is re-enacted and internalized in Darwin's concept of human nature as such.

In the above passage, as in many others throughout his work, Darwin denies that there is any absolute scale of development on which species can be ranked. The human mind is not more perfect than the instincts that direct animal behaviour; each is a response to different contingencies. As we have seen, however, Darwin in other passages asserts without apparent reservation that it requires a "far higher" mental organization to *learn* skills like language, as human beings must, than to inherit them as instincts, as birds, in Darwin's view, inherit the ability to sing. Versions of these irreconcilable positions coexist throughout Darwin's evolutionary writings, and constitute a major crux in their interpretation.[3] For his own framing of the problem, we turn to a final passage from the notebooks:

> It is absurd to talk of one animal being higher than another. – We consider those where the [cerebral structure] most developed, as highest. – A bee
> [intellectual faculties]
>
> doubtless would where the instincts were (189; notebook B 74).[4]

With some degree of irony, Darwin here represents speciation as determining consciousness as well as structure. Human beings' intellects and the instincts of bees are equally adapted to their respective conditions of existence; their different adaptations, however, produce a difference in consciousness, by which both bees and human beings make their specific organization into a universal standard. Even after Darwin develops the theory of natural selection to account for adaptation, as we shall see in *On the Origin of Species*, he consistently represents the power that produces useful adaptations as also producing a kind of blindness or misrecognition. Here, in what amounts to a miniature beast fable, Darwin rebukes the anthropocentrism that makes a specifically human faculty into a universal standard by arguing that, given their different faculties, bees would have a different standard. The bees' consciousness in this argument is surely something of a heuristic fiction – but then so is that of human beings themselves, even though Darwin identifies it as his own with an underlined "we." For in the first sentence of the note, Darwin flatly contradicts what he says in the second that "we" consider to be true. The whole point of the analogy between human beings and bees is to show that the error it denounces is not just an absurd view held by some people, but is rather proper to humans as a species. As a human being denouncing a characteristic human error, Darwin occupies an ambiguous position. At once inside and outside an illusion, his position in this note is divided by a contradiction homologous with others we have discussed in this section: on the one hand, a position of blindness that he defines by analogy to the blindness to its own motives of a creature acting by instinct; on the other hand, a position of knowledge, somehow located outside the history of material bodies and dispositions that limits the first. The problem of this difference within the Darwinian observer remains in *On the Origin of Species*, no less because in that work the development of human beings is almost excluded from explicit consideration.

II.

Because of the absence of a transcendental or theological ground for natural selection, I have claimed, Darwin's theory is necessarily self-referential. In his writing on the subject, in consequence, it is often difficult to distinguish between scientific claims that refer to the natural world and epistemological or historical claims regarding the conditions under which that world is available to knowledge. In this section, we will examine the dialectical treatment of vision and images in *On the Origin of Species*, a text in which, both as an object and as a medium of perception, the eye is a stumbling-block to the argument. Paradoxically, as we will see, it is the eye's perfection that most conceals the process of modification by which Darwin argues it was formed.

The apparent perfection of some instincts and structures in animals and the difficulty of understanding how they could have come into being by a process of gradual modification is one of the major obstacles to his theory that Darwin discusses in *Origin*. Of such structures, the one on which he spends the most time is the eye:[5]

> To suppose that the eye, with all its inimitable contrivances ... could have been formed by natural selection, seems, I freely confess, absurd in the highest possible degree. Yet reason tells me, that if numerous gradations from a perfect and complex eye to one very imperfect and simple, each grade being useful to its possessor, can be shown to exist; if, further, the eye does vary ever so slightly, and the variations be inherited, which is certainly the case; and if any variation or modification in the organ be useful to an animal under changing conditions of life, then the difficulty of believing that a perfect and complex eye could be formed by natural selection, though insuperable to our imagination, can hardly be considered real. (140)

In this passage, the principal obstacle to recognizing the action of natural selection is the imagination. This idea appears elsewhere in *Origin*, and also recurs at the close of the discussion of the eye, where Darwin writes of anyone who has followed his argument that "his reason ought to conquer his imagination; though I have felt the difficulty far too keenly to be surprised at any degree of hesitation in extending the principle of natural selection to such startling lengths" (141). This cluster of references to the imagination as incapable of grasping natural selection around Darwin's discussion of the eye suggests that he conceives imagination primarily as a faculty of visualization. The logic of natural selection can be grasped by reason, but its operation is not amenable to visual representation. The eye is thus an obstacle to Darwin's argument in a double sense: not only does its perfection make it difficult to understand how it could have been gradually evolved, but it also produces a picture of the world in which evolution remains hidden.

The eye's reliability is in fact called into question throughout *Origin*. The dialectical form of Darwin's argument runs: the better the eye, the more deceptive. On the one hand, he founds his argument for the power of natural selection on the demonstrated possibility of modifying species under cultivation by selective breeding: "The great power of this principle of selection is not hypothetical. It is certain that several of our eminent breeders have, even within a single lifetime, modified to a large extent some breeds of cattle and sheep" (27). The principles followed by breeders provide a model for Darwin's understanding of natural selection, moreover, because they principally operate not by crossing different breeds, nor by inbreeding, but by selecting for variations in a particular trait. To succeed in this requires a trained eye:

> If selection consisted merely in separating some very distinct variety, and breeding from it, the principle would be ... obvious ... ; but its importance consists in the great effect produced by the accumulation in one direction, during successive generations, of differences absolutely inappreciable by an uneducated eye – differences which I for one have vainly attempted to appreciate. Not a man in a thousand has accuracy of eye and judgment sufficient to become an eminent breeder. (27)

The selective power of the breeder's eye thus provides Darwin with a paradigm for the power of natural selection itself as his argument unfolds in the opening chapters of *Origin*.

On the other hand, the very accuracy of the breeder's eye, its training in the recognition of differences, prevents it from recognizing the similarities among the breeds it works on as evidence of kinship:

> [A]ll the breeders of the various domestic animals and the cultivators of plants, with whom I have ever conversed, or whose treatises I have read, are firmly convinced that the several breeds to which each has attended, are descended from so many aboriginal species. Ask, as I have asked, a celebrated raiser of Hereford cattle, whether his cattle might not have descended from long-horns, and he will laugh you to scorn. I have never met a pigeon, or poultry, or duck, or rabbit fancier, who was not fully convinced that each main breed was descended from a distinct species ... Innumerable other examples could be given. The explanation, I think, is simple: from long-continued study they are strongly impressed with the differences between the several races; and though they well know that each race varies slightly, for they win their prizes by selecting such slight differences, yet they ignore all general arguments, and refuse to sum up in their minds slight differences accumulated during many successive generations. (25)

Darwin's point here is to urge modesty on readers who, unlike the expert breeder, recognize that different domestic varieties can descend from a single species – but yet deny the possibility of the transmutation of species. But there is a remarkable dialectical irony in Darwin's using as his negative example of blindness to transmutation the very authorities he also uses to show its possibility.

Given this irony and the idea elsewhere in the book that the visual image is an obstacle to understanding that reason must conquer, it is not surprising that *On the Origin of Species* is the most sparsely illustrated of Darwin's books. It contains only one figure, the schema of divergent evolution that Darwin termed a "great Tree of Life" (see Figure 2.2). Nor should we be surprised, given the equivocal meanings of the eye and the visual image throughout *On the Origin of Species*, that the book gives its single figure several divergent explanations.

As Darwin emphasizes, the figure of the tree was not unique to his work: "The affinities of all the beings of the same class have sometimes been represented by a great tree. I believe the simile largely represents the truth" (59). Darwin reworks a figure from the existing literature because what he means it to illustrate is, in part, the current state of knowledge:

> It is a truly wonderful fact – the wonder of which we are apt to overlook from familiarity – that all animals and all plants throughout all time and space should be related to each other in group subordinate to group, in the manner which we everywhere behold – namely, varieties of the same species most closely related together, species of the same genus less closely and unequally related together, forming sections and sub-genera, related in different degrees, forming sub-families, families, orders, subclasses, and classes. (99)

Darwin's figure thus illustrates in the first instance a taxonomy characterized by varying degrees of affinity between organisms of the same class. To the extent that it represents the state of knowledge before Darwin wrote,[6] it represents abstractions: relations of affinity that could be understood as arising from a concept or transcendental schema.[7] Darwin's theory gives these affinities a material and historical existence by recognizing them as relations of biological kinship. In his theory, the subordination of group to group refers literally to a chronological and genealogical sequence of events, and not to a purely conceptual subordination.

Given that Darwin's argument here aims to transform a conceptual schema into a historical one, however, his own version of the tree is strikingly abstract and lacking in historical referents. Darwin does not, for instance, choose to speculate about affinities between members of any actual class of organism. The eleven letters at the base of his diagram may represent any eleven species whatsoever having varying degrees of affinity with each other, as may all of the letters designating their descendants. Later in *Origin*, moreover, Darwin argues that the relations among genera have the same formal properties as those among species, and illustrates his argument by referring back to the figure from earlier in the book, suggesting now that the eleven letters from A to L be thought of as referring to genera rather than to species. As this retrospective revision of his earlier explication shows, Darwin understands the relations of affinity represented by his figure as formally similar at different scales; and since these relations are determined by the unfolding of kinship relations in time, the fact that the figure can be viewed at different scales means that it can be viewed as representing different periods of time. Darwin is explicit about this: "The intervals between the horizontal lines in the diagram, may represent each a thousand generations, but it would have been better if each had represented ten thousand generations" (91). Darwin's sense of constraint in the interpretation of his own figure here is odd, and is belied a few pages later when he expands its temporal scale of reference: "In the diagram, each horizontal line has hitherto been supposed to represent a thousand generations, but each may represent a million or a hundred million generations, and likewise a section of the successive strata of the earths [*sic*] crust including extinct remains" (96). On the one hand, scaling up the period of time the figure encompasses emphasizes its nature as a representation of a concept rather than of particular historical events and organisms. On the other hand, the analogy between the figure's horizontal lines and those of geological strata in the earth's crust reattaches the figure to the material world, implying that its spatial articulation of time has a geological prototype and that the tree itself might actually picture the spatial relations among fossils found at different depths.

Recalling the scepticism toward images that we have noted elsewhere in *Origin*, however, we need to wonder whether here, too, the image, the figure of the tree in its pictorial dimension, might be an obstacle to be overcome in grasping what for Darwin is the reality of his theory. And indeed, a few paragraphs before suggesting that the horizontal lines in his figure "may represent" geological strata, Darwin tells us that they are "imaginary, and might have been inserted anywhere, after intervals long enough to have allowed the accumulation of sufficient variation" (92). Even here, Darwin downplays the purely virtual and heuristic character of his figure's representation of temporal articulation. In the figure, the horizontal lines mark the moments at which variation occurs in the different lineages it traces. But in fact, Darwin's theory clearly represents variation as occurring *continuously*. Not only could the horizontal lines that locate the moment of variation appear "anywhere"; in strict logic they should be *everywhere*, covering the entire figure. To produce a schema representing natural selection's operation through time, Darwin must exclude much of that operation from representation, transforming a continuous process into a series of discontinuous events. This transformation is emphasized by his representation of the appearance of new variations as occurring not only *punctually* but also *simultaneously* on every branch of the tree. Nothing in his theory suggests that new variations appear in this coordinated way; that his figure shows them doing so is an effect of its representation of natural selection as an event transecting an otherwise empty homogeneous time.

One way to understand the exclusions in Darwin's figure is by considering its representation of intermediate forms between existing classes. On one axis, the figure denies the extinction of these forms, while on the other their extinction is presupposed. On its vertical axis, the whole point of the figure is to represent simultaneously the progenitors of a class of organism and their generations of modified successors. In this simultaneous representation, the extinction of earlier forms is ignored. On its horizontal axis, however, the blank spaces between the figure's lines and points are constituted by extinction. When he returns to a new discussion of the figure in his chapter on "Mutual Affinities of Organic Beings," Darwin shows how extinction "has played an important part in defining and widening the intervals between the several groups in each class." Though extinction has only separated groups originating from shared descent, he writes, nonetheless, "if every form which has ever lived on this earth were suddenly to reappear ... it would be quite impossible to give definitions by which each group could be distinguished from all other groups, as all would blend together by steps as fine as those between existing varieties" (317–18). As an array of points and lines, the tree of life represents Darwin's concept of how species diverge as they evolve from a common origin. It appears, though, on a blank ground that, in the form of space between the figure's lines, represents what it excludes: the extinction of the intermediate forms without which species would not exist.

The dialectical relation between what is visible and what is invisible is a recurrent topic in *On the Origin of Species,* and structures some of the work's best-known passages:

> We behold the face of nature bright with gladness, we often see superabundance of food; we do not see, or we forget that the birds which we see idly singing round us mostly live on insects or seeds, and are thus constantly destroying life; or we forget how largely these songsters, or their eggs, or their nestlings, are destroyed by birds and beasts of prey; we do not always bear in mind, that though food may now be superabundant, it is not so at all seasons of each recurring year. (50–1)

There is nothing in this passage's description of the natural world that explains or requires its pervasive distinction between what can be represented by sight or memory and what remains invisible or forgotten. The distinction belongs not to the object but to the subjects of knowledge, the "we" who see only the "face" of nature while Darwin's text works to make us aware of its relation to other parts that remain unseen. This division within the knowledge conveyed by Darwin's work also structures the tree of life figure, in which it is again allegorized by a difference between what is visible and what is invisible. Here, there is no external object of representation; as I have noted, Darwin's figure does not refer to any particular period of time or class of living things. The figure rather represents a concept, one whose internal divisions it stages allegorically by the negative but mutually constitutive relation of figure to ground. This staging, though, is only one instance of a general pattern in *Origin* by which the power of vision or of visual representation is repeatedly linked to a failure to see or to an exclusion from representation. In its broadest meaning, this pattern registers in the form of Darwin's work the impossibility of any point of view affording a unified or totalizing concept of natural selection.

Where or when might such a point of view be located? The point of Darwin's illustration is to represent the emergence of new classes of organism by the selection over time of favourable variations. To do so, he presents a figure in which the process of variation is shown as occurring in a sequence of presents – a linear series of *nows*, graphically rendered by horizontal lines linking events supposed to occur at the same moment. As Darwin allows, these presents are imaginary. This would be as true for the figure's topmost line as for any other, even though this one does double duty as both an element in the diagram and as its external border. Not only are the presents represented by the figure imaginary, so too is the present it occupies. Though in all other respects the diagram represents a schema rather than any specific dates or temporal intervals, its upper line marks its intersection with the historical present. Darwin says as much in chapter 13 when he writes that "the present day" is represented by "the uppermost horizontal line" (310). In this one instance where it represents a specifiable historical moment, though, the diagram radically limits the amount of information it conveys. At its uppermost limit, the schematic representation of the secular history of variation and extinction disappears, and the diagram is left to represent only the variously related types of organism that that history has produced, variation and extinction themselves being impossible to know as events with a specific historical location.

III.

In the first section of this essay, I read passages from Darwin's notebooks to argue that his longstanding preoccupation with the topics of unconscious memory and automatism is the temporal projection of an epistemological problem. The problem is to distinguish between thinking about natural selection and unconsciously undergoing its effects, and it arises because the same body does both.

To stabilize the distinction between his own consciousness and the objects about which he thinks, Darwin tends to present the latter as bereft of consciousness, like old Elspeth in Scott's *The Antiquary*. Nonetheless, the difficulty of distinguishing between the subject of knowledge and its object – and more generally between knowing and not knowing – remains acute in *On the Origin of Species*. The puzzle of Darwin's attribution of agency to nature and natural selection has been extensively discussed.[8] An especially strange aspect of that puzzle is the way *Origin* projects onto nature responsibility, sometimes for Darwin's discovery of natural selection, but more often for his precursors' failure to discover it. In the book's conclusion, as Darwin reviews the evidence he has marshalled, he writes that "Nature may be said to have taken pains to reveal, by rudimentary organs and by homologous structures, her scheme of modification, which it seems that we wilfully will not understand" (353). Homologous and rudimentary structures, however, show modification by contrast; they demonstrate the natural "scheme of modification" by showing the affinity of organisms that have in other respects been rendered dissimilar by natural selection. Nature could thus with as much reason be seen to conceal "her" scheme as to reveal it – and, indeed, Darwin suggests as much in his chapter on the imperfection of the geological record, in which he argues that because new species are least likely to be formed in geological eras of subsidence, when fossil beds are laid down, nature "may almost be said to have guarded against the frequent discovery of her transitional or linking forms" (216). Underlying the apparent contradiction between these two passages is the dialectical logic of Darwin's theory of development, which aims to account both for positive data, such as homologous traits in related species, and also for negative, such as the limitations of the geological record. The contradiction only arises when Darwin projects his own theory, as a product of consciousness, onto a personified nature. The result is that the data the theory explains, instead of being its cause, are imagined as its effects. Nature, that is to say, is imagined as *already knowing* what Darwin knows, and as acting with intent either to reveal or to conceal her knowledge.

The dialectical structure of Darwin's thought does more than appear in *On the Origin of Species* as projected onto a personified nature; it pervades the work and makes it extraordinarily difficult to characterize the relation of Darwin's argument to what one might term his sources and authorities. As we saw in the last section, for instance, Darwin's information regarding the plasticity of animal and plant varieties under selection came largely from breeders and cultivators. An important part of the drama of his work arises from the confrontation it stages between establishment science and the unofficial knowledge Darwin obtained from these sources. As well as providing his authorities on the subject of artificial selection, though, we have seen that the breeders and cultivators he cites also serve Darwin as examples of how knowledge can coexist with and depend on blindness. And, finally, notwithstanding the importance of the encounter between official and unofficial knowledges to Darwin's argument, we recall that the effectiveness of artificial selection does not appear in *Origin* principally as evidence that Darwin's scientific peers have something to learn from animal breeders and horticulturalists, but to provide an analogy for what he goes on to say about the effectiveness of natural selection. This analogy, to which we will return below, thus provides yet another example of how *Origin* tends to represent objects of knowledge, like nature, as doubles of the subjects who know them.

In the first part of this essay, I argued that Darwin's notebooks invoke the figure of unconsciousness in order to stabilize these distinctions – between knowing and not knowing; between subject and object – that his theory threatens to collapse. The term reappears, though with significantly different associations and meanings, in *On the Origin of Species*. In the later parts of the work, which I will consider first, Darwin uses it to characterize not material evidence of development unknowingly conveyed from the past, but another kind of evidence for his theory whose bearers also do not know it as such. This evidence appears in the work of Darwin's precursors in natural history. In his chapters on morphology, Darwin returns to his earlier discussion of taxonomy, observing of "the grand fact in natural history of the subordination of group under group" that from "familiarity" it "does not sufficiently strike us" (304). His task here is thus not to present new knowledge, but to "strike" his readers with knowledge they already possess. This task Darwin describes elsewhere as one of taking ideas upon which naturalists already unconsciously base their work and raising them to consciousness. Thus he argues that the "rules and guides" for classification followed by the "best systematists" have led them "unconsciously" to use the "element of descent ... in grouping species under genera, and genera under higher groups" (313). Or he writes that in defining principles of classification, such as that which disregards the functional value of a trait for classification but rather considers whether it appears unchanged in a great number of forms or in invariable correlation with another trait, naturalists have shown themselves to be "unconsciously seeking" a genealogical classification, "not some unknown plan of creation, or the enunciation of general propositions" (308–9).

Darwin's argument in *Origin* thus concerns not only the history of species, but also the history of thought. He is able to use existing ideas about taxonomy as evidence for his theory because he views these ideas not as mere errors but as unconscious truths, presented in the inverted form that Marx characterized as ideological. When Darwin describes his precursors as "unconsciously seeking" a genealogical classification, he assumes that a genealogy showing the development of species in secular time is the reality of which the idea of "an unknown plan of creation" appears as a distorted representation. This pattern is familiar to us from Darwin's Tree of Life figure, which adapts a schema designed to express purely conceptual relationships and shows that, when properly interpreted, it actually represents a historical process. The manner of interpreting his precursors that Darwin adopts in these examples was, moreover, not limited to their ideal representations of relations between different forms: we also find it exemplified in his response to Richard Owen's concept of the vertebrate archetype. Owen's main field was vertebrate morphology, and he was the preeminent naturalist in Britain during the 1840s and 50s, when he and Darwin were on cordial terms. His major work during this period was on the homologies between different vertebrate skeletons; this work led him to postulate what he termed the vertebrate archetype, a basic skeletal plan of which he held that all existing and fossil vertebrates were variations.[9] Owen followed Cuvier in accepting the fossil evidence of extinction, and also in believing that places in the natural order vacated by extinct species were filled by successive new creations; like his peers in the early Victorian scientific establishment, such as William Whewell and Charles Lyell, however, he rejected absolutely the idea that new species could come into being by descent with modification or by any other mechanism of transmutation. His vertebrate archetype thus had no material historical existence. Rather, its recurrence as the basic pattern of which every historical vertebrate skeleton is a variation demonstrates, for Owen, the existence of a divine mind possessing foreknowledge of each of the variations that the archetype has made possible (Owen 86).

Owen presented his final version of the archetype in lectures published in 1849 as *On the Nature of Limbs*. In his copy of this work, Darwin wrote what seems like an admiring note: "I look at Owen's Archetype as more than ideal, as a real representation as far as the most consummate skill and loftiest generalization can represent the parent form of the Vertebrata" (qtd. in Ospovat 146). In light of Darwin's theory, Owen's "ideal" becomes "real," in the same way that in *On the Origin of Species* the traditional figure of the tree of life becomes a "simile" that "largely speaks the truth" (99) and the metaphorical language naturalists use in referring to the skull "as formed of metamorphosed vertebrae" turns out to apply "literally" (322–3).

Darwin's habit of finding in the work of earlier naturalists unconscious representations of the historical reality described by his theory can be understood in part as evidence of real intellectual indebtedness and affinity.[10] As it appears in *On the Origin of Species*, it has also been read as an attempt to disarm criticism by minimizing the break Darwin's theory makes with existing work in the field. Owen, it should be said, was not disarmed; his sneering and obscurantist review of *Origin* was among the most damaging it received, and it ended at a stroke the two men's friendship.[11] Nor, in my view, was Owen wrong to see Darwin's work as making a break with his; the distinction between the two is at once as fundamental and as difficult to characterize as that between Marx and Ricardo.[12]

My aim here is not to address this problem in intellectual history, but rather to read a logic internal to Darwin's work. To the intellectual-historical problem, however, my reading contributes evidence that the difficulty of Darwin's relation to his precursors is only one instance of a more general difficulty in his thinking of distinguishing between knowledge and non-knowledge of species transmutation. The result is that the origin of descent with modification as a theory is as impossible to specify as the origin of a species, or as the other events to which, using terms such as adaptation and selection, Darwin's theory refers.

We began by arguing that the fascination Darwin reveals in his notebooks with unconscious memory and inherited ideas was the historical projection of an epistemological problem, the problem being that of establishing a relation between the subject who recognizes natural selection and the one who undergoes it. Instincts, Darwin believes, appear where reason once was; this belief makes it possible to represent the task of reason in the present as one of restoring a version of itself that has been lost in the past. Nothing in Darwin's theory requires him to understand instincts as the inherited remainders of ideas that have become unconscious. That this understanding might be overdetermined by the broader structure of Darwin's thought is suggested by its recurrence in inverted form as a way of representing his relation to his precursors: as well as reading evolutionary remainders like instincts as unconscious ideas, Darwin also reads ideas as unconscious representations of evolutionary remainders. In both cases, the theory of descent is understood as a raising to consciousness of what had been unconscious.

Darwin's theory itself takes a rarely noted detour through the concept of unconsciousness. As is well known, Darwin represents natural selection, and seems to some extent to have come to understand it, by analogy with the practice of scientific breeding, which he termed artificial selection. In between his discussions of artificial and natural selection in *Origin*, though, Darwin interposes as a third term what he calls *unconscious* selection (29–33). Unconscious selection, like artificial selection, is effected by human agency; the two kinds of selection differ, however, in that unconscious selection has no intended result and, indeed, occurs without any intention at all:

> At the present time, eminent breeders try by methodical selection, with a distinct object in view, to make a new breed or sub-breed ... But, for our purpose, a kind of Selection, which may be called Unconscious, and which results from every one trying to possess and breed from the best individual animals, is more important. Thus, a man who intends keeping pointers naturally tries to get as good dogs as he can, and afterwards breeds from his own best dogs, but he has no wish or expectation of

> permanently altering the breed. Nonetheless, I cannot doubt that this process, continued during centuries, would improve and modify any breed ... Slow and insensible changes of this kind could never be recognized unless actual measurements or careful drawings of the breeds in question had been made long ago, which might serve for comparison. (29)

In unconscious selection, then, the unconsciousness refers less to the act of selection than to its effects over time. Darwin illustrates this with historical examples. The first shows how unconscious selection can operate without the kind of future object envisioned by breeders who practice artificial selection:

> The pear, though cultivated in classical times, appears, from Pliny's description, to have been a fruit of very inferior quality. I have seen great surprise expressed in horticultural works at the wonderful skill of gardeners, in having produced such splendid results from such poor materials; but the art, I cannot doubt, has been simple, and, as far as the final result is concerned, has been followed almost unconsciously ... But the gardeners of the classical period, who cultivated the best pear they could procure, never thought what splendid fruit we should eat; though we owe our excellent fruit, in some small degree, to their having naturally chosen and preserved the best varieties they could anywhere find. (31)

A second historical example follows, giving evidence that unconscious selection has proceeded in the past, while remaining unrecognized even in retrospect:

> A large amount of change in our cultivated plants, thus slowly and unconsciously accumulated, explains, as I believe, the well-known fact, that in a vast number of cases we cannot recognize, and therefore do not know, the wild parent-stocks of the plants which have been longest cultivated in our flower and kitchen gardens. If it has taken centuries or thousands of years to improve or modify most of our plants up to their present standard of usefulness to man, we can understand how it is that neither Australia, the Cape of Good Hope, nor any other region inhabited by quite uncivilized man, has afforded us a single plant, worth culture. (31)

In these remarkable passages Darwin resembles Wordsworth when, in poems like "Tintern Abbey" and "Michael," he shows us a nature haunted by the remains of culture and history. In the natural world around them, Darwin tells his readers, they see without recognizing the wild parent-forms that have been transformed to produce culture, by a historical process that itself remains unrecognized. Even the antithesis between nature and culture on which this idea rests, though, is deconstructed in the next paragraph, which argues that Europeans think of nature as different at home and in the antipodes because they do not recognize that in Europe nature is already a culture that does not know itself as such.

When he shows how, by unconscious selection, human agency can modify species without intending to do so and without recognizing its work in retrospect, Darwin is preparing the way for his description of natural selection, as he is in his demonstration of the power of artificial selection. The analogy between the two kinds of human agency that have modified species over time and the agency of natural selection is stated in the rhetorical question with which Darwin opens his treatment of the latter: "As man can produce and certainly has produced a great result by his methodical and unconscious means of selection, what may not Nature effect?" (65). Darwin's analogy between selection by "Nature" and selection by human agency requires him, having demonstrated the principle of selection, to show first that it can operate without intent and second that it need not be teleological. Though the argument is not made explicit, this is what he achieves by introducing unconscious selection as a third term between artificial and natural selection. Without any intention of doing so, by unconscious selection human actors modify species over time in ways that serve their interests. To make the analogy complete, having produced a concept of agency without intentionality, Darwin then goes on to imagine one without interests. Natural selection works in the same way as selection by human agency – but it has no interests of its own to serve: "Man selects only for his own good; Nature only for that of the being which she tends" (65). The concept of natural agency in this argument is produced by negation: by deleting the traits of intention and interest that constitute the subject who acts in artificial and unconscious selection, Darwin theorizes a form of agency without a subject, or with the no-subject he personifies as Nature. In the most schematic reading of his argument, Darwin establishes the possibility of this kind of negative concept by his use of the mediating term "unconscious," which supplies a model for the work of the negative in everything that follows.

Mediating third terms, though, often destabilize the arguments in which they appear. As we have seen, by introducing the concept of unconscious selection, Darwin radically undermines the antithesis between nature and culture upon which rests the entire distinction between human and natural agency in the selection and modification of living forms. Unconscious selection is defined by its blindness to its own agency; it thus has a standpoint from which it is impossible to distinguish between the agency of human beings and that of nature. This standpoint of blindness is actually the correct one; it can be shown to be the unacknowledged standpoint adopted in *On the Origin of Species*. For human beings *are,* in Darwin's view, part of nature. And unconscious selection in the broadest sense *is* natural selection: of *all* the players in the infinitely complex system of mutually dependent forms of life Darwin describes, it could be said that they are mutually and unconsciously selecting one another. Darwin distinguishes between natural selection and selection by human agency with the claim that only the former acts purely for the benefit of the organism under selection; but the dialectical logic of his own argument makes this distinction unsustainable. If we consider his example of the improvement of pears by unconscious human agency, as it were, from the pears' point of view, it turns into the story of a fruit becoming adapted to human tastes so as to give itself a competitive advantage. The concept of unconscious selection disallows the privileging of the human point of view by making it as much an interpretative construct as any other. And without such a privilege, the evolution of the pear becomes indistinguishable from that of any fruit that evolves by making itself progressively more palatable to the organisms that feed on it. Darwin himself discusses as a kind of natural selection the mutual adaptation of plants and the insects that pollinate them, each of which benefits itself by becoming fitted to the needs of the other (*Origin* 71–4).

The concept of unconscious selection thus provides a way of understanding human agency as a *part* of natural selection. Why, we need to ask, does Darwin use it to distinguish the two? His decision not to write about human beings as objects of natural selection in *On the Origin of Species* is well-known, and widely understood as arising from a wish to present his theory in the least controversial way possible. Less easy to understand, but perhaps more fundamental, is the book's non-discussion of human beings' agency in natural selection. I have argued in this essay that Darwin's work represents knowledge of natural selection as constitutively incomplete because faculties shaped by natural selection will necessarily tend to conceal its operation. Natural selection cannot be fully known from within; I thus argued that in Darwin's notebooks, the work of knowing is divided between, on the one hand, subjects who embody unconscious ideas of the past in the form of hereditary instincts and, on the other, the disembodied scientist in whom these ideas are raised to consciousness.

To understand the figure of unconsciousness in Darwin's later writing, I now want to suggest, we need to turn this view of it as the projection of an epistemological problem upside down. I've argued that what Darwin in *On the Origin of Species* terms natural selection is conceived by analogy with, but also as the negation of, forms of selection involving human agency. Only by this negation is Darwin able to produce natural selection as an object that can be comprehended from outside. From the moment it incorporated human agency, natural selection would cease to exist as such. Instead of being an object available to knowledge, it would be a practice freighted with libidinal and political investments. This has indeed been the history of the interpenetration of *bios* and *techne* in the century-and-a-half since Darwin's book. But, as I've argued, they were never really separate. What Darwin in *Origin* terms unconscious selection is indistinguishable from natural selection except from the standpoint of anthropocentrism. It is not too much to say that the historical interpenetration of human agency and natural selection constitutes the unconscious of Darwin's ideas, just as a certain material and historical reality constituted the unconscious truth his work revealed in the ideas of his precursors. We know this not only from the instability of the concept of unconscious selection in *Origin* itself and the ease with which it can be shown to coincide with its apparent antithesis, but also because of the recurrent appearance throughout Darwin's work of figures of unconscious agency, most often when he discusses habits that have become hereditary in the form of instincts. In these discussions, he represents forms of agency that cannot recognize themselves as such, and being as they are the traces or remainders of past history, cannot properly be represented *as present* at all. These figures, I suggest, allegorize in Darwin's writing traits of natural selection that elude representation as such: notably, its invisibility *in the present* as a process that becomes recognizable only in retrospect, and even then not in its actual effects, but only in the unassimilated remainders that its operation has left behind. Among these figures is a character who haunts Darwin's earliest writing on natural selection, Elspeth Mucklebackit from Scott's *The Antiquary*. With her yarn, spindle, and unceasing work, Elspeth already appears in Scott not only as a figure typifying a specific period in past history, but also as an allegory of something like destiny as a general and transhistorical agency. In Darwin she not only exemplifies history's power to shape human action and consciousness in particular ways, but also, and more uncannily, the alienated and misrecognized form under which human agency in history appears in any imaginable present.

NOTES

1 Darwin's interest in Elspeth may have been related to her anxiety-dispelling revelation that the hero's birth had *not* been incestuous. Emma Wedgwood, whom Darwin married in January 1839, was his first cousin; later, Darwin worried about the effect on his children of their parents' consanguinity. The effect of inbreeding was a central topic in his researches into heredity; for its connection to his own life, see Browne 1:282. For a discussion of Darwin's marriage and his views on heredity in the broader context of Victorian marital conventions and ideas about incest, see Kuper 83–103.

2 In the sixth edition of *Origin*, Darwin altered this sentence to assert only that effects of habit are "in many cases" subordinate to those of natural selection in the formation of instincts. See Darwin, *Origin* 382.

3 For a history of the idea of progress in the Victorian era in which Darwin's theory is understood as presenting a broadly non-progressive and non-teleological view of evolution, see Bowler. The contrary view of Darwin is maintained by Richards.

4 This entry must have been made before the end of September 1838. The entries cited above from Notebooks M and N also belong to the second half of 1838.

5 Darwin's attention to the problem of the eye is an implicit response to its use as a principal exhibit in the argument for design mounted in William Paley's *Natural Theology* 19–41.

6 When he adopts a hierarchical classification of groups under groups, Darwin distinguishes his theory of classification from that of Lamarck. For the latter, the classification of organic forms was made by human beings and has no basis in nature: "La nature n'a rien fait de semblable; et au lieu de nous abuser en confondant nos oeuvres avec les siennes, nous devons reconnoitre que les *classes*, les *ordres*, les *familles, les genres*, et les *nomenclatures* a leur égard, sons des moyens de notre invention, dont nous ne saurions nous passer, mais qu'il faut employer avec discretion" (20) ["Nature has not made anything of the kind, and instead of deceiving ourselves by confusing our work with hers, we should recognize that classes, orders, families, and genera, together with their nomenclatures, are tools of our invention, which we could not manage without, but which must be used with discretion" (my translation).] Lamarck's classification, moreover, does not use groups under groups, but rather consists of a single series of organic life arranged in order from lower to higher forms.

7 Darwin's vague assertion that the figure of a tree has "sometimes" been used to represent the affinities among different forms of life obscures its specific reference to idealist and anti-Lamarckan theories of life's ascent current in Britain during the 1840s. For an account of these, see Desmond 276–372. To compare with Darwin's figure, here is a passage from the anatomist and follower of Coleridge J.H. Green's 1840 Hunterian Oration, *Vital Dynamics*:

> We might perhaps venture to symbolize the system of the animal creation as some monarch of the forest, whose roots, firmly planted in the vivifying soil, spread beyond our ken; whose trunk, proudly erected, points its summit to a region of purer light, and whose wide-spreading branches, twigs, sprays, and leaflets, infinitely diversified, manifest the energy of the life within. In the great march of nature nothing is left behind, and every former step contains the promise and prophecy of that which is to follow. (qtd. in Desmond 369–70)

8 Recently, as part of their argument against the Darwinian concept of natural selection, Fodor and Piattelli-Palmarini have argued that Darwin cannot get rid of the language of agency in his account because the idea of selection is incoherent without the idea of a selector – so that the idea of selection itself must be rejected. See *What Darwin Got Wrong* 95–138. For an account of Darwin that takes his personifications seriously as evidence for an idealist reading of natural selection, see Richards 514–54. For a deconstructive argument that this problem in Darwin's

argument arises from the necessarily metaphysical language in which it is expressed, see Beer 53–4, 67–73.

9 Showing his conviction that the relations among different expressions of the archetype are ideal rather than genealogical, Owen suggests that modifications of the type unknown on earth might exist on other planets. See Owen 83–4.

10 For an account of how Darwin's theory was shaped by developments in morphology and taxonomy during the 1830s and 1840s, see Ospovat 115–200.

11 Owen's review was published anonymously in the *Edinburgh Review* in April 1860. For a narrative of his relations with Darwin, and of the break following the publication of *Origin*, see Browne 2:110–12.

12 Marx produced a theory of fetishism to explain his relation to bourgeois political economy. Darwin produced no such historical metatheory. For discussion of the historical problem of Marx's relation to the labour theory of value as it was formulated in classical political economy, see Rowlinson, *Real Money* 100–16.

WORKS CITED

Beer, Gillian. *Darwin's Plot: Evolutionary Narrative in Darwin, George Eliot, and Nineteenth-Century Fiction*. London: Routledge & Paul, 1983. Print.

Bowler, Peter J. *The Invention of Progress: The Victorians and the Past*. Cambridge, Mass.: Basil Blackwell, 1989. Print.

Browne, Janet. *Charles Darwin: A Biography*. 2 vols. New York: Knopf, 1995, 2002. Print.

Darwin, Charles. *Charles Darwin's Notebooks, 1836–1844: Geology, Transmutation of Species, Metaphysical Enquiries*. Ed. Paul H. Barrett et al. London; Ithaca, N.Y.: British Museum of Natural History; Cornell UP, 1987. Print.

Darwin, Charles. *The Descent of Man, and Selection in Relation to Sex*. 1879. Ed. Adrian Desmond and James R. Moore. London: Penguin, 2004. Print.

Darwin, Charles. *The Expression of the Emotions in Man and Animals*. 1872. Ed. Paul Ekman. Oxford: Oxford UP, 1998. Print.

Darwin, Charles. "Letter to Emma Wedgwood January 6-7 1839." Nov. 13 2011. http://www.darwinproject.ac.uk/letter/?docId=letters/DCP-LETT-484.xml. Web. 29 May 2017.

Darwin, Charles. *On the Origin of Species*. Ed. Gillian Beer. Rev. ed. New York: Oxford UP, 2008. Print.

Darwin, Charles. *The Origin of Species: A Variorum Text*. Ed. Morse Peckham. Philadelphia: U of Pennsylvania P, 1959. Print.

Desmond, Adrian J. *The Politics of Evolution: Morphology, Medicine, and Reform in Radical London*. Chicago: U of Chicago P, 1989. Print.

Fodor, Jerry A., and Massimo Piattelli-Palmarini. *What Darwin Got Wrong*. London: Profile Books, 2011. Print.

Kuper, Adam. *Incest & Influence: The Private Life of Bourgeois England*. Cambridge, Mass.: Harvard UP, 2009. Print.

Lamarck, Jean Baptiste. *Philosophie Zoologique, Ou Exposition Des Considerations Relatives a L'histoire Naturelle Des Animaux*. Paris: Dentu, 1809. Print.

Ospovat, Dov. *The Development of Darwin's Theory: Natural History, Natural Theology, and Natural Selection, 1838–1859*. Cambridge: Cambridge UP, 1981. Print.

Owen, Richard. *On the Nature of Limbs: A Discourse*. 1849. Ed. Ron Amundsen. Chicago: U of Chicago P, 2007. Print.

Paley, William. *Natural Theology*. London: Printed for R. Faulder by Wilks and Taylor, 1802. Print.

Richards, Robert J. *The Romantic Conception of Life: Science and Philosophy in the Age of Goethe*. Chicago: U of Chicago P, 2002. Print.

Rowlinson, Matthew. "Foreign Bodies: Or, How Did Darwin Invent the Symptom?" *Victorian Studies* 52.4 (2010): 535–59. Print.

Rowlinson, Matthew. *Real Money and Romanticism*. Cambridge: Cambridge UP, 2010. Print.

Scott, Walter. *The Waverly Novels*. 25 vols. Edinburgh: William Paterson, 1885–7. Print.

PART TWO

Romantic Temporalities

Chapter Four

Deep Time in the South Pacific: Scientific Voyaging and the Ancient/ Primitive Analogy

NOAH HERINGMAN

I. Temporalities

Deep time is a surprisingly young concept, a product of modern geology and evocative popular science writing. In 1987 Stephen Jay Gould praised "deep time" as "a beautifully apt phrase," paying tribute to John McPhee six years after McPhee coined the phrase in his book *Basin and Range* (Gould 2). Gould and others have adopted McPhee's expression along with his definition of deep time as a marker of the unbridgeable gap between geological and historical time scales, between the earth's gradual changes over millions of years and the rapid changes occurring in even a century of human history. Gould shows that the concept of this gap is older than McPhee's expression, although, while it was formulated roughly around the mid eighteenth century, it is still young in historical terms. Mobilizing the same gap for evolutionary theory, Charles Darwin pointed out that the evolution of species maps more readily onto geological than human time. More recently, other writers have used "deep time" metaphorically to refer to human cognitive processes or to the lifespan of monuments, among other topics. Geology textbooks trace the idea as far as James Hutton's *Theory of the Earth*, originally published in 1788 (Putnam 4), and there seems to be a strong consensus that deep time depends on the emergence of modern geology. In this essay, however, I want to propose another historical version of deep time, namely the attempt by eighteenth-century explorers and historians to envision human prehistory and human origins through art. Like evolutionary biology, archaeology today stands on a geological foundation, dealing systematically with the earth's youngest strata. But the recognition of geological time itself may owe something to the European perception of other living cultures as primitive. By comparing Pacific islanders, among others, to the peoples of ancient Europe, explorers translated cultural difference into an abyss of time. In this context, the term "deep time" captures an unfamiliar aspect of the conventional trope of exploration as "time travel": although it is an ethnocentric trope, it also unsettles the history of the species as radically as the "time revolution" later unsettled geology.

Early art historians and Pacific explorers brought expanding temporal frameworks to bear on differing "varieties of the human species," as John Reinhold Forster called them in his *Observations Made During a Voyage Round the World* (1778). Operating under the aegis of "travel as science," both Forster and antiquaries such as Pierre Hugues d'Hancarville saw themselves as "exploring the past," a method formalized by Joseph-Marie Degérando in 1800: "the philosophical traveller, sailing to the ends of the earth, is in fact traveling in time" (qtd. in Fabian 7). Johannes Fabian takes up this substitution of time for space in *Time and the Other*, locating these philosophical travellers of the Enlightenment in the middle of a process that generated "naturalized Time" out of secular time. Anthropology became a modern discipline, Fabian argues, by deriving a strict teleology of progress from evolutionary theory, a development that he terms "intellectually regressive" (16) as compared to experimental Enlightenment practices of "temporalization." Philosophical travellers such as Forster anticipated the modern synthesis, however, by developing the "comparative method" of situating cultures in time, which depended on the presupposition that "dispersal in space reflects ... sequence in time" (12). "*Primitive* being essentially a temporal concept," it implied that the traveller's point of origin was not only geographically central but historically advanced (18). For Fabian, "Typological Time" remains more important in anthropology than "physical" or absolute time (23) because it permits the distinction between traditional and modern societies: "savagery exists ... in *their* Time, not ours" (75).

The philosopher Quentin Meillassoux, by contrast, insists that absolute time (as measured by carbon dating or stellar spectroscopy) is a "time of science" from which "humanity is absent" (26). What I am arguing here is that the current spatial metaphor of "deep time" for nonhuman time recalls the prior temporalization effected by Enlightenment discourses on the primitive. These discourses have well-established implications for the origins of "racial science"; d'Hancarville's and especially Forster's pursuit of primitive origins involved incipient racial distinctions, though these were framed within a mono- rather than a polygenetic paradigm of human origins.[1] My somewhat different concern is to show how their conflation of time and space – their "horizontal stratigraphy," in Fabian's apt phrase (75) – promoted an empirical attention to artefacts and to customs and manners that joined civil with natural history and disrupted the dominant model of uniform "stages" of human civilization.

Because of the long vista that it seemed to open on both natural and human history at the moment of its European discovery in the 1760s, the Pacific island of Tahiti has been described as "a foundation stone of the romantic movement" (Beaglehole 1:xciv–xcv). J.C. Beaglehole's description is not so far removed from the more conventional explanation that derives the Romantic movement from the French Revolution. George Forster, John Reinhold's son and fellow Pacific explorer, also achieved distinction as a revolutionary in the 1790s. The French and American Revolutions transformed the Enlightenment views of human society that inspired them, and likewise Pacific exploration became a testing ground for the philosophical theories of Jean-Jacques Rousseau, Adam Ferguson, and many others concerning the origin and progress of society. These Enlightenment philosophers were already armchair travellers: both Rousseau and Ferguson cited descriptions of Native American societies from travel narratives to illustrate their model of the early "stages" of social development. Another would-be historian of humankind, Charles de Brosses, coined the term "fetishism" to describe West African religious practice, based solely on his reading of travel narrative. Though these philosophers did not travel around the world, their books did, and their ideas merged with the observations of real travellers, particularly the philosophical travellers who took part in the voyages of Bougainville and Cook, including the Forsters. By the time that numerous Pacific exploration narratives began to appear in print in the 1770s, the influence of these two modes of travel became thoroughly reciprocal. Travellers became prone to imagining Pacific Islanders as primitive people who resembled what Europeans had been at an ancient, preliterate stage of their history, and philosophers such as Denis Diderot eagerly incorporated details from these narratives into new conjectural histories.[2]

Voyagers charted their course to "places apparently remote in time," in Neil Rennie's words (1), under the influence of these conjectural histories as well as another body of thought with longstanding connections to travel: neoclassicism. The period's histories of ancient art made a signal contribution to the practice of marking cultural difference as distance in time: many voyagers shared with early art historians, or antiquaries, the focus on artefacts and the conception that primitive art is "wrought out of nature," as the Earl of Shaftesbury had claimed (qtd. in Décultot 53). The propensity to "temporalize difference," in Nicholas Thomas's succinct formulation, led voyagers to imagine their encounters with islanders as glimpses of "remote antiquity," or what we would call prehistory, but the vocabulary available to them came from their study of recorded antiquity, particularly ancient Greece (Forster 422n9). European explorers compared the native peoples they encountered with Greeks and other ancient Europeans virtually from the beginning of the Age of Exploration in the sixteenth century. By the time of Cook's voyages, much more was known about classical antiquity, and the prejudice in favour of classical aesthetics was stronger than it has ever been before or since. The fascination of ancient art produced the earliest attempts at art history and archaeology, and these more empirically driven histories, too, made use of analogies their authors found in travel narratives. These antiquarian authors were primarily armchair travellers, though some of them visited archaeological sites: Winckelmann and d'Hancarville, for example, were strongly influenced by the recent discoveries at Pompeii and Herculaneum.

European explorers between the Enlightenment and Romanticism saw themselves entering a new order of time in the South Pacific. An anthropological kind of deep time arose from the shock of recognition and the simultaneous reaction that made these voyagers want to suggest that they were encountering only faint traces of themselves, a human past so ancient as to be almost forgotten. Pompeii and Herculaneum had raised new questions about the once-familiar classical past, fostering the kinds of tactics also brought to bear on Pacific encounters: attention to everyday life, speculation on human prehistory, and especially the contemplation of artefacts as a medium of cultural empathy.[3] My argument for deep ethnographic time describes a prehistoric turn in the understanding of human time, distinct from the historical turn in the study of nature famously described by Michel Foucault in 1966 as "the breaking up of the great table," the displacement of "static" natural history by biological concepts around 1800 (275). Paolo Rossi has argued that a "dark abyss of time" was opened even earlier when historical methods were applied to fossils and strata, and more recently Gould and Martin Rudwick (182) have also explored the historical turn in natural science.[4] These are different kinds of arguments, but they all track a movement of chronological organization from human history into the study of nature. The natural history of Pacific populations, however, had the opposite effect of disrupting *human* temporality. The explorers' perverse insistence on the antiquity of living cultures and artefacts not only flattered the presumption of European superiority, but also promoted a prehistoric turn in the understanding of human time itself, challenging the short, accepted chronology of the human species without explicit recourse to geological or to what Fabian terms "naturalized Time."

Darwin still recalled the voyages of Captain Cook to describe his encounter with the "savages" of Tierra del Fuego (Darwin 1:263–4), shortly before he applied the new geological principles of Charles Lyell in his more famous encounter with recently evolved species on the Galápagos Islands.[5] Similarly, James Hutton incorporated earlier cosmogonies in his analysis of "primitive rocks" before his uniformitarian view was systematized (by Lyell and others) to support a geohistorical understanding of everything from stratigraphic sequence to the archaeological periods of human prehistory. Darwin's "savages" and Hutton's "primitive" are traces of a rift within human time itself – my subject in this essay – that preceded the absolute distinction between human time and "nature's own history" (Rudwick 348). Ethnographic "time travel" in the Enlightenment not only restructures the horizontal space of geography, as Rennie, Thomas, and others have noticed; it also anticipates early geology's restructuring of vertical space in the pursuit of primitive rocks.

II. Artefacts

Writing in 1768, just months before Cook set sail on his first circumnavigation, the Italian antiquary Octavien de Guasco insisted on the analogy between ancients and "natives" in his history of ancient sculpture:

> Nothing could be more appropriate for a correct idea of the state of the arts, customs, and usages of the ancients, as yet little civilized, than to point out the practice of peoples who in our own times still live in a state of barbarism. This should be a rule for antiquaries always to keep before their eyes, in order to judge intelligently the customs, usages, and the monuments of earliest antiquity, because the same situation, the same needs, the same deficiencies in the arts produce the same ideas, the same practices, according to the climate.[6]

Guasco traces the history of sculpture from natural objects, such as sacred trees or standing stones, through carvings with a limited number of human features, all the way up to the Greco-Roman marble statues that were the most celebrated objects in European museums. He wrote this history with a fifty-volume collection of voyage narratives by his side, and followed his own rule so well that every chapter in the book carries multiple footnotes referring to this collection. These references range so widely, however – from Japan to Latin America to West Africa in a single chapter – that it is not as easy to correlate specific ideas and practices with specific situations or climates as he appears to claim in this passage. Moreover, Guasco is a scriptural literalist, so the Hebrew Bible provides a distinct point of origin for antiquity as he understands it. According to Guasco, sculpture and the other arts originate as a form of religious expression, and any polytheistic religion, whether ancient or modern, has degenerated from an original monotheism (9–11).[7] Though he is intellectually more conservative than other philosophical travellers, Guasco's vexed commitment to monogenesis and his uneasy identification of apparent lack of progress with degeneration pose problems that also preoccupied younger scholars such as J.R. Forster.

Within the larger group of antiquaries concerned with ancient religion and the origin of the arts, the writer who borrowed the most from Guasco was Pierre François Hugues, aka Baron d'Hancarville. D'Hancarville occupied the opposite end of the spectrum on religious belief, using his analysis of ancient religion to promote relativism and scepticism. While d'Hancarville borrowed Guasco's ideas in the second half of his work, Guasco or his publisher also borrowed an image from volume 1 of d'Hancarville's *Collection of Etruscan, Greek, and Roman Antiquities*, a sprawling, richly illustrated history of ancient art that appeared in four volumes between 1766 and 1776.[8] Two of his engravings (Figures 4.1–2) capture much of the common ground that he and Guasco shared,

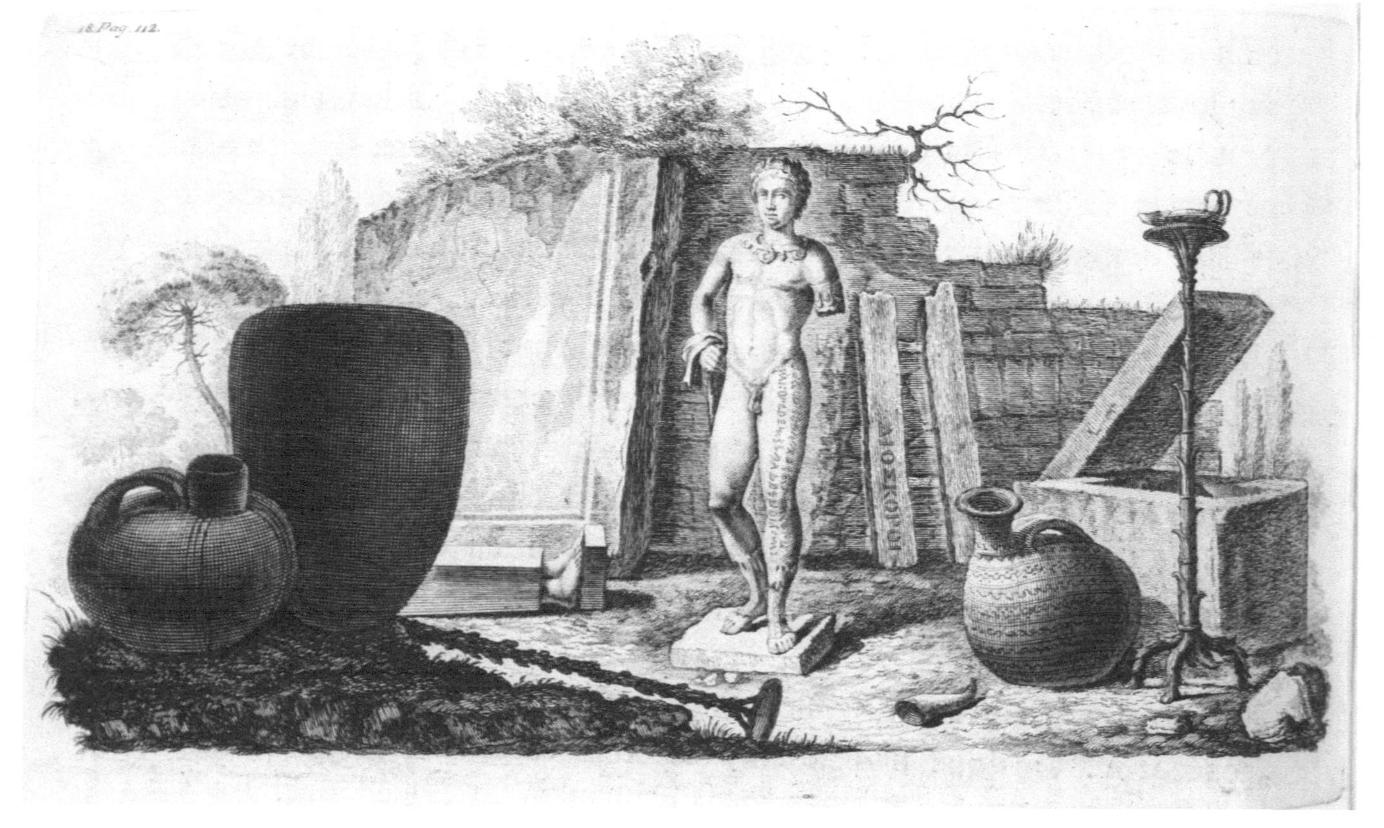

Figure 4.1. Unsigned engraving. *Collection of Etruscan, Greek, and Roman Antiquities* [*AEGR*], Vol. I, p. 112 (Chapter III headpiece [English text]). Courtesy of Margaret M. Bridwell Art Library, University of Louisville.

Figure 4.2. Unsigned engraving. *Collection of Etruscan, Greek, and Roman Antiquities* [*AEGR*], Vol. I, p. 113 (Chapter III headpiece [French text]). Courtesy of Margaret M. Bridwell Art Library, University of Louisville.

documenting some of the earliest stages in the evolution of the arts by means of artefacts that emulate natural objects.

The two engravings together make up d'Hancarville's synoptic exhibition of the kinds of useful and, increasingly, ornamental objects created in response to what Guasco calls "deficiencies in the arts." A revised version of the second engraving appears on Guasco's title page. D'Hancarville points out the formal allusions to natural materials in many of these artefacts, allusions that he describes as "conserving" the prehistory of their fabrication. So the jug in the left hand corner of Figure 4.2, for example, preserves the form of the ostrich egg from which such vessels were originally fashioned, and the cow horn in the foreground of Figure 4.1 recalls the natural history of the drinking horn. The Roman bronze candlesticks (Figure 4.1) include the nubs of branches to recall the prehistoric moment when they were made from canes or young trees stripped of their smaller branches. The faint human profile on the wall in the background (Figure 4.1) alludes to the birth of painting, attributed to the legendary Maid of Corinth who traced her lover's shadow in charcoal on the wall of a cave. The statue with Greek characters engraved on its left leg shows the fourth stage of the emergence of sculpture (Figure 4.1), preceded (in Figure 4.2) by the more "primitive" Egyptian statue without articulated limbs, the "term" or column showing only the head and feet, and finally the standing stone or "symbolic column" evoked by the horizontal log. The log also alludes to the origin of architecture, which originally deployed cut logs before stone columns were devised in the same form, a transformation signalled by the far end of this log/column with its carved necking (cp. d'Hancarville I:177). D'Hancarville develops four elaborate stages out of Guasco's rudimentary stadial sketch (Guasco 1), and although there are four stages, it is notable that he does not correlate them with the four stages of barbarism, pastoralism, agriculture, and commerce familiar from Adam Smith and conjectural history. Though less rigorous than modern archaeology, it is nonetheless d'Hancarville's empirical attention to artefacts that produces a time scale differing markedly from the dominant stadial model.

In the absence of a geologically based archaeology, d'Hancarville's exhibition must pose the question of human origins through artefacts from recorded antiquity, most noticeably Egyptian, Greek, and Roman times. The neoclassical pursuit of Italian antiquities, however, extended to semi-legendary indigenous populations predating classical times, which also created categories for the reception of new South Seas artefacts. D'Hancarville spent six years in Naples, and had some experience of the archaeological sites near Vesuvius and the newly rediscovered temple complex at Paestum. Maybe a kind of archaeological instinct led him to include one genuinely prehistoric artefact, the egg-shaped jug from the first engraving, which survives and has been identified as an Italic *askos* made between 2800 and 2400 BC (Jenkins and Sloan 234). When the Italian antiquary Giovanni Giovene described a cache of even older artefacts found in the Kingdom of Naples in the 1780s, he compared them explicitly to the South Pacific artefacts brought back to Europe by Captain Cook. The Neolithic jadeite axes from this site, in Giovene's view, precisely resembled stone tools that were collected on Tahiti and passed on to a collector in Giovene's local network.[9] D'Hancarville's patron and collaborator, Sir William Hamilton, also referred to Cook's voyages in a natural history of volcanoes that was modelled on the scholarship of the Italian connoisseurs. Hamilton made the point that these stone tools from Tahiti were made of volcanic rock, and thus supported his thesis that volcanic eruptions were geologically formative events not only in antiquity but across the earth and across time.[10]

III. Voyages

Maria Toscano points out that Giovene and his colleagues in the Neapolitan Republic of Letters considered themselves "naturalist-antiquaries" and demonstrates their influence on Hamilton. Toscano's formulation helps to make sense of the collecting that was done on Cook's voyages as well. Those who collected stone tools and other artefacts in the South Pacific included Joseph Banks, who is cited here by Hamilton, and others on Cook's first voyage, as well as the Forsters, father and son, who served as naturalists on Cook's second voyage. Many of the armchair travellers who theorized human origins debated whether cultural difference should be attributed to differing rates of progress or to degeneration. Guasco, for one, viewed non-European peoples categorically as degenerated (12, 22). Forster senior, the first Pacific voyager who was thoroughly familiar with these debates, saw evidence of both progress and degeneration in his broad historical sketch of Pacific peoples, recognizing on the strength of his first-hand knowledge that the evidence was not sufficient to support either hypothesis categorically. Forster argued that empirical evidence of the diverse populations in the Pacific brought the various so-called "stages" of civil society into focus, and also proved that existing scholarship had no authority for its hypotheses concerning (in modern terms) racial difference or the human evolutionary process:

> The History of mankind has often been attempted, [but] ... None of these authors have ever had the opportunity of contemplating mankind in this state [of original simplicity], and its various stages from that of the most wretched savages, removed but in the first degree from absolute animality, to the more polished and civilized inhabitants of the Friendly and Society Isles. Facts are the basis of the whole structure [i.e., Forster's book]. (9–10)

In Forster's argument, the "most wretched savages," exemplified somewhat conventionally by the people of Tierra del Fuego (now southernmost Argentina), reached that state by degeneration, for which Forster assigned the environmental cause of migration from a warmer into a colder climate. Unlike some anthropological explanations that postdate evolution and genetics, Forster's does not take the most apparently primitive people to be at the earliest evolutionary stage. He argues for environmentally driven transformation of human varieties over long periods of time, and compares some living populations to ancient ones, both European and Pacific. Customs and manners, he argues, provide the only evidence concerning human antiquity in cultures without written records. This evidence from customs and manners, along with biogeography, informs his conclusion that "the warm tropical climates seem to have been originally the seat of the human race" (342).

Focusing mainly on Tahitian society, like his predecessors, Forster argued that Pacific "arts and sciences" – including their tools – represented a creative response to the natural setting of the islands and had evolved a great distance from the ancient practices of the population that originally migrated to the island. He also believed, however, that Tahitian religion had degenerated from an ancient Asiatic cult, traces of which survived the process of migration and resettlement. In other words, Forster vacillates between a long and a short time scale in his history of Pacific peoples. This contradictory temporalization organizes Pacific space by means of "heterochronism" – to borrow a term from Foucault's "Of Other Spaces" – as well as "allochronism," Fabian's term for the separation between "savage" and ethnographic time. In some areas Forster recognizes that change is slow, an insight that favours speculation about remote human origins and provides an alternative to the proto-racial distinctions he makes elsewhere (172–90). In other cases, such as religion, he attempts to trace Polynesians (for example) to living Malay peoples. At the same time, however, Forster adopts the habit of other educated voyagers such as Banks and Bougainville, regularly comparing Pacific islanders to Greeks and other ancient peoples.[11] This comparative gesture, though it seems to flatten cultural distinctions, also makes some reckoning with human origins unavoidable. One end result is a diachronic turn in the understanding of preliterate societies that precedes the geologically fixed hard dating established by modern archaeology.

Forster's analysis of the different Tahitian "arts and sciences" varies greatly with the subject matter, and does not always support his own general premise that knowledge is cumulative: "The more a tribe or nation preserved of the ancient systems, and modified or adapted them to their particular situation ... the more improved, civilized and happy must that tribe or nation be" (196). While this principle holds in the areas of religion and fine arts, Forster takes a different approach to "mechanical arts" like textile manufacture, emphasizing local adaptation and the use of natural materials, just as Guasco and d'Hancarville do in their theories of the origin of art:

> They beat the cloth with a square instrument of heavy wood, called *toa* ... Their dyes are very fine and bright, and would deserve more attention if they were lasting: the red dye requires a good deal of labour and care in preparing it; the fruit of a small fig called *mattée* (*ficus tinctoria*) affords a small drop of milky juice, when it is broken off from the tree; this juice is carefully gathered in a clean cup of coconut shell, and ... [t]hey soak it in the leaves of the *etoù*, or *cordia sebestena*, which imbibe the milky juice, and soon tinge it of the finest crimson imaginable. (276)

Among the various natural materials detailed here, the coconut shell-cup, in particular, recalls the vessels made of eggs and horns in d'Hancarville's engravings. In the case of music and poetry, Forster is more strongly influenced by conjectural history and neoclassicism:

> The least happy occurrence in life is sufficient to inspire them with a high degree of glee, which sets their whole body in motion: they begin to frisk and DANCE, this makes cadenced or measured breathing necessary; if in this situation man wishes to communicate his ideas to the by-standers, he will naturally give his words that kind of measure or cadence, which he has adopted with his breathing, this, together with the voice of exultation may be considered the first origin of singing and MUSIC [which in turn] gives rise to POETRY. (284)

The ambivalence already apparent in his account of "mechanical arts," registered in the form of a doubt concerning the permanence of the red dye, becomes more pronounced in this openly paternalistic and idealizing narrative of the fine arts. At the same time, however, their primitive genius Hellenizes the "Taheiteans," whose spontaneous "verses ... are always delivered by singing, in the true antient Greek style" (286).

The contradictions in Forster's account support his claim to be testing philosophical conjectures experimentally in the field, but they also reflect the ambivalence inherent in the discourse of primitivism itself. Empirical study of customs and artefacts readily unsettled the orderly sequence of four stages of society partly because the theorists of those stages themselves suspected or recognized that the sequence was not orderly. Nicholas Thomas points out that although both Forsters readily invoke the ancient/primitive analogy to explain violence among the Maori, their response is ultimately ambivalent, mirroring the ambivalence already inherent in stadial histories such as that of John Millar. Thomas notes that Millar made "liberty" characteristic of both early and late stages of civil society and argues that J.R. Forster shared his anxiety about the slippage from "liberty" to "license" in both stages (*In Oceania* 86). Nostalgia for "savage" liberty casts doubt on the celebration of commerce and progress in Forster as in Millar and even in more conservative theorists such as Adam Ferguson. Harriet Guest notes that "Forster was concerned to develop these theories" – Millar's in particular – "and adapt them to his novel experiences" (56). In a similar vein, but emphasizing the Forsters' use of European climate theory, David Bindman points out that "experience made things complex and unstable, challenging the formulae of climate and social life that had dominated views of the 'savage' world" (149–50).[12] Richard Lansdown expands this broader European frame of reference by tracing the "bipolar vision" of "cultural" and "chronological" primitivism from classical antiquity down to the Enlightenment, emphasizing "the depth of Rousseau's ambivalence" as a major influence on the voyagers (69–70). Thus both new empirical evidence and instability in the theories themselves made it far from a "simple matter," in Thomas's words, "for Europeans to apply their prior notions of savage or primitive life" to Pacific peoples (71).

Forster's last example departs completely from his Rousseauvian commonplaces about the origins of poetry, attributing a local and unique origin to Pacific geography, astronomy, and navigation, which represent (for him) the highest achievement of these groups:

> The inhabitants of the islands in the south sea have made very considerable navigations in their slight and weak canoes; navigations which many Europeans would think impossible to be performed, upon a careful view of the vessels themselves ... [Therefore it is] probable that the inhabitants of these isles were the inventors of their own astronomy and geography: and if they had strength of mind sufficient to enable them to invent sciences which require accurate observations, and a remarkably strong sagacity; why may we not think them equally capable of being the inventors of the whole cyclus of their knowledge. (318–20)

Here the semi-obligatory condescension of the voyage narrative gives way before an example that implies a "deep," quasi-evolutionary time scale. Philibert Commerson, the naturalist on Bougainville's voyage, pointed to the same body of autochthonous knowledge to argue that the Tahitians were "a primitive people" in the literal sense that they did not migrate to the island from elsewhere (Lansdown 84). Forster, although he may have known Commerson's letter (published in 1769), saw himself as the first to gather empirical evidence for the inquiry into human origins, and his findings were inevitably inconsistent as a result.

Forster sees improvement and degeneration in the same cultures, and alternates between imitative and original explanations of cultural practices. His focus on the arts remains constant, however, and continually prompts an extension of the time scale that unsettles the stadial assumptions behind the comparison of native and ancient peoples: "It is the work of ages," he writes, "to bring the mind of a whole nation to maturity ... a few years cannot bring on a material change among them" (200). The empirical study of populations and artefacts by explorers such as Forster, inspired by antiquarianism, disrupted the progressive model of four stages of human history and revealed profound variations among "primitive" populations, along with evidence for much longer, open-ended historical processes behind the development of native technologies, arts and customs. The reciprocal relationship of ethnographic voyaging and early archaeology fostered exploration beyond the written record, opening the domain we now call prehistory.

IV. Monuments

Forster is more conservative in the area of religion, in which more innovative work was done by other explorers as well as antiquaries. At the same time, he only spent four weeks on Tahiti, and so his emphasis on Tahiti as the most advanced civilization of the Pacific is sometimes supported more by his reading and presuppositions than by his fieldwork. Forster had very little firsthand exposure to Tahitian religious practice, and other explorers' accounts must be used to contextualize his claim that "they are ... still in the infant state of humanity, not yet ripened to the use of argument and reason in religious matters" (323). The *Endeavour*, in which Cook performed his first voyage, remained anchored off Tahiti for three months, so the naturalists on board had better opportunities to make observations, even if they were not informed as thoroughly as Forster by philosophical reading. The journals from this voyage also offer several distinct approaches. The *Endeavour* journal of Joseph Banks, who collected the stone tools noticed by scholars in Italy, differs symptomatically from that of the illustrator Sydney Parkinson, who was employed by Banks to make a visual record of the full spectrum of plants, animals, people, and artefacts. The classically educated Banks, for example, favoured classical analogies, comparing Tahiti repeatedly to the legendary Greek region of Arcadia (Hawkesworth II:120), whereas Parkinson compared the islanders to ancient Britons (23) and avidly gathered oral histories (125). Parkinson also was the first to recognize the islanders' navigational feats and to speculate about their ancient migrations.[13]

Banks and Parkinson make different kinds of observations concerning Tahitian religion and public space, but they both differ from most theorists of their time in taking these subjects seriously as fields of empirical study. This difference is most apparent by contrast to William Robertson's *History of Scotland*, which Parkinson had on board the *Endeavour*. Robertson declared that "Nations, as well as men, arrive at maturity by degrees, and the events, which happened during their infancy or early youth, cannot be recollected, and deserve not to be remembered" (I:1). Banks's analogies, reminiscent of Winckelmann's neoclassicism, do privilege the "infancy" of Greece as a way of legitimating the Pacific cultures he found particularly worthy of study. Bindman makes the same connection when he suggests that Winckelmann's Athens is just as "fragile" as the voyagers' Tahiti (150). It is partly the art historians' interest in religion and monuments that makes them, at times, more relevant to the voyage narratives – even if less directly influential – than the conjectural historians are. Robertson, setting himself against the Rousseauvian thesis of the noble savage, is more skeptical about the "infancy" of nations. His Scottish contemporaries, as I have noted, share this ambivalence to some extent, but all of them still rely on travel narrative, as Rousseau had done, for evidence concerning the early stages of civilization (Ferguson 80; cf. Meek 37–67). Fieldwork offered philosophical travelers an even better opportunity to study the ancient past empirically through the "primitive" present, but evidence from their encounters sometimes suggested both a more civilized present and a deeper past than they anticipated.

Parkinson also drew on popular writers and costume books on British antiquity, and Banks participated in the renewed attention to Gothic architecture that later became known as the Gothic revival.[14] The refashioning of the European middle ages as a kind of antiquity merits comparison with the antiquarian approach to Oceania as an example of "barbarous" postclassical culture – more recent than Greece and Rome, if not exactly close to the Pacific present – that gained scholarly currency by this refashioning. The Society of Antiquaries, which sponsored this early research, focused in the 1770s and 1780s on gathering "faithful representations," both visual and verbal, of the funerary sculpture in England's Gothic cathedrals. These material remains, as their director Richard Gough argues in *The Sepulchral Monuments of Great Britain,* offer the most reliable evidence available on "our manners, habits, arts, national taste, and style of architecture" in bygone ages. On the one hand, Gough derives some British traditions from Greco-Roman antiquity, quoting copiously in Greek and Latin (e.g., I:5–6); on the other, he uses Cook's *Resolution* narrative to corroborate Homer's account of Greek customs (I:i), displaying the reciprocal influence of different versions of antiquity. Like Parkinson, however, the illustrators involved in this project lacked both classical education and Royal Academy training, along with the prejudices that accompanied them. The Gothic revivalists insisted on preservation as the ultimate goal of studying material culture, and the same motive can be traced in different ways in Parkinson's and Banks's ethnographic work in the South Pacific. The affiliation between these projects appears clearly in an engraving of "Various Instruments and Utensils, of the Natives of Otaheite" in Parkinson's *Journal* (plate XIII), based on a drawing by Samuel Hieronymus Grimm, who created a very similar composition of ecclesiastical artefacts discovered beneath the floor of Lincoln Cathedral for Gough's book (II:lxviii).

Banks adopts the idiom of architectural antiquities in his discussion of a monument on Raiatea (or "Ulietea") that he visited on 29 June 1769. Architecture and funerary customs belong to the subset of topics, along with botany, that were of special interest to Banks and Parkinson. Banks was especially active in his investigation of "morai" or *marae*, gathering places first noticed by the British on Tahiti, who understood them primarily as "burying grounds ... [and] places of worship" (Hawkesworth II:166). As Banks and Cook explored the southwestern portion of the main island, they found two sites of architectural interest, the first a smaller burial site containing a pyramid, as well as the first piece of stone carving they had seen in ten weeks on the island. If this was impressive, the second was astonishing: "we no sooner arrivd there than we were struck with the sight of a most enormous pile, certainly the masterpeice of Indian architecture in this Island so all the inhabitants allowd. Its size and workmanship almost exceeds beleif, I shall set it down exactly." This monument, terraced and shaped like a large pitched roof (267 feet long by seventy-one wide and forty-four high), belonged to Purea, or Queen Oberea, as she was known to the British. Banks points out a curvature in the stone steps suggesting that this monument may be much older than its possessor, but he sees the workmanship as "ancient" primarily in a technological sense: "it is almost beyond belief that Indians could raise so large a structure without the assistance of Iron tools to shape their stones or mortar to join them, which last appears almost essential as most of them are round; it is done tho, and almost as firmly as a European workman would have done it" (southseas.nla.gov.au/journals/banks/17690629.html). The use here of "pile" and especially "architecture," which occurs very rarely in the Cook voyage narratives, owes something to contemporaneous antiquarian writing on Gothic architecture and on ancient monuments including megaliths.

In his engagement with this monument, Banks addresses preservation as well as material culture. His insistence on recording dimensions ("I shall set it down exactly"), and on having Parkinson sketch the sites, reflects this effort. Parkinson was, in fact, the first voyager to notice and describe the *marae* (on 6 May 1769). Subsequent discussions and illustrations of funeral rites on Tahiti, which remained central to early Pacific ethnography, incorporate these intertexts – not only Banks's and Parkinson's recorded observations, but also their associations with ancient Britons and classical and medieval antiquities – along with new observations. William Woollett's engraving of a Tahitian funeral (Figure 4.3), based loosely on a watercolour sketch by William Hodges, provides a case in point. Hodges made this sketch in August 1773 shortly after the *Resolution* and *Adventure* landed on

Figure 4.3. *A toupapow with a corpse on it, attended by the Chief Mourner in his habit of ceremony.* Engraving by William Woollet from a drawing by William Hodges. James Cook, *A Voyage towards the South Pole*, second ed. (London, 1777), vol. 1, plate XLIV [fp 184]. Courtesy of the Linda Hall Library of Science, Engineering, and Technology, Kansas City, MO.

Tahiti for the first time, as reported in George Forster's narrative (I:164). The only person present with the corpse on its elevated bier (*tupapau*), according to Forster, was a mourning woman seated like the central figure here. The "chief mourner" with his elaborate headdress, who is not present in Hodges's sketch, was evidently interpolated by Woollett from an image originally drawn by Parkinson (70), a very similar composition engraved for John Hawkesworth's *Account* (plate V). For his journal, Parkinson created a similar landscape without the figures (*Journal* plate X), but then added a separate portrait of the "priest" or *heiva* (XI) in headdress, along with a footnote describing his role as chief mourner, based on information from Banks. Woollett combines all these images, and his composite is almost surely informed as well by the same Banks journal account, as rendered in Hawkesworth's *Account* of Cook's first voyage (II:234–9) – a text often cited by both Forsters and by Cook himself.[15] Woollett's engraving for Cook's narrative, together with Parkinson's and Banks's descriptions of the "altar" and "sacrifices" employed in these ceremonies, captures something of the antiquarian spirit that informed artists' renderings of trilithons and Druids, such as C.H. Smith's *Arch-Druid in His Judicial Habit* (1815).

Hawkesworth's account of the large monument on Raiatea adds one detail from Cook's journal that escaped Banks in his enthusiasm: the ornaments on top include large carvings in wood and stone, the latter of which is broken (evoking the then-current deterioration of medieval built works) (Hawkesworth II:166). Without making a direct argument concerning their antiquity and preservation here, Banks notes that the *marae* are threatened as well by military devastation, as suggested by a coastal battlefield strewn with human bones very near the site. Banks also participated in a funeral on Tahiti, and gives a vivid subjective account of some of the same "solemnities" there (southseas.nla.gov.au/journals/banks/17690610.html). Unlike Parkinson, Banks puts some of his observations on funerary customs into a systematic concluding description, retaining some of his field notes and altering or omitting others. He incorporates historiographic reflections here as well, comparing Tahitian social organization to "the early state of the feudal laws" of Europe (southseas.nla.gov.au/journals/banks_remarks/185.html). Parkinson's more detached observations emphasize the stages of the funeral process and the construction of the sites, particularly the "sort of stone pyramid" that becomes the permanent resting place of the bones after decomposition (35).

In his concluding "Remarks" Banks argues explicitly that some of these sites are "of great antiquity," citing the use of "immensely large" stones assembled in "rough" fashion without mortar. D'Hancarville uses the same criteria to confirm the legendary antiquity of the walls of Tyrinth (I:108). The antiquarian emphasis on architecture and funerary customs, brought together so neatly in Gough's title, *Sepulchral Monuments*, contributed significantly to the "temporalizing" classification of Pacific cultures in the Cook era. Parkinson's and Banks's writings on the *marae* demonstrate this influence, and they also enact a substitution by which death becomes a major avenue for understanding the life of another culture. Fabian's concept of temporalization is appropriate here in part because the monuments and their presumed antiquity are overlaid on the existing temporality – a "coevalness" of the dead and the living – that gave meaning to the funeral rites themselves (cp. Fabian 34). Once again, temporal distance, with its attendant objectivity, takes the place of cultural difference. But at the same time, these quasi-antiquarian studies recognize the independent history of Pacific cultures in at least a rudimentary fashion, insofar as the analogy to Western antiquity becomes visible as analogy. This essay is not the place to address more recent developments in paleoanthropology, but it is worth noting the continued importance of burial sites in the reckoning of Pacific antiquity, from the forty-thousand-year-old Lake Mungo site in Australia to the Neolithic cemetery discovered in Vanuatu in 2003.

V. Histories

In the art historical domain, d'Hancarville and Guasco again stand out for connecting close ethnographic attention to religion with speculation about human origins, a concern that recedes into the background with Banks and Parkinson. D'Hancarville, in particular, adopts the premise that customs and manners provide the best evidence concerning prehistory. This is the premise that J.R. Forster brings to bear on Pacific cultures as well, though "arts and sciences" such as *tapa* (barkcloth) manufacture are more central for him than funerary and other religious customs.[16] D'Hancarville combines the close attention to religion that we saw in Banks and Parkinson with theoretical concerns more akin to Forster's. For colonial voyagers, ethnography was part of natural history, and natural history played a significant role for the historians of ancient art as well. Pliny's *Natural History* was crucial for d'Hancarville, who concentrates on ancient sources, though he faults Pliny for subordinating the history of art to the history of nature (III:iii). D'Hancarville also drew extensively on his own observations of Neapolitan customs and manners as a form of field evidence (however indirect) for his interpretations of ancient vase paintings and the religious customs he thought they depicted. Guasco owed a great part of his art history to the ethnographic observations of contemporary voyage accounts, and Friedrich Schlegel, among other contemporaries, spoke casually of the "natural history of art" as an area of study, some vestiges of which survive today in the anthropology of art.[17] D'Hancarville boasts that he will be guided by the "monuments of art itself" (IV:vii) in his voyage into the deep uncharted past, but Pliny and natural history prove hard to escape. So too does travel narrative, represented heavily in d'Hancarville's as in other histories of ancient art by Pausanias's *Description of Greece*, the only record remaining of numerous ancient works.

The traveller's field notes have a special resonance for the anthropology of art in the era of the scientific voyage narrative. Guasco makes this resonance especially clear by juxtaposing Pausanias regularly with Engelbert Kaempfer, Jakob Roggeveen, and dozens of other contemporary voyagers represented in the fifty-volume (octavo) *Histoire Générale des Voyages* that Guasco owned and duly cited in his notes in 1768.[18] Within five years, the publication of Bougainville's and Cook's voyages made even more material available. D'Hancarville uses modern travel narrative much more sparingly, and his scepticism informs some compelling disciplinary questions that self-consciously scientific voyagers such as Forster must also have begun to ask themselves: does art belong to natural history? does ethnography? do customs or artefacts provide reliable records of the period before writing? D'Hancarville sets himself apart from antiquarianism as well, or at least from its prevalent negative stereotype, by insisting on a critical use of both ancient and modern sources. Even so, he remains deeply indebted to natural history and antiquarian scholarship, and the intensely situated nature of his archaeological work – an aspect of archaeology recovered and embraced by some recent practitioners – suggests that Pliny's fusion of art and natural history may have been more than accidental after all.[19]

D'Hancarville's project of recovering history from art and myth commits him above all to a history of religion. His methodical translation of mythic time into human prehistory provides the first stage of this history, what might be called an evolutionary narrative concerning the co-adaptation of religion and art.[20] Engraved gems and other types of jewellery evolved, he argues, as increasingly portable forms of the *boetile* or god-stone, as amulets that protected the wearer against evil spirits (*mauvais Génies*) (IV:28). As with many other artefacts examined in the course of the work, these amulets are both "assuredly of the highest antiquity" (27n) and yet – in a temporalizing view of European space – the object of "a custom still practiced today in my native country" (27n), presumably the country around Nancy, where d'Hancarville was born. Ancient vase paintings, in this analysis, show us the context in which *boetiles* and other ritual objects continued to feature not just in private devotions but in public worship. The first *boetiles* were anointed with oil (here d'Hancarville borrows Guasco's illustration from Genesis 28:18) and "wrapped in bands of wool," a practice that accounts for the myth of Kronos eating a stone wrapped in cloth and taking it for the infant Zeus (26–7n). In their capacity as illustrations of traditional practices, even Greek vase paintings become monuments of a prehistoric phase in the development of art, expanding the historical domain by providing visual and formal cues for conjectures about the "infancy" of art.

By framing art as a "natural writing" and mythology as an "aesthetic religion," d'Hancarville created "a narrative of the cultural origins of art preceding historical times," as Pascal Griener has observed (64). To explain the novelty of this approach, Griener fixes specifically on archaeological images, especially the famous engraving of a partially excavated tomb at Trebbia in d'Hancarville's second volume: "they illustrate perfectly a new conception of history as the resurrection of the past and an approach to art history strictly tied to the history of religion." "Like many Enlightenment philosophers," Griener adds, d'Hancarville was secular in outlook (59), and he is sometimes compared to Rousseau in particular. Jenkins and Sloan suggest that d'Hancarville's "primitive" artist, "like Rousseau's noble savage, stands intelligent but culturally naked before us" (151).

Like the Forsters, however, d'Hancarville uses Rousseau and the other conjectural historians critically and selectively. His writing shows their influence less strongly than Winckelmann's, whose art history tends to become "a general history of ancient peoples," as Elisabeth Décultot has argued (45). Quite possibly with Winckelmann in mind, d'Hancarville makes a point of deriving the history of nations, conversely, from the history of the arts (IV:5): he repeatedly revisits the prehistoric succession of southern Italian peoples, concluding with a long excursus note revised on the basis of new evidence from their monuments (IV:73–96n). Guasco, too, takes "the steps of the human mind" for his subject and likewise insists that antiquarianism becomes philosophical when joined to the history of manners (ii, v). But for Guasco, manners should be progressive, and the history of pagan art is merely the history of superstition. Guasco premises an originary monotheism, of which all forms of polytheism or "fetishism" are merely the decadent descendants (9–11; cf. G. Forster I:170–1). He relies on the history of civil society for his notion of progress both in morality and religion (26–7), and even more on sacred history. Therefore, modern "primitive" peoples who have not yet found their way back to monotheism do not fare well in Guasco's comparative ethnography: they are "nations naissantes et sauvages" (193) and are, he argues, everywhere the same (12, 22). These ignoble savages place total reliance on their "fetishes," a term that encompasses everything from the colossal heads of Easter Island (193) to figures of the virgin in Naples (202n). Guasco's summary rejection of devotional art that "makes philosophy blush" (229) represents a horror of "superstition" that was dismissed more or less successfully by those histories of primitive art – including d'Hancarville's and that of Forster, among other voyagers – that pursued art and religion more fully into the shadowy domain of prehistory.

While the voyagers were apt to translate distance into time, the comparatively sedentary d'Hancarville translated time into distance in an extended metaphor that maps deep time, in Romantic fashion, as a journey of imagination:

> Antiquity is a vast country, separated from ours by a long interval of time; some travellers have discovered its coasts almost waste, others more undertaking have dared to push on to its very heart, where they have found but the dismal rubbish of towns formerly magnificent, and Phantoms of incredible description. My two first Volumes may be looked upon as attempts, to discover unknown lands; I have endeavoured to fix the situation of some places, but for want of instruments, not being able to do it with all the nicety I would have wished for ... I have taken measures to ... rectify the errors. (III:3)

D'Hancarville visited the coasts of Phoenicia, Etruria, Ausonia, Pelasgia, and many other quasi-historical nations that were almost as spectral as Cook's Southern Continent on the contemporary map of prehistory. His absorption in the cultural landscape of Vesuvius, while studying vases recovered from ancient tombs carved into the tuff produced by more ancient volcanoes, gives his speculations a geological ground. Just as Hamilton, by frequent repetitions especially apparent in his commentary on the plates in *Campi Phlegraei*, establishes the abundance of local instances that confirm volcanic evidence of unsuspected antiquity, so too d'Hancarville multiplies instances of artefacts that attest to the development of the arts at periods far earlier than those located by Winckelmann and other predecessors. Debates about the Easter Island heads (*moai*, also made of tuff) among Pacific voyagers similarly suggest that deep ethnographic time emerges where geology and human prehistory meet. Disagreement concerning the putative antiquity of these sculptures in voyage narratives that frequently cited each other – including those of George Forster (I:320) and La Pérouse (II:85–8) – depended as much on debates about cultural degeneration as on geological analyses of the stone heads. "Ever since" the production of these temporalizing voyage narratives, in Fabian's view, "anthropology's efforts to construct relations with its Other ... implied affirmation of difference as *distance*" (16). Deep time, so often constituted in opposition to human historical time, also has roots in the ethnographic and aesthetic experiences of the voyagers, whose "history of mankind" buckled under its efforts to fold in the histories of others.

NOTES

1 See J.R. Forster 175 for his most direct approach to this distinction. On the broader issue, see Marks ch. 1. David Bindman offers an insightful reading of both Forsters in relation to ideas about race and aesthetics (123–50, 173–81). See also Schmied-Kowarzik.

2 In a curious testament to this reciprocal exchange, Jonathan Lamb declares that "uncertain anthropology ... yields to the theories of cultural difference and change propounded first by the Forsters and then by Adam Ferguson; Henry Home, Lord Kames; John Millar; and Gottfried Herder" (77) – even though all four of these authors' treatises on the history of civil society were published *before* the Forsters returned on the *Resolution* in 1775. On this exchange of ideas between voyagers and philosophers, see also Lansdown 64–72.

3 Bernard Smith has observed that, for some explorers, the Pacific voyage was an extension of the Grand Tour of classical sites in Italy (17). For one set of archaeologically inspired conjectures on prehistory, see d'Hancarville IV:73–6n.

4 I refer primarily to Foucault's argument about Cuvier rather than the related arguments about the human sciences in the same volume. On the historical turn as it relates to early geology, see also Heringman, "'Very vain.'"

5 Thomas points out that pre-evolutionary anthropology was actually *less* likely to regard "primitive" peoples such as the Fuegians as "living exemplars of primeval stone age ways of life" (Forster xxx). The dawning of evolutionary time on the *Beagle* voyage is conventionally ascribed exclusively to the influence of Lyell on Darwin. Darwin is still a long way from evolution, and particularly human evolution, at this point. Evolutionary aesthetics today, however, shows some intriguing parallels to the voyage narratives in its linkage between art and human origins. See Dissanayake and, for a more archaeological approach, Coe.

6 Guasco continues: "There, I repeat, is the antiquarian philosophy, but how few antiquaries are philosophers!" (*De l'usage des statues* xiii–xiv; qtd. in Jenkins and Sloan 99). I adopt Jenkins and Sloan's translation of this passage from their very valuable discussion of Guasco in the context of d'Hancarville's art history.

7 Other historians of ancient art, drawing on some of the same ancient authors and monuments, also made strong connections with religion, but Guasco is unique in his strong emphasis on travel narrative as well as his insistence that all worship involving devotional objects – including popular Catholicism in Italy (202) – is an idolatrous misuse of sculpture.

8 Their disagreement on religious issues may be one reason why he took Guasco's ideas without acknowledgment for the third and fourth volume of this work. D'Hancarville felt justified in his plagiarism partly because he had expressed some of the same ideas in his first two volumes, which appeared before Guasco's book, and he almost surely noticed that Guasco recycled his engraving. See further Griener 81–2 and Heringman, *Sciences of Antiquity* 159–60.

9 "When I saw the hatchets belonging to the Tahiti islanders in the museum of Signor Poli in Naples," Giovene wrote, "I was surprised to find that they resembled exactly those from Pulo at Molfetta" (qtd. in Toscano 231–2).

10 In Hamilton's words, "all the implements of stone brought by Mess. Banks and Solander from the new-discovered islands in the South-Seas, are evidently of such a nature as are only produced by Volcanos" (I:84n).

11 For a critical reflection on the comparison to ancient Greece specifically, see George Forster I:232. D'Hancarville's literal interpretation of Greek myth as a record of prehistoric events (III:206–7n) parallels the ancient-primitive analogy deployed in voyage narratives inasmuch as their apparent Greekness confirms the "primitive" or prehistoric character of Pacific peoples.

12 I am conflating passages from the very useful introduction to Lansdown's book, an anthology of Pacific writings, in which he discusses the classical legacy (11–12) and introduces his concept of bipolar vision (16), with the introduction to his section on the "noble savage," where he develops his distinction between cultural and chronological primitivism (65) and his reading of Rousseau. Bindman and Lansdown offer a larger European framework for understanding the voyages, which is just as important – especially in the case of continental intellectuals such as the Forsters – as the Scottish Enlightenment framework emphasized by Thomas and Guest. This framework is also in play in Guasco's art historical study of "people who in our own times still live in a state of barbarism," quoted above.

13 On Banks's gentlemanly classicism, see further Joppien and Smith I:21. Parkinson was ahead of his time in suggesting that Polynesians were capable of deliberate navigation over long distances (over two thousand miles) (*Journal* 125). Cook himself rejected the possibility, and it was not taken seriously by scholars before the nineteenth century. See Durrans 151, 153.

14 Though a member of the Society of Antiquaries, Banks was not allied with the Gothic faction of that body, led by Richard Gough. He did, however, publish *An Elegy on the Demolition of the Spires of Lincoln Minster*.

15 The images in Parkinson's *Journal*, along with his plain descriptions (26, 70–1), stand out as the most detailed and exact in this whole body of work on the subject. Cook's description of a *tupapau* closely resembles George Forster's, but since no artist accompanied him in this instance, the Woollett/Hodges image was used to illustrate his description, together with his inquiries concerning human sacrifice (Cook I:184–5). Hodges's original sketch is reproduced in Joppien and Smith, vol. 2, Fig. 52A.

16 This modern-ancient trajectory is especially clear when he reconstructs the ancestral culture of Polynesians from the customs of modern Caroline Islanders (*Observations* 352–7).

17 Schlegel observed in a letter to his brother August Wilhelm on 5 April 1794 that "the history of Greek poetry is a complete natural history of the beautiful and of art, and for that reason my work is – aesthetics" (229). See also Gell.

18 Guasco specifically cites a Mexican voyage from vol. XLVIII of this edition (Guasco 59), but elsewhere he cites the quarto edition of the same text, originally published serially between 1746 and 1759 under the editorship of Abbé Prévost. He cites several other collections of voyages as well.

19 D'Hancarville himself criticized Pliny's art history as accidental (IV:119n) – an epiphenomenon of his *Natural History* – yet relied on him exclusively for what he took to be the ancient theory of art (e.g., IV:13–14n). On situated archaeology, see Tilley.

20 Though pre-evolutionary, d'Hancarville's thesis that art and ritual are coordinated adaptive behaviours is in some ways quite close to contemporary evolutionary aesthetics as practiced by Dissanayake and Coe, among others (see note 5, above). In a remarkable gloss on the "circle called mythic," d'Hancarville historicizes the period described by Proclus as extending from Uranus to Ulysses (III:206–7n). The names of these characters, like early sculpture, allegorize their essential personal or biographical traits. By pairing this metonymic "discourse" with the infant "forms" of sculpture, he recodes mythical time as human prehistory, as an evolutionary stage in the history of art. On the general problem of dating in d'Hancarville, see Jenkins and Sloan 149–55.

WORKS CITED

Banks, Joseph. *An Elegy on the Demolition of the Spires of Lincoln Minster*. London: W. Bulmer, 1807. Print.

Banks, Joseph. *The* Endeavour *Journal of Joseph Banks. South Seas: Voyaging and Cross-Cultural Encounters in the Pacific, 1760–1800.* [National Library of Australia Online Edition.] http://southseas.nla.gov.au/index_voyaging.html. Web. 13 December 2012.

Beaglehole, J.C., ed. *The Journals of Captain James Cook on His Voyages of Discovery.* 3 vols. Cambridge: Hakluyt Society, 1955. Print.

Bindman, David. *Ape to Apollo: Aesthetics and the Idea of Race in the Eighteenth Century.* Ithaca: Cornell UP, 2002. Print.

Coe, Kathryn. *The Ancestress Hypothesis: Visual Art as Adaptation.* New Brunswick, NJ: Rutgers UP, 2003. Print.

Cook, James. *A Voyage towards the South Pole.* 2nd ed. 2 vols. London, 1777. Print.

Darwin, Charles. *Journal of Researches into the Natural History and Geology of the Countries Visited During the Voyage of HMS* Beagle. 2 vols. New York: Harper and Brothers, 1846. Print.

De Brosses, Charles. *Du culte des dieux fétiches, ou Parallele de l'ancienne Religion de l'Egypte avec la Religion actuelle de Nigritie.* [Paris]: n.p., 1760. Print.

Décultot, Elisabeth. "Comment l'art est-il venu aux Grecs? Winckelmann face à Shaftesbury, Caylus, et Herder." *Klassizismen und Kosmopolitismus: Programm oder Problem? Austausch in Kunst und Kunsttheorie im 18. Jahrhundert.* Ed. P. Griener and K. Imesch. Zürich: Schweizerisches Institut für Kunstwissenschaft, 2004. 45–58. Print.

D'Hancarville [Pierre François Hugues]. *Collection of Etruscan, Greek, and Roman Antiquities.* 4 vols. Naples: F. Morelli, 1766–67 [i.e., 1768–76]. Print.

Dissanayake, Ellen. *Art and Intimacy: How the Arts Began.* Seattle: U of Washington P, 2000. Print.

Durrans, Brian. "Ancient Pacific Voyaging." *Captain Cook and the South Pacific.* Ed. T.C. Mitchell. London: British Museum, 1979. 137–66. Print.

Fabian, Johannes. *Time and the Other: How Anthropology Makes Its Object.* New York: Columbia UP, 1983. Print.

Ferguson, Adam. *An Essay on the History of Civil Society.* Ed. Fania Oz-Salzberger. Cambridge: Cambridge UP, 1995. Print.

Forster, George. *Voyage Round the World.* 1777. 2 vols. Ed. Nicholas Thomas and Oliver Berghof. Honolulu: U of Hawaii P, 2000. Print.

Forster, John Reinhold. *Observations Made During a Voyage Round the World.* 1778. Ed. Nicholas Thomas, Harriet Guest, and Michael Dettelbach. Honolulu: U of Hawaii P, 1996. Print.

Foucault, Michel. *The Order of Things: An Archaeology of the Human Sciences.* New York: Random House, 1966. Print.

Gell, Alfred. *Art and Agency: An Anthropological Theory.* Oxford: Clarendon P, 1998. Print.

Gough, Richard. *Sepulchral Monuments in Great Britain. Applied to Illustrate the History of Families, Manners, Habits, and Arts, and the Different Periods from the Norman Conquest to the Seventeenth Century.* 2 vols. in 5. London: Printed by J. Nichols, for the author, 1786–96 [1799]. Print.

Gould, Stephen Jay. *Time's Arrow, Time's Cycle: Myth and Metaphor in the Discovery of Geological Time.* Cambridge, Mass.: Harvard UP, 1987. Print.

Griener, Pascal. *Le Antichità Etrusche Greche e Romane 1766–1776 di Pierre Hugues d'Hancarville. La pubblicazione delle ceramiche antiche della prima collezione Hamilton.* Rome: Edizione dell'Elefante, 1992. Print.

Guasco, Octavien de. *De l'usage des Statues chez les Anciens. Essai historique.* Brussels: J.L. de Boubers, 1768. Print.

Hamilton, Sir William. *Campi Phlegraei: Observations on the Volcanos of the Two Sicilies.* 2 vols. Naples, 1776. Print.

Hawkesworth, John. *Account of the Voyages undertaken by the order of His present Majesty for making discoveries in the Southern Hemisphere.* 3 vols. London: Printed for W. Strahan and T. Cadell, 1773. Print.

Heringman, Noah. *Sciences of Antiquity: Romantic Antiquarianism, Natural History, and Knowledge Work*. Oxford: Oxford UP, 2013. Print.

Heringman, Noah. "Romantic Explorations of Antiquity: Customs and Manners in d'Hancarville's *Antiquités Etrusques, Grecques, et Romaines*." *Romantic Explorations: Papers from the Koblenz Conference*. Ed. Michael Meyer. Trier, Germany: Wissenschaftlicher Verlag Trier, 2010. 149–71. Print.

Jenkins, Ian, and Kim Sloan, eds. *Vases & Volcanoes: Sir William Hamilton and His Collection*. London: British Museum, 1996. Print.

Joppien, Rüdiger, and Bernard Smith. *The Art of Captain Cook's Voyages*. 3 vols. New Haven: Yale UP, 1985. Print.

Lamb, Jonathan. *Preserving the Self in the South Seas, 1680–1840*. Chicago: U of Chicago P, 2001. Print.

Lansdown, Richard. *Strangers in the South Seas: The Idea of the Pacific in Western Thought*. Honolulu: U of Hawaii P, 2006. Print.

La Pérouse, Jean-François. *Voyage de La Pérouse autour du monde*. 4 vols. Ed. M.L.A. Milet-Mureau. Paris, 1797. Print.

Marks, Jonathan. *Human Biodiversity: Genes, Race, and History*. New York: Aldine de Gruyter, 1995. Print.

Meek, Ronald. *Social Science and the Ignoble Savage*. Cambridge: Cambridge UP, 1976. Print.

Meillassoux, Quentin. *After Finitude: An Essay on the Necessity of Contingency*. Trans. Ray Brassier. London: Continuum, 2008. Print.

Parkinson, Sydney, and Stanfield Parkinson. *A Journal of a Voyage to the South Seas, in His Majesty's Ship, the Endeavour*. 1772. Adelaide: Libraries Board of South Australia, 1972. Print.

Putnam, William C. *Geology*. Fourth ed. Ed. Edwin E. Larson and Peter Birkeland. New York: Oxford UP, 1982. Print.

Rennie, Neil. *Far-Fetched Facts: The Literature of Travel and the Idea of the South Seas*. Oxford: Clarendon P, 1995. Print.

Robertson, William. *The History of Scotland During the Reigns of Queen Mary and of King James VI*. 2 vols. London: Printed for A. Millar, 1759. Print.

Rossi, Paolo. *The Dark Abyss of Time: The History of the Earth and the History of Nations from Hooke to Vico*. Trans. Lydia Cochrane. Chicago: U of Chicago P, 1984. Print.

Rudwick, Martin. *Bursting the Limits of Time: The Reconstruction of Geohistory in the Age of Revolution*. Chicago: U of Chicago P, 2005. Print.

Schlegel, Friedrich. *Kritische und ästhetische Schriften*. Ed. Andreas Huyssen. Stuttgart: Reclam, 1987. Print.

Schmied-Kowarzik, Wolfdietrich. "Der Streit um die Einheit des Menschengeschlechts: Gedanken zu Forster, Herder, und Kant." *Georg Forster in interdisziplinärer Perspektive:*

Beiträge des internationalen Georg Forster-Symposions. Ed. Claus-Volker Klenke. Berlin: Akademie-Verlag, 1994. Print.

Smith, Bernard. *European Vision and the South Pacific, 1768–1850: A Study in the History of Art and Ideas*. Oxford: Clarendon P, 1960. Print.

Thomas, Nicholas. *In Oceania: Visions, Artifacts, Histories*. Durham, NC: Duke UP, 1997. Print.

Tilley, Christopher. *The Materiality of Stone: Explorations in Landscape Phenomenology*. Oxford: Berg, 2004. Print.

Toscano, Maria. "The Figure of the Naturalist-Antiquary in the Kingdom of Naples: Giuseppe Giovene (1753–1837) and His Contemporaries." *Journal of the History of Collections* 19 (2007): 225–37. Print.

Chapter Five

Malthus Our Contemporary? Toward a Political Economy of Sex

MAUREEN N. MCLANE

Species Means Guilt.

– Bruce Andrews

... desire,
as Aristotle knew, is all
angle, and so he gave us the math
to keep track of our loves: *Number*,
he said, *has two senses: what is counted*
or countable, and that by which we count.

– Angie Estes, "Take Cover," *Tryst*

I. Thinking with Malthus

I begin with a remark of Hazlitt, who in *The Spirit of the Age* glossed an aspect of what we might call The Malthus Meme: "There is this to be said for Mr. Malthus, that in speaking of him, one knows what one is talking about" (254). But do we know what we're talking about when we talk about Malthus? This essay proposes to reopen the question of "Malthus," and of Malthusian reckoning – or rather, this essay hopes to suggest how Malthusian political economy might provide a horizon for thinking about change in structure, in historicizable yet still resonant ways. Malthus is, among many things, a Romantic theorist of change in biological populations; he is also necessarily a theorist of time – the time of production and reproduction within and across generations (and indeed across species), how this timing might be sped or slowed. What one sees in Malthus is that these populations – and the very category "population" – resist a purely biological or species definition. Malthus's work intriguingly anticipates the possibility of rethinking biological networks through information theory. Malthus also emerges as a sexual theorist worth taking seriously, one who "marks time" (to invoke the title of this volume): the time of desire and the time of labour emerge from and fold themselves into (or are truncated within) broader networks of social structure. The problem of population in Malthus is always already a problem of time as well as sociocultural formation. So while we may in some respects know what we are talking about when we talk about Malthus, it may be that "Malthus" obscures how we might still think with Malthus.

For one wittily efficient take on The Malthus Meme, consider Stephen Leacock, Canadian poet and humourist, in his Depression-era "Oh! Mr. Malthus!" (1936):

"Mother, Mother, here comes Malthus,
Mother, hold me tight!
Look! It's Mr. Malthus, Mother!
Hide me out of sight."
This was the cry of little Jane
In bed she moaning lay,
Delirious with Stomach Pain,
That would not go away.
All because her small Existence
Over-pressed upon Subsistence;
Human Numbers didn't need her;
Human Effort couldn't feed her.
Little Janie didn't know
The Geometric Ratio.
Poor Wee Janie had never done
Course Economics No. 1;
Never reached in Education
Theories of Population, –
Theories which tend to show
Just how far our Food will go,
Mathematically found
Just enough to go around.
This, my little Jane, is why
Pauper Children have to die.
Pauper Children underfed
Die delirious in Bed;
Thus at Malthus's Command
Match Supply with true Demand.

"Theories of Population," "The Geometric Ratio," "subsistence," "Supply" and "Demand" – all are keywords of Malthusian political economy, "geometric ratio" perhaps the most marked as a Malthusian phrase, designating the rate of increase of animal populations (destined to outstrip any increase in the vegetal kingdom, which according to Malthus would increase only along an "arithmetic ratio"). Malthus's concerns and his conceptual apparatus were of course those of his age, yet as Leacock's bouncing couplets attest, they were still and newly citable in the 1930s; and in several ways, *mutatis mutandis*, Malthus's concerns remain ours.

It seems timely, then, to assess whether and how Malthus might still be "good to think with," in Claude Levi-Strauss's terms (*Totemism* 89). That Malthus is considered still good to think with is suggested by any Google search, as well as by more specialized research inquiries into demography, population theory, and, indeed, political economy. An under-acknowledged Malthusian formula appears in popular TV: on *The Daily Show*, Jon Stewart casually invoked geometric versus arithmetic progression in a conversation with his guest Jim Cramer, of CNBC's "Mad Money" ("Jim Cramer"). And those contributing to more recondite conversations also invoke Malthus: the *London Review of Books* published a letter from historian Jonathan Steinberg of the University of Pennsylvania, who notes that Malthus was a sharp critic of Ricardo's commitment to "the illusion that economics could be reduced to mathematical formulae." Steinberg concludes his letter thus: "A fascinating, yet unwritten, chapter in the history of economic thought would explore how Malthus's *Principles of Political Economy* lay dormant for a century until Keynes rediscovered them in the 1920s" (6). The post-Crash (that is, post-September 2008) return to Keynes has prompted, as well, a return to some important aspects of Malthus, who, as Keynes recognized (and Steinberg reminds us), was among the earliest and astutest critics of Ricardo's theory that demand could never be deficient.

So Malthus, or "Malthus," or "Malthusianism," remains contemporary, however much the specifics of his engagement with Condorcet and Godwin (say), or his critique of the Poor Laws, or his controversial theodicy in the final chapters of his 1798 *Essay*, might be relegated to the footnoted dustbins of scholarship. Indeed, the case of Malthus is an interesting one regarding the matter not only of the history of the production of knowledge but its historicity, for if Malthus's cautions about population outstripping food supply seemed by the mid-nineteenth through the mid-twentieth centuries to be overzealous given advances in food production (not to mention birth control), in certain respects Malthus may be once again, to invoke Foucault, "in the true." As Alan MacFarlane, the Cambridge historian and anthropologist, put it in "The Malthusian Trap" (2005),

> Malthus wrote before the huge resources of energy for humankind locked up in coal and then oil became widely available. For a while, from the middle of the nineteenth century,

> it looked as if the Malthusian trap was no longer operative. A combination of science (in particular chemistry) and of new resources had made it possible to more than double production in each generation. First England, then parts of Europe, Japan and elsewhere escaped from the trap. His laws could be inverted. Population grew slowly, resources exponentially. Yet now in the early twenty-first century, as the resources reach their limits and the external costs of the massive use of carbon energy become apparent in pollution and global warming, it appears that the ghost of Malthus has arisen again. (2)

But if Malthus remains, in however vexed a fashion, our contemporary, to what extent is Malthus a Romantic, much less a Major Romantic Writer? That was the implicit polemical thrust of a 2009 MLA panel, "Major Romantic Writers," organized and chaired by Kevin Gilmartin – a panel that featured presentations on Malthus, John Stuart Mill, and Thomas Paine.[1] It seems to me conclusively demonstrated in an efflorescence of work over the past twenty-five years that Malthus was not only a major writer of the period, not only a major point of reference for such go-to Romantics as Wordsworth, Southey, Coleridge, and Shelley, but a Major Romantic Writer in his own right. Let's agree not to go down the wormhole of contended period and movement terminology; let us, for the sake of argument, assume that "Romantic writer" designates a writer in the period formerly known as Romantic – say, 1789–1832. Or 1770–1830.

Or maybe we should not concede this at all. To do so might well evacuate "Romantic" of any specificity it might yet have. That literary scholars and cultural historians working in both Romantic and Victorian periods have undertaken intense, acute, illuminating reckonings with Malthus in recent years – and here I think of Frances Ferguson, Catherine Gallagher, Mary Poovey, and Philip Connell most recently – suggests, however, that the very horizon of the literary, not to mention of Romanticism across the board, has decisively altered.[2] Writing of "the Romantics and the political economists," Catherine Gallagher "urge[s] an awareness of the unacknowledged continuity of shared premises ... Romanticism and political economy should be thought of as competing forms of 'organicism'" (73). Mary Poovey proposes Malthusian moral arithmetic as a crucial element in the transition from conjectural history to political economy.[3] Frances Ferguson has persuasively specified the logic of a "Romantic political economy," informing Wordsworth's *Prelude* as much as Malthus's *Essay* ("Malthus, Godwin" 106). And Richard Bronk, coming from another intellectual and professional formation in the City of London, has published an entire book rehabilitating not just Malthus or political economy but the very notion of "The Romantic Economist." Historian Donald Winch has incisively and extensively analysed Malthus's contribution to a discourse on "riches and poverty" – not least in dialogue with the Lake poets – and has explored Malthus more broadly as a key figure in the history of economic thought alongside Adam Smith. And for an account of Malthus disseminated, see James P. Huzel.

Given this richly suggestive work, then, one can no longer reproduce the caricatures of Malthus that the poets and writers of his and succeeding ages were prepared to do: the image of Malthus as Parson Malthus (viz. Marx), Priest of the Workhouse, Contemptor of the Poor, Complacent Supporter of Hierarchical Order, or Nattering Nabob of Sexual Negativism has given way to a richer, more nuanced, historicized picture.[4] Connell (for example) has given us a Malthus as conversant with dissident intellectual inquiry as Mary Wollstonecraft; a frequenter of Joseph Johnson's circle; a Cambridge-schooled latitudinarian who sought to offer a progressive Anglican theodicy; and ultimately, especially in revisions to and later editions of his 1798 *Essay*, a reasoned social meliorist visionary grounded in natural theology.[5] Connell's Malthus shares the project of disciplinary humanization and concern for social sympathy with Wordsworth and Southey; he turns out to be, in his way, as sex-positive as Shelley.[6]

Indeed, it was the very high valuation Malthus placed on sex – as a need as urgent as that for food – that earned him the disgust of Coleridge (as Connell notes), Southey (as Gallagher observes), Hazlitt (see his portrait in *The Spirit of the Age*), and later John Stuart Mill, among others. One can live without sex, Malthus's critics often said, but not without food.[7] About which more later. For this is true for individuals yet not for the species. Part of the scandal of Malthus's essay is its bravura albeit unsteady tacking between individual and species, between individuals as socially inscribed desiring-machines and humans as a generic animal species subject to the law of population as well as the law of desire. (And here the reader may register the first of several Deleuzian turns on the Malthusian body, considered not only as an organic, reproducing body but also as a "desiring-machine." I hope this slight anachronism will serve to defamiliarize or reframe productively the Malthusian problematic and our possible engagement with it; this Deleuzian heuristic is one Slavoj Žižek also activates in his own mediations on the problem of the body, consciousness, and the "loop of Life" in *Organs Without Bodies* [120].)

It is worth asking what the Principle of Population is a Principle *of*. On the level of explicit proposition, Malthus announced: "I think I may fairly make two postulata: First, That food is necessary to the existence of man. Secondly, That the passion between the sexes is necessary and will remain in nearly its present state" (19). If Deleuze and Guattari assert in *Anti-Oedipus* that "*There is only desire and the social*" (29; emphasis in original), Malthus would seem to assert that there is only desire and food. The elision of the social at crucial theoretical junctures, only to have it return with a vengeance at others, is symptomatic of Malthusian political economy – and indeed, we could argue, of political economy *tout court*.

We might recall Louis Althusser's handy definition of ideology as the production of obviousnesses;[8] that the desire for food and the desire for sex are equally pressing and exigent desires continues to animate the latest policy papers.[9] Malthus proposes two things: that the level of population is constrained by the means of subsistence; and that if a population outstrips its means of subsistence, it will be checked. Checks to population are, moreover, always already operating, in Malthus's view, through a complex amalgam of disease, malnourishment, natural disaster, and other miseries (the so-called positive checks) as well as, to a lesser extent, foresight (the preventive check) – to wit: celibacy and delayed marriage.

Malthus's argument proposes a theory about reproducing bodies; it is also, via "foresight," a theory of subjectivity and imagination. Malthus is one of the great, melancholy meditators on imagination and futurity. It is instructive to read Malthus not only as an incipient political economist or an early demographer, but as a theorist of the relation between what he called "mind" and "matter" – what we today might recode as "subjectivity" and "bodies."

There are several resources to help us recode Malthus; indeed, such re- or over-writing of previous discursive systems has been the hallmark of twentieth-century critique.[10] Consider Gayle Rubin's trenchant essay "The Traffic in Women: Notes toward a 'Political Economy' of Sex" (1975); Rubin follows Lacan in proposing that we "conceive of psychoanalysis as a theory of information rather than organs" (188). We might follow this through for Malthus, undertaking along the lines of the extremely productive de-biologization of Freud a de-biologization of Malthus. Or rather, we might think of Malthus's theory as a theory of information rather than one of organs or bodies or populations – a media theory, perhaps, rather than an algorithm of reproduction and death. Or perhaps it is more precise to say that, given transformations in biology, any biological theory is also, simultaneously, a theory of information.

That Malthus might be participating in a broader network of information was suggested in his own moment by none other than Thomas Love Peacock. In his *Paper Money Lyrics*, written during the economic crisis of 1825–6, one lyric features a righteous Scot named MacFungus – apparently a Malthusian as well as a fiscal conservative – who inquires:

A weel sirs, what's the matter?
An hegh sirs, what's the clatter?
   Ye dinna ken,
   Ye seely men,
Y'ur fortunes ne'er were batter.
There's too much population,
An' too much cultivation,
An' too much circulation,
That's a' that ails the nation. (104)

Other verses on this topic include the "Lament of the Scotch Economists on the Extinction of One-Pound Notes," the "Chorus of the Northumbrians on the Prohibition of Scotch One-Pound Notes in England," and the "Chorus of Scotch Economists on a Prospect of Scotch Banks in England," a verse of which reads:

Come, sing as we've said it – Oho! Oho!
Sing "Free trade and credit" – Oho! Oho!
Sing "Scotch education,"
And "O'er-population,"
And "Wealth of the nation," – Oho! Oho!

These jokes point to, even as they enact, a broader circulation of ideas, a discursive economy, and what we might call an information network. The very proliferation of rhymes suggests the ease with which political-economical abstractions amplify themselves in an acoustic feedback loop – population, circulation, wealth of the nation. They become (at least for MacFungus) the very currency of the thinkable, the analytic of the situation. One notes, too, the persisting horizon of this analytic: the nation. Peacock intimates that what ails the nation is as much political economy – Scotch education – as "O'er-population."

Peacock's lyrics raise the question about the efficacy of information: Do these circulating concepts actually help to illuminate, much less relieve, the causes of distress in the social body? Is political economy just another instance of traffic in population? How does information move through, and constitute, actual bodies and minds and networks thereof?

To refocus Malthus via information theory and the problematic of subjectivity, Slavoj Žižek's work on "autopoiesis" might be of use. In *Organs without Bodies*, Žižek asks apropos of recent work on genetics and the problem of modelling "life": "[H]ow are we to pass from this self-enclosed loop of Life to (Self)Consciousness?" (120). As Žižek notes, theories of "autopoiesis" aim to evade the problems of "mechanism," but they introduce another set of questions instead: "The properly *materialist* problem is, How does *subjectivity* emerge in this reproductive cycle (of genes)?" (121).

Malthusian foresight marks exactly that passage – from the "self-enclosed loop of Life to (Self)Consciousness." Or rather, "foresight" marks the spot where consciousness might intervene in the very "loop of Life." For truly, despite the apparent algorithmic inexorability of Malthus's principle, what strikes this reader is Malthus's interest in imagining the potentially supervenient consciousnessness of his geometrically reproducing subjects – the consciousnesses of men, it should be said, for, aside from a few brief albeit extremely interesting forays, Malthus is overwhelmingly interested in the male subject, and in a particular kind of male subject. It is the male-as-prospective-husband who most frequently inspires Malthus's thought-experiments with foresight.

Malthus's essay is nothing if not the work of a sensitive husbandry – husbandry in all its sense: agricultural, matrimonial, economical, theological. In the logic of Malthus's essay, the divisions among humans and plants and animals are prior to the fall into gender. Malthus presupposes an anthropologic: it is precisely "the structure of the human mind" (119) that potentially distinguishes human reproductive fate, and certainly distinguishes human (or at least men's) experience from our fellow reproducers/desiring-machines on earth – plants and animals:

> [T]he effects of this check [on population] on man are more complicated. Impelled to the increase of his species by an equally powerful instinct [as that governing plants and animals], reason interrupts his career and asks him whether he may not bring beings into the world, for whom he cannot provide the means of subsistence. In a state of equality, this would be the simple question. In the present state of society, other considerations occur. Will he not lower his rank in society? Will he not subject himself to greater difficulties than he at present feels? Will he not be obliged to labour harder? And if he has a family, will his utmost exertions enable him to support them? May he not see his offspring in rags and misery, and clamouring for bread that he cannot give them? And may he not be reduced to the grating necessity of forfeiting his independence and of being obliged to the sparing hand of charity for support? (23–4)

Such a passage maps a complex dance between "the career" of sexual instinct (a transpecies career) and that of the sensitive husband; between "the career" of the generic human species and that of the consciously hesitating, individually desiring man who "in the present state of society" must ask himself even more complicated questions than would be required were he living in "a state of equality" (in which he would need to ask only whether he could provide the means of subsistence for a future family – not whether he could maintain them in the station to which they were accustomed or felt entitled). In this performance of the logic of foresight, we confront a series of escalating interrogatives. Such a passage offers a mimesis of consciousness in action, in which thinking is precisely not fucking.

The capacity for such thinking, such foresight, is not equally distributed among the population; here is one limit of "the preventive check." Malthus observes:

> The preventive check appears to operate in some degree through all the ranks of society in England. There are some men, even in the highest rank, who are prevented from marrying by the idea of the expenses that they must retrench, and the fancied pleasures that they must deprive themselves of, on the supposition of having a family.

> These kinds of considerations are certainly trivial, but a preventive foresight of this kind has objects of much greater weight for its contemplation as we go lower. (33)

If aristocrats are preoccupied with trifles and luxuries, nevertheless accession to consciousness withers once you move down the ladder to rungs below men "of education." Consider the labourer of sensibility, the lowly man of feeling, who might wish to marry his beloved: "Harder fare and harder labour he would submit to for the sake of living with the woman that he loves, but he must feel conscious, if he thinks at all, that should he have a large family, and any ill-luck whatever, no degree of frugality, no possible exertion of his manual strength could preserve him from the heart rending sensation of seeing his children starve, or of forfeiting his independence, and being obliged to the parish for their support" (35). The labourer *must feel conscious, if he thinks at all*: herein lies the crux of consciousness unevenly developed across all "ranks of society."

Though the preventive check operates "to some degree" throughout society, it also operates, or should operate, according to a class-stratified logic. "Preventive foresight" – the power of thinking futurity, of calculating costs and benefits – in fact should weigh heavier on the lower orders (though typically, according to Malthus, they refuse to feel it): the costs of not-thinking are so much higher (parish charity, social shame, malnourishment, death of kin versus the mere loss of luxuries and station that await impulsive aristocrats). Malthus swiftly moves down through "all the ranks of society in England," passing from aristocrats to the "man of liberal education" to the "sons of tradesmen and farmers" (34) to "labourers" and finally "servants" (35). Moving through the ranks of society, he moves as well through stratified *mentalités* – for as one moves "two or three steps of descent in society" one encounters that threshold "where education ends and ignorance begins" (34).

What we have here, in this "sketch of the state of society in England" (35), is a highly organized picture of socialized subjectivity, one that would do Pierre Bourdieu proud: a hierarchy of consciousness presented as "things as they are" rather than a state of affairs that might be inquired into. (This hypostasis particularly outraged Hazlitt.) It is in this socially stratified logic, not in his swerve from thinking contraception, that we might find the heart of the Malthusian *impensé*. In such a picture, the quantitative mania and its aporia display themselves, for we soon find that comparing, much less quantifying, the happinesses (or miseries) of aristocrats with those of, say, labourers, proves elusive. (As Bronk observes, such impasses when confronting "incommensurable values" [xiv, 2, 172] inevitably haunt a utilitarian calculus.) Yet Malthus is dedicated to considering, as he repeatedly says, "the mass of happiness" (41).

Malthus devotes heartfelt sentences to what one feels is his core audience as well as his self-image:

> A man of liberal education, but with an income only just sufficient to enable him to associate in the ranks of gentleman, must feel absolutely certain that if he marries and has a family he shall be obliged, if he mixes at all in society, to rank himself with moderate formers and the lower class of tradesman. The woman that a man of education would naturally make the object of his choice would be one brought up in the same tastes and sentiments with himself and used to the familiar intercourse of a society totally different from that to which she must be reduced by marriage. Can a man consent to place the object of his affection in a situation so discordant, possibly, to her tastes and inclinations? (34)[11]

Not perhaps since Burke wept over Marie Antoinette had such a chivalrous outburst coloured the pages of a sociopolitical treatise.

For truly Malthus's essay is as much a theory of moral sentiments as a quantitative analytic. And this will be no surprise to anyone who has read Ferguson's landmark work on Malthus, Godwin, and the Spirit of Solitude; or Connell's differently oriented work on Malthus as a moderately progressive Anglican. Malthus's "sketch of the state of society in England" (35) is also a stadial theory of subjectivity and its possible progress, its development in time as well as its refinement ascending through social rank. A spectre is haunting these pages, the spectre of downward mobility and discordant or broken sympathy, not only the struggle for existence. And Malthus here reveals himself as much a theorist of companionate marriage as of anything else: in his mention of "[t]he woman that a man of education would naturally make the object of his choice," and her possible distress at consorting with farmers and tradesmen, we have the very sociologic underpinning a Jane Austen novel, not to mention an Oscar Wilde comedy or a novel by D.H. Lawrence. Here, rarely, the woman emerges explicitly as a separate, gendered, desiring, socially located subject; in most of Malthus's survey of the ranks of society, it is men's foresight that is scrutinized, men's need for caution and discipline that is discussed, men's desire that is liberated or constrained.

II. Desire Interminable? Matches and Measures

Aristotle defined man as the political animal; Marx analysed man as the labouring animal; Raymond Williams construed man as the communicating being; Malthus proposes man as the desiring being. But what does an organism want? To increase its numbers. Thus Darwin, following Malthus, enjoined us "never to forget that every single organic being may be said to be striving to the utmost to increase in numbers" (*On the Origin of Species*; qtd. in Malthus, *Essay* 159). Yet for all Malthus's "actuarial terror" (to invoke Ferguson, *Solitude and the Sublime* 118), his dystopia of inexorably, mathematically reproducing bodies, what an organism wants in Malthus is not so much to "increase its numbers" as to have sex. Malthus notably does not reckon much with the desire for children per se, for desired "increase in numbers." Here one encounters the crucial elision between individual desire and a species-condition: In what sense can a population desire anything? Malthus, in the end, is more focused on singly desiring bodies – or rather, on classes of desirers: human desiring-machines organized into social structure.

Malthus's theory of desire, of sex-passion, emerges in part in response to Godwin's speculations in *Political Justice* (1793), in particular Godwin's proposal that man's desire would atrophy as he marched into his perfectible future.[12] Godwin envisioned a future time "when the earth shall refuse itself to a more extended population" (776). It is striking when and where Malthus and Godwin converge: both are equally anxious about a depleted earth. Yet Malthus argues that the earth has always already refused itself to population – that we live amid extinguished millions. Godwin's solution, in this thought experiment about the soon-to-be-refusing earth, is to propose "men ... [who] will probably cease to propagate. The whole will be a people of men, and not of children. Generation will not succeed generation, nor truth have, in a certain degree, to recommence her career every thirty years" (776). In his *Essay*, Malthus quickly disposes of this aspect of Godwin's speculation: "towards the extinction of the passion between the sexes, no progress whatsoever has hitherto been made. It appears to exist in as much force at present as it did two thousand or four thousand years ago" (19). Malthus returns to this theme in Chapter 11: "No move toward the extinction of the passion between the sexes has taken place in the five or six thousand years that the world has existed" (76). Indeed, Malthus's principle implies quantifiable desire: "The passion of the sexes has appeared in every age to be so nearly the same that it may always be considered, in algebraic language, as a given quantity" (53).

This algebraic language outlines a mathesis of desire. (Here we recall one of my epigraphs: Angie Estes on Aristotle's "math / to keep track of our loves" might be relevant.) One confronts here the question of sex-passion as part of or opposed to bodily capacity in general: Is sexual desire a special kind of desire, or just desire in general, the general-equivalent of desire or passion in terms of the potentialities of humans as "desiring-machines"? Malthus famously rejected the labour theory of value (because he rejected the abstractions of exchange value that could not be reconverted into sustenance) (see, e.g., Gallagher, "Body Versus the Social Body" 93), but his treatise does implicitly propose a complex algebra for assessing the relations of desire, labour, reproduction, and death. What Malthus does not investigate, what he does not see the need to investigate, is the sexual division of labour and the sexualized division of its fruits.

It is precisely at this juncture that Rubin's essay might intervene: offering notes on a "political economy" of sex, she calls for an analytic capable of describing the "sex/gender system" in its full ramifications – "the set of arrangements by which a society transforms biological sexuality into products of human activity, and in which these transformed sexual needs are satisfied" (159). Rubin remorselessly culturalizes Malthus's variables (as Malthus himself does throughout his *Essay*, all editions):

> The needs of sexuality and procreation must be satisfied as much as the need to eat, and one of the most obvious deductions which can be made from the data of anthropology is that these needs are hardly ever satisfied in any "natural" form, any more than are the needs for food. Hunger is hunger, but what counts as food is culturally determined and obtained. Every society has some form of organized economic activity. Sex is sex, but what counts as sex is equally culturally determined and obtained. (165)

> Marx observed that beer is necessary for the reproduction of the English working class, and wine necessary for the French. (163)

Malthus counts by food and bodies. But these are nationalized and culturalized – not consistently treated as variables in a formula. Regarding food: "The labourers of the South of England are so accustomed to eat fine wheaten bread that they will suffer themselves to be half starved before they will submit to live like the Scotch peasants. They might perhaps in time, by the constant operation of the hard law of Necessity, be reduced to live even like the lower Chinese" (54). (One thinks of Samuel Johnson's chauvinistic joke in his *Dictionary*, regarding "oats": "a grain, which in England is generally given to horses, but in Scotland supports the people.") Such a passage is reminiscent of Malthus's reckonings with the different mentalities of gentlemen of education vs. farmers vs. labourers vs. servants: what counts as a good marriage, what counts as good food, is informed by far more than mere serviceability for copulating or eating.

In a version of You Are What You Eat, Malthus proposes a calculus of happiness via comparative foodstuffs: here, the dematerialized phrase "the means of subsistence" finally achieves a material specificity – even as the minds of the anxious, potential husbands elsewhere in the essay acquire very specific socio-cognitive contents when contemplating good marriages for their rank:

> Other circumstances being the same, it may be affirmed that countries are populous according to the quantity of human food which they produce, and happy according

> to the liberality with which that food is divided, or the quantity which a day's labour will purchase. Corn countries are more populous than pasture countries, and rice countries more populous than corn countries. The lands in England are not suited to rice, but they would all bear potatoes; and Dr. Adam Smith observes that if potatoes were to become the favourite vegetable food of the common people, and if the same quantity of land were employed in their culture as is now employed in the culture of corn, the country would be able to support a much greater population, and would consequently in a very short time have it. (55)

This prospect of turning England into a nation of potato-eaters anticipates with unwittingly savage historical irony the nineteenth-century British policy toward the Irish; and it points as well to that cultural, as well as situational, specificity by which "means of subsistence" might be procured. Climate and land quality and a host of other factors affect food production, this aside from (though contributing to) longstanding and hard-to-transform foodways.

One thinks of Wallace Stevens's poem "Frogs Eat Butterflies. Snakes Eat Frogs. Hogs Eat Snakes. Men Eat Hogs." The food chain is not, however, simply a series of links; it is also a series of material substitutions and transubstantiations – operating along a logic more metaphoric than metonymic. Consider the famous Malthusian set-piece of the fatted calf, in which Malthus observes that years ago, cattle were thin, spindly affairs raised on "waste lands," rarely fatted (107). As the cost of preparing cattle for the market decreased, higher quality butcher's meat became desirable and sellable, allowing farmers to invest in fattening up their cattle, and encouraging them as well to convert land that might be devoted to agriculture to pasturage. But "the present price will not only pay for fatting cattle on the very best land, but will even allow of the rearing of many on land that would bear good crops of corn. The same number of cattle, or even the same weight of cattle at the different periods when killed, will have consumed (if I may be allowed the expression) very different quantities of human subsistence. A fatted beast may in some respects be considered, in the language of the French economists, as an unproductive labourer" (107).[13]

Malthus's musing on the fatted calf as an unproductive labourer cannot but remind us of his cautious strictures on aristocrats' love of trifles, as well as his animadversions on the overvaluing of manufactures over agriculture. (The biblical precedent here offers another ramifyingly intriguing, ironical dimension, which Malthus the clergyman obviously meant to evoke: fatted calves in the modern period signify prodigality and bad policy, rather than a reward for the return of the prodigal son.) In another famous passage, Malthus suggests that the labourer who makes lace is similarly to be considered an "unproductive labourer," as he "will have added nothing to the gross produce of the land: he has consumed a portion of this gross produce, and has left a bit of lace in return" (111).[14] The lace-maker, the fatted calf, the gentlemen keeping his horses for pleasure – all these are figures of unproductive labour, "productive" understood quite materially in Malthus as "productive of the means of subsistence," not "productive of exchange value," much less of surplus value. Following Malthus's own logic, one might begin to calculate the repression of human population (to use his diction) in favour of fattened cattle or lace for "the vanity of a few rich people" (111); one notices, too, that such passages must function theologically and typologically as stories of misdirected worship, fatted calves (not to mention lace) the false idols of Malthusian political economy. In this story, though Malthus will not fully tell it, cattle eat men.[15] And cattle eat – repress – men in part by displacing corn. And what they eat are not men in general but men "of the common people" – of the lower social ranks.

Malthus's improving eye surveys a landscape of absences: the fields that might have been dedicated to corn and not cattle; the millions that might have lived; the lands that might yet be devoted to potatoes instead of corn. Alongside his relentless commitment to "things as they are" – that Malthusian as well as Godwinian phrase – lurks the negative pressure of things as they were, might have been, or may yet be. Sceptical of what he considered wild conjectures, he was nevertheless a ceaseless conjecturer. Demolishing some obviousnesses – for example, that populousness was a good in itself – he continued to traffic in others: that contraception was vice; sexual passion largely constant; desire heteronormative.

Malthus's meditations lead him to an extraordinary counting of the missing, a calculus of absence. "In every State in Europe, since we have first had accounts of it, millions and millions of human existences have been repressed from this simple cause [positive checks]; though perhaps in some of these States, an absolute famine has never been known" (56). (Here one thinks of Amartya Sen's remarkable work on the missing millions of girls in India and China – "missing" because of sex-based positive or preventive checks:[16] the systematic undernourishing or killing of female infants creates its own gendered imbalances within populations Malthus typically preferred to leave, considered as children, ungendered.) The checks to population are always already operating, Malthus insists, and have been throughout recorded history. Malthus argues contra Condorcet that "the period when the number of men surpass their means of subsistence has long since arrived, and that this necessary oscillation, this constantly subsisting cause of periodical misery, has existed ever since we have had any histories of mankind, does exist at present, and will for ever continue to exist, unless some decided change take place in the physical constitution of our nature" (59–60).

Malthus, in the end, maps for us a political economy of population, not a "principle" of population. His treatise outlines a system of insufficient mediations – insufficient distances among desire, sex, and children; between working and eating; between eating and labouring: "The labouring poor, to use a vulgar expression, seem always to live from hand to mouth. Their present wants employ their whole attention, and they seldom think of the future" (40). Malthus points here to the problem of time, the time of and in thought, as well as the time of and in labour. His "hand-to-mouth" figure partakes of a repeated history of bodily mapping, of hand-figuration, from Locke through Smith's infamously invisible hand; in Malthus, we get a conjoined figuration of labouring hands and mouths – the poor who "live from hand to mouth" (this not least a figure of eating). We are presented as well with the spectre of the man who might "be reduced to the grating necessity of forfeiting his independence and of being obliged to the sparing hand of charity for support" (35); we encounter gentlemen with horses' reins implicitly in hand; we read of anxious prospective husbands offering hands in marriage. If unchecked, the "hand to mouth" life of the poor will lead to dependence on "the sparing hand" of charity: Malthus's critique of poor relief involves not least a figure of certain hands withdrawn, other hands forestalled. In terms of bodily mapping, too, we find here instead of Deleuze's and Guattari's "body without organs" an insistently figured body with hypertrophied hands and mouths: for all the reproductive pressure on population, this is a text far more orally than genitally fixated, the optimally albeit regretfully closed hands of parish charity-dispensers checking the perpetually open, hungry mouths of the labouring poor.

Considered from the perspective of a system, the problem of population as Malthus outlines it is in part a problem of time – a collapse of time between crucial nodes in a productive/reproductive network. Malthus envisions a feedback loop so nearly immediate that unless checks intervene, desire moves to sex moves to children in a figurative instant. Foresight is the great mental condom; and here Malthus anticipates some aspects of the theologico-pedagogical complex known as Abstinence Education. For certainly Malthus, like all good utilitarians, calls for a pedagogy – in his case, a pedagogy of prudence, an inculcation of consequences, a disciplinary humanization: "A man who might not be deterred from going to the ale-house from the consideration that on his death, or sickness, he should leave his wife and family upon the parish might yet hesitate in thus dissipating his earnings if he were assured that, in either of these cases, his family must starve or be left to the support of casual bounty" (41). Malthus calls for abstaining from sex outside marriage, yes – a call he need not even overtly make, as it falls under the obviousness of virtue, under the strictures of vice; he also, much to John Stuart Mill's later disgust, seems to be incapable of calling for continence within marriage (an admittedly perverse call for an Anglican divine to make). Sociologist Arland Thornton notes, in a recent paper on population theory, that for Malthus, marriage rates and fertility rates were so correlated that he was "sometimes using marriage rates as indicators of fertility rates" (10). Any breaking of the link between sex and reproduction, or that between marriage and sex, was for Malthus *verboten* – not unthinkable, but immoral. These are precisely the links that the twentieth century so flagrantly broke, and the means by which the systemic relations of sex and power in the family world system (to invoke Göran Therborn's work in *Between Sex and Power*) have transformed themselves in so-called developed nations.

It's no surprise that a founding text of political economy should waver between the "is" and the "ought," and that its formulations of both "is" and "ought" should come in for sustained ideological critique for over two centuries. Malthus offers us a diagnostic and still-relevant inquiry – not so much in its reckoning with population in relation to the means of subsistence (though this will continue to be a crux for development economics and global political economy and discussions of distributive justice) as in its networking of what he called "body," "mind," and the body's "wants." Malthus's profound respect for "the wants of the body" and not only for the body is notable; so too his interest in what he called "the structure of the human mind" (119). If his essay seems all too often to devolve into preoccupations with the structure of the gentleman's mind, or the nature of labourers' mindlessness, well, these assessments we still have with us, inasmuch as poverty, community, consciousness, and agency remain critical concerns for a possible politics. The coming community, in Giorgio Agamben's terms, will be broader than Malthus could fully allow himself to see – though he repeatedly pointed to the anthropological and, indeed, the transhuman, interspecies feedback loops involved in any analytic of the principle of population; his was, in the end, a study of human populations, but grounded in a more general inquiry into the conditions of reproducing life, animal and vegetal.

Malthus's salutary preoccupation with the earth and with lands, what they might sustain and produce, how they could be territorialized by cattle or corn or rice, remains politically trenchant in this era of genetically modified foods and sustainability discourse. He wavers between what a Heideggerian would call a preoccupation with earth and with world. Thinking along these lines in *The Human Condition*, Hannah Arendt listed "life itself, natality and mortality, worldliness, plurality, and the earth" as among "the conditions of human existence" (11). We might say that Malthus is weakest in thinking plurality, though he quite scrupulously presents it: man as multiple, variously affiliated, organized into communities national, ranked, and otherwise grouped.

The question of plurality, that "we" are among many, and that there are many kinds of "we," returns us to the tropologics and politics of number, something that preoccupied Romantic-period writers across the spectrum of commitment and activity. Here Marjorie Levinson's recent neo-Spinozan work on the Romantic morpho-politics of number, form, the multitude, and multiplicity might illuminate our path.[17] In these essays, Levinson invokes Georg

Cantor's work on set theory and other works on the history of mathematical thinking, and she follows these in productively distinguishing "counting" from "matching" (and ordinality from cardinality).[18] When Wordsworth hails "the host of dancing daffodils" – "Ten thousand dancing in the breeze" – or when he asks of rejuvenated Man in *The Prelude* (1805), "Why is this glorious Creature to be found / One only in ten thousand?" (12.87–8), he is not counting actual thousands or tens of thousands (or truly counting the "one only" found). When Shelley in his "Ode to Liberty" invokes "many a swarming million" of unliberated man (34), when he wonders whether "earth can clothe and feed / Amplest millions at their need" (246–7), when in *The Triumph of Life* he likens people on the "public way" (43) of life to "numerous ... gnats" (46), each one "one of the multitude" (49), "borne amid the crowd as through the sky / One of the million leaves of summer's bier" (50–1), he is not counting actual millions – no: Wordsworth's ten thousands and Shelley's recurring millions are – like Malthus's geometrically proliferating populations – figures of multitude, of undifferentiated multitude; these are not numeric counts, but rather figures marking a theoretical aspiration toward a differentiated, unpressured multiplicity.[19]

In the end, Malthus is not about counting; he is about matching. In "Take Cover," a recent poem by Angie Estes, the poet invokes Aristotle as a way of clarifying this double sense of "number": "*Number* / he said, *has two senses: what is counted / or countable, and that by which / we count.*" As mentioned before, Malthus counts by food and bodies. He also counts food and bodies (particularly in the revised and expanded editions of his *Essay*, chock full of charts and data to support what looks to be a mathesis – a mathematic system predicated on calculable representation).[20] Malthusian counting – or rather, his matching – poses a number of questions, not least: What set(s) are we in? National populations? Social ranks or classes? Gendered tranches of reproducers? Fertile heterosexuals? Transnational migratory flows? Global populations? Malthus himself wavers as he establishes his sets: food and sex-passion, at first unmarked in his laws, are increasingly specified, territorialized: we encounter the peculiarly anemic nature of sex-passion in North American Indians (e.g.), and the problem of population in England vs. China vs. Antiquity.[21] Malthus tacks among global and national and social and individual flows, and this is both a strength and a weakness of his analysis. Apparently biological law and political-juridical formation will not easily align themselves in his analyses, even as the gentleman's concerns will not match the labourer's.

Malthus and his contemporaries were preoccupied with humans as a species (viz. *Frankenstein*), but also with humans in intensely local and potentially transformable communities (viz. Southey's and Coleridge's dream of the Pantisocracy). Malthus's invocation of the earth and its fertility – derived from Robert Wallace, but drawing of course on a much longer tradition – points to this quintessentially Romantic question of the possible matching of earth and mind in a humanly worlded world. Malthus is, after all, a theorist of happiness, and of what kinds of matches – sexual, economic, etc. – are productive of happiness. Thus I'll conclude with a later Major Romantic Writer, Wallace Stevens, and his "Auroras of Autumn" (420):

> A happy people in an unhappy world –
> it cannot be [...]
>
> Turn back to where we were when we began:
> an unhappy people in a happy world.

NOTES

1 Kevin Gilmartin's invitation to join this panel provided the occasion for thinking (again) about Malthus: Jon Mee presented on Thomas Paine, Frances Ferguson spoke on John Stuart Mill, and I spoke on Malthus. This essay benefitted immeasurably from their conversation and responsiveness, and from the comments of Charles Rzepka. I would like to thank as well the editor of this volume, Joel Faflak, for his keen engagement and shepherding. Best thanks as always to Laura Slatkin.

2 The work of Gallagher, Poovey, and especially Ferguson informed my own earlier inquiry into the relations among Romantic poetry, population theory, imagination, futurity, and species-logic. See McLane. I was at that time unaware of Connell's or Winch's illuminating work. Recent and ongoing work by Ron Broglio, James P. Huzel, Richard Bronk, and Winch, e.g., further confirms this return to Malthus. Murray Milgate and Shannon C. Stimson explore in another key the complex tradition of political economy, from Smith through Malthus and beyond – a tradition quite different from its polemical seizure by neo-classical economics.

3 Poovey further notes that Malthus's "use of numbers" helped to transform political economy from a moral science (as Malthus still considered it) to the amoral, "dismal" one more familiar to us (xxiii).

4 Marx perfected this strain of caricature, mistakenly characterizing Malthus (cheerfully married Anglican cleric and father of several children) as a celibate parson. See his famous footnote to "The General Law of Capitalist Accumulation" in *Capital* (note 3), included in Malthus, *Essay* 159–61.

5 See Connell 13–62.

6 See Gallagher, "The Body Versus the Social Body," for Malthus as "vindicator of the rights of the body" (88), especially its sexual rights.

7 See Gallagher, "The Romantics and the Political Economists," on Southey's revulsion against "Malthus's sensualism" (104). For a sample of Mill's antipathy, see his *Principles of Political Economy*, excerpted in Malthus, *Essay* 154–5.

8 Althusser, "Ideology and Ideological State Apparatuses": "It is indeed a peculiarity of ideology that it imposes (without appearing to do so, since these are 'obviousnesses') obviousnesses as obviousnesses, which we cannot *fail to recognize* and before which we have the inevitable and natural reaction of crying out (aloud, or in the 'still small voice of conscience'): 'That's obvious! That's right! That's true!'" (172; emphasis in original).

9 See, for example, the essay, "Is Food the New Sex?" by Mary Eberstadt who, like Malthus himself, invokes "the two things without which human beings cannot exist: food and sex" – she takes this as a given, not as variables themselves subject to inquiry.

10 Sustainability discourse is only one obvious route for the recoding of Malthusian concerns – neo-Keynsianism (as previously noted) another.

11 Frances Ferguson first brought this passage to my attention, in her "Malthus, Godwin, Wordsworth, and the Spirit of Solitude."

12 For an extended meditation on the twenty-some-year Malthus-Godwin controversy, and its impact on the work of Mary as well as Percy Shelley (as well as its broader dispersal in work throughout the period, e.g., that of Hazlitt and Peacock), see McLane 84–7, 100–8, 109–18, 163–79.

13 For evidence of the ongoing contention over cattle as (nationalized) food, and cattle vs. human populations in competing national markets, see the images in the appendix, by Joao Pina, as well as Barrionuevo.

14 See Gallagher, "The Body Versus the Social Body," for an extended, trenchant meditation on these very passages (95–7).

15 And here I align with Gallagher's bravura reading of the fatted calf passage in "The Body Versus the Social Body": "The biological economy envisioned here is one in which cattle 'eat' men" (97).

16 See, for example, Sen.

17 See Levinson's papers "Bounded Infinities" and "Clouds and Crowds," and her recent article, "Of Being Numerous." In "Of Being Numerous," Levinson finds, via Spinoza and the cloud-taxonomist and theorist Luke Howard and the mathematician Georg Cantor, a "model of singularity that is already multiple, diverse, and dynamically continuous with its environment" (635) – something Wordsworth and, after him, George Oppen explored in their poetry. Levinson understands her method as a historical "conjunctural" method, our contemporary moment of "material and informational connectedness" "as that which lights up the holisms of the past" (636). Our "post-organic" moment, that is, allows us to see what was always already there, in Wordsworth, and in other Romantic phenomena.

18 See, for example, Levinson, "Of Being Numerous," 651–2, as she pursues her illuminating reading of Wordsworth's "I wandered lonely as a cloud," which reading prompts my meditation here.

19 This "matching" logic of Wordsworthian number is further borne out when one considers the differences between the 1805 and 1850 *Prelude*: in Book 12, Wordsworth continues, "What one is / Why may not many be?" (88–9) – lines that become, in Book 13 of the 1850 *Prelude*, "What one is, / Why may not millions be?" (88–9). This pivoting between the many and the millions points precisely to this shadowy territory of multitudes massed into a set, not cardinally numbered and arithmetically counted. For a more extended meditation on Shelley's calculus of "millions" and his analytic of multitude, see McLane 197–201.

20 See Poovey for a brilliant account of Malthus's *Essay*, and his revisions, as one key site where "the meanings of numerical representation were reworked at the end of the eighteenth century" (278–95). Poovey further notes that "the revisions he made to his *Essay* robbed theological utilitarianism of its providential and ethical dimension because the numbers he used supported a thesis that made it all but impossible to argue, as theological utilitarians did, that whatever is, is right" (283). She observes as well that Malthus's "populations" (and other aggregates like "the poor") were not fully statistical populations, but something proto-statistical, closer to populous masses than any numerical count (287). Poovey's thoughts on number thus chime intriguingly with Levinson's more recent inquiries.

21 For Malthus's aside on North American Indians and their "less ardent" passion "between the sexes," see Chapter 3 (27–8). And for comparative assessment of different periods of society and contemporary nations, see, e.g., chapters 3–7 (27–56).

WORKS CITED

Althusser, Louis. "Ideology and Ideological State Apparatuses." *Lenin and Philosophy and Other Essays*. Trans. Ben Brewster. New York: Monthly Review, 1971. 121–76. Print.

Andrews, Bruce. "Species Means Guilt." *Postmodern American Poetry: A Norton Anthology*. Ed. Paul Hoover. New York: Norton, 1994. 532–4. Print.

Arendt, Hannah. *The Human Condition*. Chicago: U of Chicago P, 1989. Print.

Barrionuevo, Alexei. "A Market for Beef Ambles Across a Border." *The New York Times*. 15 December 2009. http://www.nytimes.com/2009/12/16/world/americas/16uruguay.html. Web. 22 May 2012.

Broglio, Ron. "Docile Numbers and Stubborn Bodies: Population and the Problem of Multitude." Romantic Numbers (Gallery Two), Romantic Circles Praxis Series (April

2013). 25 August 2013. http://www.rc.umd.edu/praxis/numbers/HTML/praxis.2013.broglio.html. Web. 2 November 2014.

Bronk, Richard. *The Romantic Economist: Imagination in Economics*. Cambridge: Cambridge UP, 2009. Print.

Connell, Philip. *Romanticism, Economics, and the Question of "Culture."* New York: Oxford UP, 2001. Print.

Deleuze, Gilles, and Felix Guattari. *Anti-Oedipus: Capitalism and Schizophrenia*. Trans. Robert Hurley, Mark Seem, and Helen R. Lane. Minneapolis: U of Minnesota P, 1990. Print.

Eberstadt, Mary. "Is Food the New Sex? A Curious Reversal in Moralizing." *Policy Review* 153 (February/March 2009). Hoover Institution, Stanford University. http://www.hoover.org/research/food-new-sex. Web. 28 March 2010.

Estes, Angie. "Take Cover." *Tryst*. Oberlin, OH: Oberlin College P, 2009. 62–3. Print.

Ferguson, Frances. "Malthus, Godwin, Wordsworth, and the Spirit of Solitude." *Literature and the Body: Essays on Populations and Persons*. Ed. Elaine Scarry. Baltimore: Johns Hopkins UP, 1988. 106–24. Print.

Ferguson, Frances. *Solitude and the Sublime: Romanticism and the Aesthetics of Individuation*. New York: Routledge, 1992. Print.

Gallagher, Catherine. "The Body Versus the Social Body in the Works of Thomas Malthus and Henry Mayhew." *The Making of the Modern Body: Sexuality and Society in the Nineteenth Century*. Ed. Catherine Gallagher and Thomas Laqueur. Berkeley: U of California P, 1987. 83–106. Print.

Gallagher, Catherine. "The Romantics and the Political Economists." *The Cambridge Companion to English Romantic Literature*. Ed. James Chandler. Cambridge: Cambridge UP, 2009. 71–100. Print.

Godwin, William. *Enquiry concerning Political Justice and its Influence on Modern Morals and Happiness*. 3rd ed. 1798. Introd. Isaac Kramnik. New York: Penguin, 1985. Print.

Godwin, William. *Of Population: An Inquiry concerning the Power of increase in the Numbers of mankind being an answer to Mr. Malthus' Essay on that Subject*. 1820. New York: Augustus M. Kelley, 1964. Print.

Hazlitt, William. "Mr. Malthus." *The Spirit of the Age: or, Contemporary Portraits*. London: Bentley, 1825. 251–76. Print.

Huzel, James P. *The Popularization of Malthus in Early Nineteenth-Century England: Martineau, Cobbett and the Pauper Press*. Modern Economic and Social History. Burlington, VT: Ashgate P, 2006. Print.

"Jim Cramer Battle." *The Daily Show*. 11 March 2009. http://www.cc.com/video-clips/fttmoj/the-daily-show-with-jon-stewart-exclusive---jim-cramer-extended-interview-pt--1. Web. 17 June 2017.

Johnson, Samuel. "Oats." *Dictionary of the English Language*. London: J. and P. Knapton, 1755. np. http://www.bl.uk/learning/images/texts/dict/transcript1395.html. Web. 4 June 2017.

Leacock, Stephen. "Oh! Mr. Malthus!" *Hellements of Hickonomics in Hiccoughs of Verse Done in our Social Planning Mill.* New York: Dodd, Mead, 1936. https://tspace.library.utoronto.ca/html/1807/4350/poem1283.html. Web. 2 November 2014.

Levinson, Marjorie. "Bounded Infinities: Unending Romanticism." MLA Panel. "The Ends of Romanticism." Philadelphia, PA. 28 December 2009. Private Communication.

Levinson, Marjorie. "Clouds and Crowds, Solitude and Society: Revisiting Romantic Lyric." Plenary Lecture at "Romanticism and the City." The 2009 International Conference on Romanticism. New York City. 6 November 2009. Private communication.

Levinson, Marjorie. "Of Being Numerous," *Studies in Romanticism* 49:4 (Winter 2010): 633–57. Print.

Levi-Strauss, Claude. *Totemism.* Trans. Rodney Needham. Boston: Beacon P, 1963. Print.

MacFarlane, Alan. "The Malthusian Trap." Draft of article to be published in the International Encyclopedia of the Social Sciences. 2nd ed. Written August 2005. http://www.alanmacfarlane.com/FILES/pop.html. Web. 2 November 2014.
http://www.alanmacfarlane.com/FILES/pop.html. Web. 2 November 2014.
MacFarlane, Alan. *Thomas Malthus and the Making of the Modern World.* Ebook only available at: http://www.alanmacfarlane.com/FILES/ebooks.html. Web. 2 November 2014.

Malthus, Thomas Robert. *An Essay on the Principle of Population.* Ed. Philip Appleman. New York: Norton, 1976. Print.

Malthus, Thomas Robert. *Parallel Chapters from the First and Second Editions of An Essay on the Principle of Population by T.R. Malthus 1798: 1803.* New York: MacMillan, 1895. Print.

McLane, Maureen N. *Romanticism and the Human Sciences: Poetry, Population, and the Discourse of the Species.* Cambridge: Cambridge UP, 2006. Print.

Milgate, Murray, and Shannon C. Stimson. *After Adam Smith: A Century of Transformation in Politics and Political Economy.* Princeton: Princeton UP, 2009. Print.

Peacock, Thomas Love. *Paper Money Lyrics and other Poems. Poems and Plays.* Vol 7. *The Halliford Edition of the Works of Thomas Love Peacock.* 8 vols. Ed. H.F.B. Brett-Smith and C.E. Jones. New York: AMS P, 1967. Print.

Pina, Joao. "Cows Outnumber Uruguayans More than Three to One in a Country Where about 80 Percent of the Land is Used for Cattle Grazing." *The New York Times,* 16 December 2009. http://www.nytimes.com/imagepages/2009/12/16/world/16uruguay_CA0.html. Web. 22 May 2012.

Pina, Joao. "Preparing Meat for Processing in Montevideo. Big Beef Investors Are Betting on Uruguay's More Market-Friendly Policies." *New York Times,* 16 December 2009. http://www.nytimes.com/imagepages/2009/12/16/world/16uruguay_CA1.html. Web. 22 May 2012.

Poovey, Mary. *A History of the Modern Fact: Problems of Knowledge in the Sciences of Wealth and Society*. Chicago: U of Chicago P, 1998. Print.

Rubin, Gayle. "The Traffic in Women: Notes on the 'Political Economy' of Sex." *Toward an Anthropology of Women*. Ed. Rayna R. Reiter. New York: Monthly Review P, 1975. 157–210. Print.

Sen, Amartya. "More Than 100 Million Women Are Missing." *New York Review of Books* 37.20 (20 December 1990). http://www.nybooks.com/articles/1990/12/20/more-than-100-million-women-are-missing/. Web. 2 November 2014.

Shelley, Percy Bysshe. "Ode to Liberty." *Shelley's Poetry and Prose*. 2nd ed. Ed. Donald H. Reiman and Neil Fraistat. New York: Norton, 2002. 307–15. Print.

Shelley, Percy Bysshe. *The Triumph of Life*. *Shelley's Poetry and Prose*. 481–500. Print.

Steinberg, Jonathan. "Economic Illusions." Letter. *London Review of Books* 31.23 (3 December 2009): 6. https://www.lrb.co.uk/v31/n23/letters. Web. 2 November 2014.

Stevens, Wallace. "The Auroras of Autumn." *The Collected Poems of Wallace Stevens*. New York: Vintage, 1982. 411–21. Print.

Therborn, Göran. *Between Sex and Power: Family and the World, 1900–2000*. New York: Routledge, 2004. Print.

Thornton, Arland. "Robert Malthus, the Developmental Paradigm, Reading History Sideways, and Family Myths." https://developmentalidealism.org/pubs/rhs/related-working-papers.html. Web. 2 November 2014.

Winch, Donald. *Malthus: A Very Short Introduction*. Oxford: Oxford UP, 2013. Print.

Winch, Donald. *Riches and Poverty: An Intellectual History of Political Economy in Britain, 1750–1834*. Cambridge: Cambridge UP, 1996. Print.

Winch, Donald. *Wealth and Life: Essays on the Intellectual History of Political Economy in Britain, 1848–1914*. Cambridge: Cambridge UP, 2009. Print.

Wordsworth, William. *The Prelude, 1799, 1805, 1850*. Ed. Jonathan Wordsworth, M.H. Abrams, and Stephen Gill. New York: Norton, 1979. Print.

Žižek, Slavoj. *Organs without Bodies: Deleuze and Consequences*. New York: Routledge, 2004. Print.

PART THREE

Goethe and the Contigencies of Life

Chapter Six

Structure and Advancement in Goethe's Morphology[1]

GÁBOR ÁRON ZEMPLÉN

In my garden's care and favour
From the East this tree's leaf shows
Secret sense for us to savour
And uplifts the one who knows.
Is it but one being single
Which as same itself divides?
Are there two which choose to mingle
So that each as one now hides?
As the answer to such question
I have found a sense that's true:
Is it not my songs' suggestion
That I'm one and also two?

– Goethe "Ginkgo Biloba"[2]

The essential value of Goethe's scientific contributions ... is closely related to organization and form, for it is precisely its form that prevents the content of individual parts from being torn from the mysterious whole.

– Wilhelm von Schütz[3]

Linné's Taxonomy and Goethe's Morphology: Grasping and Pursuing

In 1822 the *Göttinger Anzeigen* printed an anonymous review of a botanical textbook by the Marburg Professor Georg Wilhelm Franz Wenderoth. The reviewer was a young lecturer in Göttingen, Ernst Meyer. He differentiated two approaches concerning the study of organisms. One can "pursue the plant in its living metamorphosis as a 'something' capable of existence only in regulated alteration." But one may also "wish to grasp it as something constant, and therefore dead, in one or several widely separated scientific situations." The review stresses that the choice is crucial. There is a clear dichotomy between the two approaches, yet the reviewer is not applauding the one and condemning the other – a typical argumentative strategy – but instead highlights a difference in the safety of the two approaches: "Whoever declares himself with Linné for the latter method, takes the safer course. However, once we have ventured into the cycle of metamorphosis, we may no longer hesitate or even turn back" (BW 114; FA 24:535).

Goethe republished paragraphs of the review contrasting his approach with that of Linné in the fourth issue of the first volume of his *Morphologische Hefte.* Editing the short-lived journal is a clear testimony that Goethe aimed to popularise his morphology. His commentary on the review, *Increasing Difficulty in Botanical Instruction* [*Erschwerter Botanischer Lehrvortrag*], outlined not only the attractiveness but also the difficulty of the method: "That it is difficult to deal by this method with didactic or even dogmatic aspects of the subject is no secret for us who understand the value of the method" (BW 115).[4] This is a puzzling description: Goethe's method "pursues" rather than "grasps"; it is somehow difficult to teach; and unlike Linné's method, morphology is cyclical. To elaborate on these points, Goethe invited Meyer in the next issue of the journal (Vol. 2.1) to respond to his essay "Problem."

Linné was an important author, influencing both Goethe and Meyer. Goethe recalls that in his youth he took Linné's *Terminolgy* [*Termini botanici*], bound together with his *Fundamentals* [*Fundamenta botanices*], "into the highways and byways ... the active, happy days ... those pages opened up a new world to me." He studied daily Linné's *Philosophy of Botany*, "thus advancing farther and farther in ordered knowledge, attempting to acquire as far as possible all that might procure for me a more general view of this broad realm" (BW 153–4).[5] Meyer explicitly defended Linné in his response [Erwiderung], and dissolved the tension he set up in the review between Linné's successful taxonomic classification and Goethe's approach.

A gifted and keen observer, religiously inspired and poetically active, Linné made little name with his hymns to the spring countryside outside Sweden, but did succeed with his system of classification. The method had its critics by the 1780s, however. As Goethe states:

> We repeatedly heard it said that this science of botany which we were so assiduously pursuing was by and large only a nomenclature, a system based on counting – and not very accurate counting at that; that it could satisfy neither the reason nor the imagination, and that it could achieve no satisfactory results. In spite of this objection we confidently pursued our way, which indeed promised to take us far enough into the science of plants. (BW 156)[6]

Linné presupposed that nature made no leaps; his system, arranging natural objects, held species to be constant and unchanging. The key to classifying them depended on *divisio* and *denominatio.* The observables required grouping and arrangement as well as the labels that differentiated them. "In this way," as Sten Lindroth argues, "Linnaeus narrowed down the field of botany greatly. He was monumentally one-sided – everything other than nomenclature and classification was scarcely accorded the rank of science" (26). His systematic biology required enormous knowledge of existing forms to give the right names to plants. Morphological work, however, also took development into account. As Meyer wrote: "we must pursue the course of development from the first utricle whence fungus and alga, as well as the seed of the highest plant, emerge" (BW 114; FA 24:535).

Goethe described morphology in his "Problem" as an approach to study plants in their form and formation, and started a dialogue with a general problem that stands in the way of systematization: "Natural system – a contradiction in terms. Nature has no system; she has, she *is* life and its progress from an unknown centre toward an unknowable goal. Scientific research is therefore endless" (BW 116).[7] In the *Goethezeit,* polarities were *en vogue.* Like many others, Goethe also systematically established and dissolved them. He explains how the problem of unfounded systematization can be overcome with the help of metamorphosis:

> The concept of metamorphosis is a highly estimable gift from above, but at the same time a highly dangerous one ... It leads to formlessness, destroys knowledge, disintegrates it. It is like centrifugal force and would lose itself in the infinite if a counterweight were not provided. I am referring to the specification force [Spezifikationstrieb], that tenacious capacity for persistence inherent in whatever has attained existence, a centripetal force. (BW 116)[8]

The essay ends by highlighting two deviations where scientific research can go amiss, and these correspond to the forces discussed earlier: "all our efforts must be in the direction of eavesdropping on the methods of Nature herself, so that we may prevent her from becoming obstinate over enforced prescriptions, and yet not be deterred from our purpose through her arbitrary behaviour" (BW 118).[9]

Meyer took up the challenge in the *Response* [*Erwiderung*]. The young follower mirrors the *topoi*, style, and philosophy of the master. Concerning the dangers, Goethe focuses on "enforced prescriptions," and Meyer highlights that our demand for a natural system appears to lie beyond human understanding, yet the demand is upheld, and thus the *Spezifikationstrieb* restrains the flow of Nature.[10] He reflects on Goethe's enterprise, using the language game of morphology by pointing out the *regressus*: "already the effort to dissolve the contradiction [inherent in 'natural system'] is a natural drive" that cannot be fully satisfied (FA 24:585). Even behind the attempt to pursue, there is an attempt to grasp. The "young friend" had already advanced so far on the path Goethe held in esteem that giving parts of his botanical collection to Meyer seemed unnecessary, as he explained in a letter to Sartorius in 1822.[11]

A contemporary reader feels that he is caught up in the thicket of nineteenth-century *Naturphilosophie*; a few lines on art and symbolism, and a longer paragraph on significant actions and deeds of the individual, surround the discussion on the inherent conflict of nature and system. The medley is typical of the ageing Goethe's relentless publishing activities, and Meyer's style suggests that the language-game of morphology is multi-player, that there is a Denkkollektiv[12] with which Goethe can discuss important botanical content: the examples, like the genera Rosa and Erica, are heteromorphous botanical taxa, which with their many forms and unclear species-boundaries posed well-known challenges to Linné's system of classification. Goethe's morphology offered a Romantic alternative, a new and modestly popular approach to the study of living forms.

There is clearly a tension between understanding morphology in a narrow sense, as a non-Linnean enterprise in botany, with some followers, and partial transmission to many fields of research, and the much broader claim that it offered a very peculiar (and infectious) approach to observation and theory-construction. In Goethe's "Problem" the cycle of metamorphosis is also referred to as a way of life, a way of understanding the alternations of life:

> [A]n idea cannot be demonstrated empirically, nor can it actually be proved. An individual not in possession of it, will never catch sight of it with his physical eye. The individual who does possess it, easily trains himself to look beyond outer appearances, although returning to reality, after this diastole, to reorient himself. It is possible that he might follow this alternating procedure throughout his life. (BW 115)[13]

The diastole is described cryptically as a gaze directed behind the appearances, yet in the text we do not learn anything about the systole, the contraction, the counterpart of the organ being filled with blood, life-force, or what you will; the exposition is elliptical.[14]

Is morphology just a narrow field, among many in the ever-changing landscape of science, or a way of life, a generalizable and implementable approach? In a posthumously published manuscript on morphology's relationship with physiology, Goethe delineates eight sciences. Among these, morphology is the penultimate discipline, defined as the "[c]onsideration of form both in its parts and as a whole, the conformities and deviations, apart from all other considerations" (BW 88; FA 24:364). Although morphology is a "separate science only through definition" and "is everywhere considered the hand-maiden of physiology," on the normative/

declarative level the definition leaves much more room: "morphology should include the theory of form, formation, and transformation of organic natures" (BW 88; FA 24:364). Goethe understands morphology both as an autonomous science that can utilize other sciences like physics or chemistry, intertwining with various research programs, and as a unique (by definition) approach to forms: studying conformities [*Übereinstimmungen*] and deviations [*Abweichungen*]. In principle, the latter part fits any domain in which conformities and deviations occur. Morphology as a language-game was both a vehicle to convey meaning, including theoretical concepts, descriptions, and comparisons of specimens, and a maze of vague and picture-like metaphors that pursued the living forms, yet – by definition? – ever failed to grasp them.

Goethe's work ranges from comparative anatomical studies of various animal and plant types to a reinvestigation of prismatic colour-phenomena. The morphological approach pops up in diverse loci in Goethe's huge corpus, through various referents, descriptions, and definitions, and metamorphosed into essays, books, or just short marginalia and reflections. These texts outline a theory of scientific language and changes thereof, and even a theory of observation exemplified by domain-specific applications tailored to optimize epistemic effort at a time when historicity was gaining significance in many fields of research. The peculiarity of Goethe's enterprise is the way the static, structural, and stable are connected with the dynamic, changing, and evolving.

Many elements of Goethe's later research agenda appear in the earliest texts, but there is clearly a historical development to his thought, as is apparent from his comments a few years after the joint publications with Meyer on another anonymous text: "The piece of writing in question was given to me from among the papers of our late beloved Duchess Amalia. It is in handwriting well known to me, that of a person upon whose services I often drew in the eighties" (BW 244–5). Just as Meyer's thought was agreeable, so were these ideas: "I cannot recall actually writing these remarks, but they do agree with the ideas occupying my mind at that time" (BW 245). The fragment *Die Natur* showed "inclination toward a kind of pantheism," where "an inexplorable, undefined, humorous, self-contradictory entity is visualized at the base – a playful jester, one to be taken nevertheless in bitter earnestness" (BW 245).[15] It was full of contradictions, comparisons, contrasts, dichotomies, and dilemmas – paradoxes even: "She seems to stake everything on individuality, yet sets small value on the individual" (BW 242); or "Nature is even the unnatural. Those who cannot see her everywhere will not see her clearly anywhere" (BW 243).[16] The piece is clearly inspired by the Orphic hymn "To Nature" (Arber 120).

Half a century later, Goethe criticizes the piece: "The composition lacks the consummating concept of two of Nature's activating forces: polarity and progression"

(BW 245). Goethe establishes dichotomies, but using a different language: "Polarity is a property of matter insofar as we conceive of it as material; progression is a property of spirit, insofar as we conceive it as spiritual. The first is in continual attraction and repulsion, the latter in constant upward striving" (BW 245).[17] His commentary abounds in clustered images from various domains, multiple correspondences, and similes that establish multidimensional links between tenor and vehicle. Lucretius used a similar technique to communicate Greek philosophy to a less-educated Latin audience, deploying their less-advanced language to describe key notions and at the same time to persuade. Yet the polarities that permeate Goethe's text are quickly de-dichotomized and dissolved into polarities of intertwining natures:

> But since matter never exists without spirit, and spirit never without matter, matter is capable of advancing and spirit has the power to attract and repulse. We have an analogy in the fact that only an individual who has analyzed sufficiently is in a position to do the thinking prerequisite to synthesis, and only one who has sufficiently synthetized, is in a position to make a reanalysis. (BW 243)[18]

The analogy with human thinking anticipates the essay's crescendo, a self-laudatory outburst characteristic of the ageing Goethe: "If one recalls the splendid development of this idea, through which all natural phenomena have gradually been linked together for the human intellect, and if one then carefully rereads the essay here referred to, one can smilingly contrast the comparative, as I have called it, with the superlative achieved, and rejoice in fifty years of progress" (BW 245).[19]

The superlative, advanced Goethe writes differently. The driving concepts (*Triebrad*) of nature are polarity and *Steigerung*, best conveyed as progression/enhancement/evolution. They leave their traces in language use: matter advances (enhancement), while spirit can attract and repulse (polarity). Analysis (separation) and synthesis (connection) require one another (as non-exclusive polarities) for the development (evolution/advancement) of thinking to take place. They are also the keys to understanding Goethe's scientific approach on many levels, from observation to theory-construction.

Scientific Research Using Polarity and Progression

In Goethe's scientific research, the search for polarities and progressions can serve as both a didactic and a heuristic device. In the domain of natural philosophy, Goethe undertook a series of experiments to demonstrate regularities and variations (Marcum). Whereas Newton established his theory with ingenious glassworks (Schaffer), Goethe's attack on Newton's theory of white light and colour used not only optical gadgets – a lens, a prism – but also the beholder's own eye. In his early research of the 1790s, Goethe provided readers of his *Contributions to Optics* [*Beyträge zur Optik*] with prisms in order to follow his description of subjective prismatic experiments (twenty years later, he took special care to supply readers with printed plates for both *Contributions* and his 1810 *Theory of Colours* [*Farbenlehre*]).

As Jonathan Westphal notes, "the crucial claim made by Goethe, which is at the centre of his polemic against Newton, [is] that (as we would say) colour is an edge-phenomenon" (9). When one looks at black-and-white strips through a prism, two coloured fringes appear: Goethe's experimental series establishes conformities and deviations, building up in the reader a near-sensory expectation of the key structural elements of his explanation: polarity and enhancement. Inspecting the fringes, we see that they are symmetrical, containing thicker red and blue and thinner yellow and violet bands. Polarity is apparent between warm and cool colours, the two pairs of thinner and thicker coloured bands. Describing the ways in which light interacts with darkness, white with black, shows that "without a boundary ... no colours appear. That is, the boundary condition is fundamental" (Sepper 222). The coloured bands spread out as we move away from the prism, make the strip thinner, or use a prism with greater refractive angle. When the fringes meet, new colours appear, green for the white strip on a dark background, peach blossom (like magenta) for the black strip on the white background. Enhancement in bandwidth results in the overlapping of the two coloured bands, and a new polarity of complementary colours emerges: the green of the white strip (visible in Newton's spectrum) opposed to the extra-spectral red of the black strip, absent from Newton's colour wheel. The new colours spread farther, and extinguish the two colours that gave birth to them: the yellow and the blue in the case of the white strip, the violet and the red in the case of the black strip.

Goethe used his experimental series as a research tool in his exploratory research (Ribe and Steinle) to establish the polarity and progression of the phenomenal domain, and it also served as a didactic and heuristic tool. In his "Confessions of the Author" ("Konfession des Verfassers"), published at the end of the historical part of the *Theory of Colours*, it becomes clear that Goethe endeavoured to develop his earlier work into a more systematic study of colours, by looking for polarities and advancement in phenomenal domains not covered by his early work.[20]

He uses the same explanatory scheme in his 1790 *Metamorphosis of Plants* [*Versuch die Metamorphose der Pflanzen zu erklären*], and just as in his optical writings, polarity and progression also help the linguistic portrayal of phenomena. In this popular – and, even today, readable – treatise, Goethe describes the different organs of the plant, starting from seed leaves, through stem leaves, to the formation of calyx, corolla, the staminal organs, the style, and finally the fruit. Following a brief introduction, he stresses the effect of surroundings on development, and explains the changes in the form of the different organs in terms of changes in the "sap" (BW 42, §30). Goethe incorporates "sap," an observational term not uncommon to eighteenth-century physiology (Portmann), as a theoretical term into his explanatory framework using "polarity" and "progression." As a material substrate, "sap" is gradually refined, and this changing quality is visible in the plant as its organs develop in temporal and spatial succession through a series of "contractions" and "expansions." An account using polarities as explanatory crutches supplements the advancing/progressive series. Goethe was "convinced that with some practice it would not be difficult to account for the diversified forms of flowers and fruits in this manner. To be sure, the conceptions established above – of expansion and contraction, compression and anastomosis – would have to be manipulated as expertly as algebraic formulae, and would have to be applied in the right places" (BW 72, §102).

Goethe (de)limited morphology to differentiate conformity from deviation, yet these terms point to comparisons, stable points of reference. Once an even vague and intuitive grasp of some progression (formation and transformation, and underlying pattern) is attained, a classification of phenomena might follow. That is why dealing with the subterranean parts of the plants was an "unjust demand [Unbillige Forderung]" for Goethe:

> I was not concerned with [the root] at all, for what had I to do with an organ which takes the form of strings, ropes, bulbs, and knots, and – thus limited – manifests itself in such unsatisfying alternation, an organ where endless varieties make their appearance and where none advance. And *it is advance solely that could attract me, hold me, and sweep me along my course.* Let everyone go his own way. Let him, if he can, look back upon forty years of accomplishment, such as the Good Genius has granted me. (BW 118; emphasis added)[21]

Polarity and enhancement are theoretical assumptions behind much of Goethe's thought (Amrine, "Metamorphosis"; Hegge), and his historically significant scientific achievements use explanatory schemes that share structural similarities. The conformities are established through the concatenation of phenomena, and ostensive gestures from this array of visual arguments point to special forms. Of these, the most significant are the archetypal form, the archetypal phenomenon, teratological (monstrous) examples, and examples/exemplars that facilitate model building.

Maintaining the existence of an archetypal form (*Urform*) can be used to distil the conformities of a given phenomenal domain, or even to set the boundaries of that domain, as an entry from Goethe's Italian wanderings shows, written after he visited the Botanical Garden in Padua:

> Here where I am confounded with a great variety of plants, my hypothesis that it might be possible to derive all plant forms from one original plant becomes clear to

me and more exciting. Only when we have accepted this idea will it be possible to determine *genera* and *species* exactly. So far this has, I believe, been done in a very arbitrary way. At this state of my botanical philosophy, I have reached an impasse, and I do not see how to get out of it. The whole subject seems to me to be profound and of far-reaching consequence.[22]

Travelling south in Italy, Goethe becomes increasingly confident that he finds the *Urpflanze,* as he writes to Herder: "The Primal Plant is going to be the strangest creature in the world, for which Nature herself shall envy me. With this model and the key to it, it will be possible to go on forever inventing plants and know that their existence is logical; that is to say, if they do not actually exist, they could, for they are not the shadow phantoms of vain imagination, but possess an inner necessity and truth."[23]

The fragments of Goethe's diary testify to the stages of his plant morphological research. The *Urpflanze* was a historical ancestor first, later an underlying scheme, the "plantness" of a plant, but is dropped after a flash of insight: "It came to me in a flash that in the organ of the plant which we are accustomed to call the leaf lies the true Proteus who can hide or reveal himself in vegetal forms. From first to last, the plant is nothing but leaf, which is so inseparable from the future germ that one cannot think of one without the other."[24] The concept of the *Urpflanze* helped the discovery process. Searching for the underlying form in the myriad variations was a successful research heuristics that facilitated Goethe's theory-construction, and informed his observation: "Because [variations] may be grouped under one concept, it gradually became clear to me that the concept could also be valid in a higher sense: a challenge which hovered in my mind at that time in the sensuous form of a supersensuous plant archetype. I traced the variations of all forms as I came upon them" (BW 162).[25] *The Metamorphosis of Plants* contains no reference to the *Urpflanze* concept, and the ageing Goethe reflects, "how, quite naively, I first conceived the idea of plant metamorphosis" (BW 166).[26]

Not only are conformities displayed, but hidden polarities are also exposed in the archetypal phenomenon (*Grund- und Urphänomen*) (HA 13:367, FL-DT, §174), another of Goethe's (in)famous concepts. In Goethe's effort to establish links between the world of objects and the world of the subject, between *explanandum* and *explanans, how* he describes phenomena didactically helps *what* he describes. As he explains the archetypal phenomenon of his *Theory of Colours* "[o]n the one hand we see light or a bright object, on the other, darkness or a dark object. Between them we place turbidity and through this mediation colours arise from the opposites; these colours too are opposites, although in their reciprocal relationship they lead directly back to a common unity" (SA 12:195; FL-DT, §175). The archetypal phenomenon

Ginkgo biloba.

Dieses Baums Blatt, der von Osten
Meinem Garten anvertraut,
Giebt geheimen Sinn zu kosten,
Wie's den Wissenden erbaut.

Ist es Ein lebendig Wesen,
Das sich in sich selbst getrennt,
Sind es zwey die sich erlesen,
Daß man sie als Eines kennt.

Solche Frage zu erwiedern
Fand ich wohl den rechten Sinn,
Fühlst du nicht an meinen Liedern
Daß ich Eins und doppelt bin.

d. 15. S. 1815

Figure 6.1 "Ginkgo biloba," poem by Goethe, 15 September 1815. Original (fair copy) in Goethe Museum, Düsseldorf.

is both ideal and real, both symbolic and concrete,[27] and as such links the phenomenal and the theoretical, the object- and the language-domains of science.

Teratological, monstrous, or irregular specimens are didactically significant deviations, seen as examples of retrogression. In §5 of *Metamorphosis of Plants,* Goethe distinguishes three types of metamorphoses: regular, irregular, and accidental.

His aim is to explain regular metamorphosis in the essay, yet it is irregular metamorphosis that gives the key "to distinguish clearly what otherwise we are allowed only to conjecture. It is by this procedure that we have the best prospect of attaining our purpose" (BW 32, §7).[28] This type is also called *retrogressive,* as Nature here "takes a step or two backward," and this enables Goethe "to bring to light what the regular type keeps hidden from view" (BW 32 §7).[29] Retrogressive (teratological) examples highlight the atypical, as when a flowering plant returns to a vegetative state (a recurrent example in Goethe's botanical writings is the perfoliate rose). In many cases Goethe merely registers seemingly irregular observations, as in the case of *Bignonia radicans* (HA 13:127). However, he often attempts to pursue the phenomenon linguistically, as in the poem cited at the beginning of the chapter on the leaf of an ancient tree, the *Ginkgo biloba.* Examples of the typical and atypical might come from any domain under investigation (specific skulls in osteology, granite or the magnet in geology), and in some cases they are developed into full-blown "theoretical" constructs, in other instances only facilitate theory construction.

The anomalous image-producing properties of Icelandic spar, for example, inspired Goethe to develop the concept of a *double image*: "Why should the Medium not be able to bring forth a double image through a cause that is unknown to us"?[30] The unpublished draft from October 1793 is directed against Newton's concept of diverse refrangibility. The tentative idea triggered by the double-refraction or polarization of the image is more fully developed in the *Theory of Colours,* where Goethe conjectures about the existence of a double image, and a special subcategory, the "auxiliary image" or *Nebenbild,* used as a bridge-concept between theories (Zemplén). In a later recollection, his description of the mineral resembles the Ginkgo poem: "I possess a peculiar specimen of this mineral with very remarkable properties. Holding the spar close to the eye, when one steps back from the object, immediately two side-images appear to the left and the right, which, depending on the orientation of the clear rhombohedron, appear sometimes singly, sometimes doubly."[31]

Why would Goethe develop this strange concept from an observed irregularity? The didactic part of the *Theory of Colours* is structured much like a *scala naturae,* leading from the most transient colours belonging to the eye itself (physiological colours), through to the increasingly less transient physical colours, to fixed chemical colours. The part on physical colours starts with the chapter on dioptric colours, which appear when light, darkness, and colourless transparent or translucent media interact (FD-DT, §143). The first class of dioptric colours introduces the archetypal phenomenon's basic polarity, light and shadow. The medium serves for enhancement, giving rise to the yellow (red) sun and the blue (at night black) sky. The explanatory model developed here is a kind of medium-modification, like the one Aristotle proposed in his *Meteorologica.*

As opposed to his medium-modificationist account, Goethe gives a boundary-modificationist account in the second class of dioptric colours, revisiting his earlier prismatic games with coloured fringes in the *Contributions to Optics*. The *Contributions,* as we have seen, operated with coloured bands, like many of the pre-Newtonian theories of colour in the seventeenth century. The auxiliary image connects the archetypal phenomenon (medium-modification) and the edge-phenomena in prismatic experiments (boundary-modification). The second class of dioptric colours can be deduced (and explained) from the first class. The two classes of dioptric colour phenomena can have a unified explanation with the help of an additional concept, the *Nebenbild.* The auxiliary image allows Goethe to reduce disparate phenomena into a common archetype.[32] The earlier research is thus incorporated in the new, its phenomena and results subsumed under the archetypal image: morphological research can be recursive.

Historical Development and the Spiral Tendency

Hidden in the chapter on the "Medieval period" in the historical part of the *Farbenlehre*, Goethe develops crucial notions of his historiography: "[A]ll that we have of the materials of history, what we have developed individually of the historical, is transmitted simultaneously, will only be the commentary to the previously expressed."[33] Goethe's language use is playful. He calls the period a *Zwischenzeit*, a term used to refer to the entertaining break between two action-packed scenes in theatrical performances. This *Zwischenzeit* is a hiatus (*Lücke*) in the ongoing evolution of human thought, a period in which no significant development in science took place, an irregular time of retrogression. Goethe the intellectual cartographer excuses himself for introducing his own conjectures into this hiatus:

> [The] earlier geographers ... who created the map of Africa, where mountains, rivers and cities were missing, used to draw an elephant, a lion, or some sort of monster into the desert, without being admonished due to this approach. One will thus also not admonish us, when we insert a few reflections into the great gap, where the exciting, alive, progressive science leaves us, and to which we will return in future.[34]

The rest of Goethe's history of colour-theories is a commentary on this discussion of the forces that act on the scientist, and that shape the evolution of science. In all its manifestations, "the conflict of the individual in the immediate experience with the mediated transmission, is actually the history of the sciences."[35] Transmitted knowledge is authority: "[W]hen we speak of transmission/tradition, we are immediately asked to speak of authority. Because specifically examined, then each authority is a type of transmission."[36] Authority is opposed to ever-changing

Nature, to experience and perception. The one detailed example of Goethe's chapter is the life and works of Roger Bacon (1214–94), which point to the regularity and lawfulness of a teratological period. Into the hiatus of development Goethe inserts the individual example of Bacon, who "feels deeply the conflict/engagement which he has to take on/pass with nature and with transmission."[37] Bacon's style of thinking thus "bring[s] to light what the regular type keeps hidden from view."

The morphological method informs Goethe's historiography, the forces that shape the individual thinker and science in general. For Goethe, knowing the structure of the explanation directs observation, and conformities and deviations disclose polarities and progression. Observing progression, gradual development throws light on irregularities, and these irregularities are used to connect the constantly changing phenomenal and the peculiar syntax and semantics of morphology. The structure is also thus a narrative device.

By reflecting on his botanical research, Goethe marks his place in the narrative. He makes clear that his metamorphosis conforms to regular, natural development of someone who is able to balance the forces acting on a scientist:

> When my essay, printed forty years ago in German, with its ingenious explanation of the laws of plant formation, became better known in Switzerland and France, people were extremely astonished to find that a poet, who normally occupied himself with moral phenomena and specifically those associated with feelings and power of imagination, could turn for a moment from his path and in a cursory study achieve such an important discovery.
>
> It is to combat this mistaken idea that the present essay has been written; it is intended to make clear how I found opportunity to devote a great part of my life with interest and passion, to nature studies.
>
> It was not through extraordinary intellectual gifts, not through momentary inspiration, not unexpectedly or suddenly, but through logical effort that I arrived at such satisfying result. To be sure, I might have complacently accepted the high honour people intended to pay to my sagacity, or at least have taken some secret pride in it. However, as it is equally harmful in scientific pursuits to rely on either experience or reason exclusively, I consider it my duty to record the event for serious investigators just as it occurred, historically accurate, though not in complete detail. (BW 165)

Goethe's advanced scientific works build on his earlier research, and he describes the stages of this development using grammatical categories: comparative and superlative. Yet he is equally aware that language fails fully to grasp this development: "[A] language is really only symbolic, only figurative, and the objects are never immediate, but only reflected."[38] As such, the bidirectional language game of morphology balances between the phenomenological and the reductionist

– systematic, in constant formation and transformation in the various *loci* of the Goethe-corpus.

Goethe in his last years studied plants and pursued his insight on the spiral tendency, which, as he writes, is "most strikingly exhibited in terminations and conclusions" (BW 130). He subsumed under a new polarity the previous concepts of his plant morphology, and wrote about two systems, vertical and spiral, in which "[n]either of the two systems can be imagined alone; they are ever and eternally one; and in complete equilibrium they produce the most perfect vegetation" (BW 132). The vertical system is responsible for "the durable, eventually solidifying, and permanent parts," while the "spiral system is the developmental, reproductive and nourishing." This latter is "temporary and almost independent of the vertical; operating in excess, it is soon exposed to ruin, and perishes; joining the vertical, it fuses with it to form a lasting union as wood or some other solid." Goethe describes the vertical as the "virile sustaining principle of growth" and the spiral "as the actual reproductive life principle" (BW 129).

The essays on the spiral tendency typify how Goethe's later work reflects back on his earlier writings. Return implies refinement, and, like his optical research, a descriptive improvement. In his treatise "The Spiral Tendency in Vegetation" (1830), Goethe states: "We had to assume in vegetation a general spiral tendency, by means of which, in combination with a vertical force, all plant structures, all plant formations, are completed according to the law of metamorphosis" (BW 129).[39] Yet the visual argument of the early morphological work reappears unchanged: "In our investigation of dicotyledons we encounter a conflict between the vertical tendency, whereby the successive development of stem leaves and buds is fostered in sequence, and the spiral system, whereby the fructification is to be completed. A perfoliate rose provides a splendid example." The ostensive gesture used forty years ago now supports a different explanatory framework, exemplifying the theoretical structure of evolving polar opposites:

> When we see that the vertical system is definitely male and the spiral definitely female we will be able to conceive of all vegetation as androgynous from the root up. In the course of the transformations of growth the two systems are *separated*, in obvious *contrast* to one another, and take *opposing* courses, to be *reunited* on a higher level. (BW 145; emphasis added)

A new polarity emerges from the enhancement of the polarity, and as polarity and progression survive transformation, morphology can be used incrementally, recursively, and universally.[40]

The method enables Goethe to display his own development as a process of metamorphosis. The self-analysis follows the pattern of other morphological writings. The mechanism of self-formation through the autobiographer's eye abounds in spirals, and this allows Bernard Kuhn to analyse Goethe's most detailed autobiographical text, *Poetry and Truth*, with recourse to "The Spiral Tendency in Vegetation." He observes that the persistent explanatory framework becomes a motive of the autobiography's narrative structure, and stipulates that an essay in the natural sciences provides the model for conceptualizing and representing the uniquely historical and dynamic self: "Teleologically driven, Goethe's spiral thought nonetheless insists on the impossibility of ever fully reaching an end-state or *telos*" (Kuhn 109). The nature of both linear and spiral development is that in each epoch the individual can believe he has witnessed progress.[41] Goethe thus explains how his scientific understanding improved, how his thinking became *clearer*. Meyer states that Goethe's thoughts are *clear*, while he is too verbose. A clear state, a development to a clearer state. Yet observing progression and change implies some uniform, underlying "true Proteus."

The empirical success of Goethe's plant morphology relied on an explanatory term that *connected* the various plants and plant-organs. Goethe wanted the reader to de-conceptualize, to uncover the thing that morphs into life, into a stem leaf or a sepal, or anything, as it permeates all of these. To connect the variations of a domain and thus capture their unity, he writes in the *Recapitulation*, "we might equally well say that a stamen is a contracted petal, as that a petal is a stamen in a state of expansion; or that a sepal is a contracted stem leaf" (BW 77, § 120). The exact terms matter little, and until a new word is formed, only existing ones can acquire the new meaning. In Meyer's review, the node had the potential for growth, development, and change, the ability to transform into many plant organs, bringing to the fore the underlying similarity of all plants and the various plant-organs of an individual plant. "We may not derive the higher organs of the plant from the root and stem but solely and singly from the node, from which the root and stem have also developed. And in contemplating the plant as a whole, we must consider it not simply as an individual, but ... each node [has] under certain circumstances the power of individual growth" (BW 114).

Darwin's reference to morphology in his *On the Origin of Species* (1859) – after discussing classification and difficulties of the "natural system" – uses another term, the leaf: "It is familiar to almost every one, that in a flower the relative position of the sepals, petals, stamens, and pistils, as well as their intimate structure, are intelligible in the view that they consist of metamorphosed leaves, arranged in a spire" (417). The original assumption of finding the underlying *Typus* was a successful research heuristics that uncovered the node or archetypal leaf producing the various organs of a plant. The hypothetical/undifferentiated node *does* conform to some underlying *Urform*, and this vindicates the original theoretical assumption. Even though the *Urpflanze* was supra-individual, and the node/

leaf was infra-individual, they referred to innumerable perceptible forms with the power of individual growth and captured them in a concept linking the direct and singular observations.

The discovery justifies the method, and the justification facilitates extension of the method. Goethe's "explorative experimentation" endorses a method of comparison, a search for similarities, experimental series, for laws and regularities, and is acutely sensitive to the irregular, teratological, pathological. Spinozist inclinations (Amrine, "Goethean Intuitions") are ideal for developing a phenomenological science of nearly everything: species change just as our concept of species changes; constancy and change, the normal and the abnormal, constitute both regular and irregular development. One can assume conformity while searching for variations in order to describe the gradual unfolding of science, a narrative on the rhythmic nature of historical change, from (imagined) freedom to the "sceptre of an imposed authority."[42]

In his history of science, Goethe uses a structure syntactically isomorphic to many of his scientific works in order to investigate a retrogressive period, to look at the normal exemplar in the abnormal period to display how progressive (regular) and regressive (irregular) forms of thinking result from the use of the same faculty: "[W]hat one usually calls superstition, has developed from an improper usage of mathematics." Bacon's thinking is essentially mathematical, and Goethe generalizes the driving force uncovered in the retrogressive period to all ages and individuals. The mathematical way of thinking is in itself neither good nor bad, but rather the source of both accepted and rejected science, even pseudoscience, as "all these nuisances take their cursed reflection from the most clear of all sciences, its obscurity from the most exact."[43]

In Goethe's thought, there is fruitful tension between the "autonomy" of morphology and the fact that it forms hybrids with various disciplines or approaches. Hybridity indicates its capacity for metaphor, but there is an archetypal structure behind the various empirical attempts to describe life itself, not unsettled by its own systematicity, but rather maintained and enhanced. Commenting on his research, Goethe notes that "a decisive *aperçu* is to be regarded as an inoculated disease: One does not get rid of it till one has fought the disease through."[44]

Tracing Goethe's concept-use allows one to reconstruct his theory-building practice, where conformities and deviations are established through observation, and concepts dichotomize and de-dichotmize the phenomenal. Although form is static, while formation and transformation presuppose the temporal, the explanatory schemes – from plant morphology to prismatic colours – have stable features. Their structural family resemblance – building on polarity and enhancement – establishes a peculiar syntax that informs observation and concept-formation.

A fundamentally developmental perspective connects Goethe's research on plants, colours, and science, on the fixed and on the dynamic. His "gift from above" is to describe the process of creation by creating an array of works manipulating One (directed) and Two (opposing yet complementing). Advancement ("rectification of sap") and polarity (prismatic edge-colours) in the early works matured into incorporation in his "Theory of Colours" and reached equilibrium and termination in "The Spiral Tendency." The intensive intertextual play with the phenomenal adds to the growing empirical foundation; the structure leaves its traces in both the theory and the description, the polarities relate to the empirical domain and inform the linguistic domain. The essence of polarities is displayed in the archetypal phenomenon, and polarities are essential to the linguistic description.

This resembles a very ancient methodology. As Lucretius's transfusion technique shows, the description of nature has much in common with the nature of the description:

> for the same beginnings constitute sky, sea, earth, rivers, sun, the same make crops, trees, animals, but they move differently mixed with different elements and in different ways. Moreover, all through these very lines of mine you see many elements common to many words, although you must confess that lines and words differ one from another both in meaning and in the sound of their soundings. So much can elements do, when nothing is changed but order; but the elements that are the beginnings of things can bring with them more kinds of variety, from which all the various things can be produced. (cited in Garani 13)

Lucretius popularized Epicurean atomism, which was to be revived in corpuscular and mechanical philosophies of the Early Modern Period, but for Goethe the "beginnings of things" were not abstract classes of entities, like Newton's corpuscles (adding "forces" to the mechanical philosophy), or Linné's species (creating an "artificial" system), or Kuhn's paradigms (tuned down to the speciation of coexisting "lexicons").

Not unlike Linnean taxonomy, morphology similarly strives toward autonomy, but its peculiar concept-formation provides very different entry-points, where language meets the world. Linné's categories used scholastic logic to label similarities (definition, genus) and differences (differentia, species), but morphology does not fundamentally rely on the species concept as it captured series via exposing links, directions, and tensions. Linné looked for discriminating traits in the structured forms (natural, observable referents like the number of pistils) and gave intensional definitions of thousands of species, while Goethe developed an alternative (transformational) study of plants, pursuing the inner dynamics of deliminated domains and giving non-natural (linguistic) referents that were often intermodal (like warm and cool colours). A systematic study without nomenclature, counting, and countless bounded abstract entities.

Morphology offered a recursive structure, an empirical approach, and outlined a heuristics of discovery. Polarity and enhancement are relational concepts that facilitate the empirical work, and unearthing the singular observations help theory-construction. Exemplars (e.g., Roger Bacon in the Medieval period) can display the forces (e.g., authority and experience) under discussion, and some observations can be linguistically refined into bridge-concepts (like the auxiliary image). The explanatory terms (vertical-spiral; expansion-contraction) create geometrical or phenomenal spaces, accounting for both regular (progressive) and the seemingly irregular (regressive) forms. This strange method that aims to pursue rather than to grasp permeates observations and concepts, structure and narrative, interweaves discovery and justification, heuristics and explanations. "Every crystallization is a realized kaleidoscope" (HA 12:370; "Maxims and Reflections" 37), Goethe writes, and his attempts to satisfy both reason and imagination are analysed in over ten thousand written works (Amrine, *Goethe in the History of Science*). The innumerable shades and shapes of the traces of morphology are like hues and shimmers intensifying and complementing each other. For Wilhelm von Schütz in 1821 it gave both Aristotle (light) and Plato (soul) (FW 24: 528). Darwin mentioned Goethe as a worthy forerunner, an "extreme partisan," in a footnote of "An Historical Sketch," added to later editions of *The Origin of Species* (1859), and the journal *Nature* started its first issue (4 November 1869) with Goethe's Orphic aphorisms on nature, the enduring influence of which Thomas Henry Huxley took care to admit:

> When my friend, the Editor of NATURE, asked me to write an opening article for his first number, there came into my mind this wonderful rhapsody on "Nature," which has been a delight to me from my youth up. It seemed to me that no more fitting preface could be put before a Journal, which aims to mirror the progress of that fashioning by Nature of a picture of herself, in the mind of man, which we call the progress of Science. (10)

The sociable Huxley, illustrious X-Club member, included parts of Goethe's commentary on "Die Natur" and made a pun on Goethe's self-description:

> Forty years have passed since these words were written, and we look again, "not without a smile," on Goethe's superlative. But the road which led from his comparative to his superlative, has been diligently followed, until the notions which represented Goethe's superlative are now the commonplaces of science – and we have super-superlatives of our own. (10)

Morphology was partially transmitted and elliptically contained in the super-superlative Victorian science of particles, forces, evolution, and specia(liza)tion. In the journal the diligent scientist lamented on the difficulties of translating German poetry into English, mentioned Goethe's discovery of the intermaxillary bone (1786), but made no reference to what the ageing poet proposed as the keystone to his nature studies, the essence of morphology, polarity and progression, the two-letter unity, an alphabet that could travel across domains to seek and explore advancement, to uncover and dissolve structure.

NOTES

1 I wish to thank the Rotman Institute of Philosophy at Western University for the financial support that made the writing of this paper possible, and to acknowledge the support of TÁMOP - 4.2.2.B-10/1 – 2010–0009 and OTKA K 72598. Hereafter, citations from Goethe's works are given according to the following abbreviations: FA (*Sämtliche Werke*); HA (*Goethes Werk*); LA (*Die Schrifte zur Naturwissenschaft*); SA (*Goethe's Collected Works*); BW (*Goethe's Botanical Writings*). I cite the original German in footnotes, except where key terms are introducd in the main text, and use the translation I find most appropriate where different versions exist, as in the case of FL-DT (*Zur Farbenlehre, Didaktischer Teil*). I thank Angela Borchert for the translations from the historical part of the *Theory of Colours* (Zur Farbenlehre, FL-HAT, LA 1:6). Discussions with Joan Steigerwald and Joel Faflak's helpful comments were much appreciated.

2 Dieses Baums Blatt, der von Osten
Meinem Garten anvertraut,
Gibt geheimen Sinn zu kosten,
Wie's den Wissenden erbaut.
Ist es Ein lebendig Wesen,
Das sich in sich selbst getrennt?
Sind es zwei, die sich erlesen,
Dass man sie als eines kennt.
Solche Frage zu erwidern,
Fand ich wohl den rechten Sinn.
Fühlst du nicht in meinen Liedern,
Dass ich Eins und doppelt bin. (HA 2:66)

English translation from *Poems of the West and East* (260–1) [Fig. 1].

3 Excerpts of Schütz's work were reprinted in Goethe's *Morphologische Hefte* I/4 (BW 194; FA 24:528).

4 "Wie schwer es sei auf diesem Wege für Didaktisches oder wohl gar Dogmatisches zu sorgen, ist dem Einsichtigen nicht fremd" (FA 24:535).

5 "Unter solchen Umständen war auch ich genötigt, über botanische Dinge immer mehr und mehr Aufklärung zu suchen. *Linnés Terminologie,* die *Fundamente,* worauf das Kunstgebäude sich stützen sollte, *Johann Geßners Dissertationen* zu Erklärung Linnéischer *Elemente,* alles in Einem schmächtigen Hefte vereinigt, begleiteten mich auf Wegen und Stegen; und noch heute erinnert mich ebendasselbe Heft an die frischen glücklichen Tage, in welchen jene gehaltreichen Blätter mir zuerst eine neue Welt aufschlossen. Linnés *Philosophie der Botanik* war mein tägliches Studium, und so rückte ich immer weiter vor in geordneter Kenntnis, indem ich mir möglichst anzueignen suchte, was mir eine allgemeinere Umsicht über dieses weite Reich verschaffen konnte" (HA 13:153; "Die Verfasser teilt die Geschichte seiner Botanischen Studien mit").

6 "Wir mußten öfters hören: die ganze Botanik, deren Studium wir so emsig verfolgten, sei nichts weiter als eine Nomenklatur, und ein ganzes auf Zahlen, und das nicht einmal durchaus, gegründetes System; sie könne weder dem Verstand noch der Einbildungskraft genügen, und niemand werde darin irgendeine auslangende Folge zu finden wissen" (HA 13:155).

7 "*Natürlich System* ein widersprechender Ausdruck. Die Natur hat kein System, sie hat, sie ist Leben und Folge aus einem unbekannten Zentrum, zu einer nicht erkennbaren Grenze" (FA 24:582).

8 "Die Idee der Metamorphose ist eine höchst ehrwürdige, aber zugleich höchst gefährliche Gabe von Oben. Sie führt ins Formlose; zerstört das Wissen, lost es auf. Sie ist gleich der vis centrifuga und würde sich ins Unendliche verlieren, wäre ihr nicht ein Gegengewicht zugegeben: ich meine den Spezifikationstrieb, das zähe Beharrlichkeitsvermögen dessen was einmal zur Wirklichkeit gekommen" (FA 24:583).

9 "Unsere ganze Aufmerksamkeit muss aber darauf gerichtet sein, der Natur ihr Verfahren abzulauschen, damit wir sie durch zwängende Vorschriften nicht widerspenstig machen, aber uns dagegen auch durch ihre Willkür nicht vom Zweck entfernen lassen" (FA 24:584).

10 "Wir begegnen hier einem zweiten Widerspruch, der dem ersten völlig analog ist, doch so, daß beide in umgekehrtem Verhältnis zueinander stehen. In der Forderung eines natürlichen Systems scheint der menschliche Verstand seine Grenzen zu überschreiten, ohne doch die Forderung selbst aufgeben zu können. Ein Beharrlichkeitsvermögen in der Natur scheint den Strom des Lebens hemmen zu wollen; und doch ist in ihr etwas Beharrliches, der unbefangene Beobachter muß es anerkennen" (FA 24:586).

11 "So haben Sie z.B. einen Doktor Ernst Meyer bei sich in Göttingen, welchem ich seinem Teil meines Nachlasses durch eine Schenkung unter den Lebendigen zu übergeben nicht nötig habe, da er auch ohne dies auf dem Wege, den ich schon längst für den rechten halte, fortschreitet" (FA 24:1121, 26.9:1822).

12 I use Ludwik Fleck's term for "thought collective," even though I focus on reconstructing the thought of a single individual of the collective. To cite another example of contact, Goethe recalls that his friend and colleague, Professor C.W. Büttner, "endeavoured to arrange the plants according to families, advancing from the simplest, almost invisible rudimentary manifestations to the most complex and devious. He was fond of exhibiting an outline ... greatly to my edification and satisfaction" (BW 156).

13 "Hier möcht ich nun nach meiner Weise noch folgendes anfügen: die Idee ist in der Erfahrung nicht darzustellen, kaum nachzuweisen, wer sie nicht besitzt, wird sie in der Erscheinung nirgends gewahr: wer sie besitzt, gewöhnt sich leicht über die Erscheinung hinweg, weit darüber hinauszusehen und kehrt freilich nach einer solchen Diastole, um sich nicht zu verlieren, wieder an die Wirklichkeit zurück, und verfährt wechselsweise Wohl so ein ganzes Leben" (FA 24:535).

14 Goethe's phenomenolgical starting point enabled him to use Newtonian concepts, much like chemists of his day: "Thus, in an attempt to study the laws whereby life is given to organic nature ... a force was ascribed to this life for purposes of discourse; and this force could be, indeed had to be, assumed, because life as a whole expresses itself as a force that is not contained within any one part" (BW 90; HA 13:126).

15 "Jener Aufsatz ist mir vor kurzem aus der brieflichen Verlassenschaft der ewig verehrten Herzogin Anna Amalia mitgeteilt worden; er ist von einer wohlbekannten Hand geschrieben, deren ich mich in den achtziger Jahren in meinen Geschäften zu bedienen pflegte. // Daß ich diese Betrachtungen verfaßt, kann ich mich faktisch zwar nicht erinnern, allein sie stimmen mit den Vorstellungen wohl überein, zu denen sich mein Geist damals ausgebildet hatte" (HA 13:48). After the anonymous piece appeared in the *Tierfurter Journal*, Goethe denied authorship in 1783 in a letter to Knebel (HA 13:576). Some suggested Tobler as composer.

16 "Sie scheint alles auf Individualität angelegt zu haben und macht sich nichts aus den Individuen" (HA 13:45). "Auch das Unnatürlichste ist Natur. Wer sie nicht allenthalben sieht, sieht sie nirgendwo recht" (HA 13:46).

17 "Die Erfüllung aber, die ihm fehlt, ist die Anschauung der zwei großen Triebräder aller Natur: der Begriff von *Polarität* und von *Steigerung*, jene der Materie, insofern wir sie materiell, diese ihr dagegen, insofern wir sie geistig denken, angehörig; jene ist in immerwährendem Anziehen und Abstoßen, diese in immerstrebendem Aufsteigen" (HA 13:48).

18 "Weil aber die Materie nie ohne Geist, der Geist nie ohne Materie existiert und wirksam sein kann, so vermag auch die Materie sich zu steigern, so wie sichs der Geist nicht nehmen läßt, anzuziehen und abzustoßen; wie derjenige nur allein zu denken vermag, der genugsam getrennt hat, um zu verbinden, genugsam verbunden hat, um wieder trennen zu mögen" (HA 13:48). Note that the English translation uses analysis-synthesis, where the original German builds on the polarity of separation-connection (trennen-verbinden).

19 "Vergegenwärtigt man sich die hohe Ausführung, durch welche die sämtlichen Naturerscheinungen nach und nach vor dem menschlichen Geiste verkettet worden, und liest alsdann obigen Aufsatz, von dem wir ausgingen, nochmals to mit Bedacht; so wird man nicht ohne Läcḫeln jenen Komparativ, wie ich ihn nannte, mit dem Superlativ, mit dem hier abgeschlossen wird, vergleichen und eines fünfzigjährigen Fortschreitens sich erfreuen" (HA 13:49).

20 "Wenn ich nun auf diese Weise das Grundlose der Newtonischen Lehre, besonders nach genauer Einsicht in das Phänomen der Achromasie, vollkommen erkannte, so half mir zu einem neuen theoretischen Weg jenes erste Gewahrwerden, dass ein entschiedenes Auseinandertreten, Gegensetzen, Verteilen, Differenzieren, oder wie man es nennen wollte, beiden prismatischen Farbenerscheinungen statthabe, welches ich mir kurz und gut unter der Formel der Polarität zusammenfasste, von der ich überzeugt war, dass sie auch bei den übrigen Farben-Phänomenen durch geführt werden könne" (FL-HT 424).

21 "So auch mit der Wurzel, sie ginge mich eigentlich nichts an, denn was habe ich mit einer Gestaltung ‘zu thun, die sich in Fäden, Strängen, Bollen und Knollen und bei solcher Beschränkung, sich nur im unerfreulichen Wechsel allenfalls darzustellen vermag, wo unendliche Varietäten zur Erscheinung kommen, niemals aber eine Steigerung und diese ist es allein die auf mich auf meinem Gange nach meninem Beruf an sich ziehen, festhalten und mit sich fortreißen konnte. Gehe doch jeder ebenmäßig seinen Gang und schaue auf das was er leistete in vierzig Jahren bescheiden zurück, wie uns ein guter Genius zu tun vergönnt hat" (FA 24:654, Weimar, 27 June 1824).

22 "Hier in dieser neu mir entgegentretenden Mannigfaltigkeit wird jener Gedanke immer lebendiger, daß man sich alle Pflanzengestalten vielleicht aus einer entwickeln könne. Hierdurch würde es allein möglich werden, Geschlehter und Arten wahrhaft zu bestimmen, welches, wie mich dünkt, bisher sehr willkürlich geschieht. Auf diesem Punkte bin ich in meiner botanoschen Philosophie steckengeblieben, und ich sehe noch nicht, wie ich mich entwirren will. Die Tiefe und Breite dieses Geschäfts scheint mir völlig gleich" (HA 11:60; Padua, 27 September 1786).

23 "Die Urpflanze wird das wunderlichste Geschöpf von der Welt, um welches mich die Natur selbst beneiden soll. Mit diesem Modell und dem Schlüssel dazu kann man alsdann noch Pflanzen ins Unendliche erfinden, die konsequent sein müssen, das heißt, die, wenn sie auch nicht existieren, doch existieren, doch existieren könnten und nicht etwa malerische oder dichterische Schatten und Scheine sind, sondern eine innerliche Wahrheit und Notwendigkeit haben. Dasselbe Gesetz wird sich auf alles übrige Lebendige anwenden lassen" (HA 11:324, Neapel, 17 May 1787).

24 "Es war mir nämlich aufgegangen, daß in demjenigen Organ der Pflanze, welches wir als Blatt gewöhnlich anzusprechen pflegen, der wahre Proteus verborgen liege, der sich in allen Gestaltungen verstecken und offenbaren könne. Vorwärts und rückwärts

ist die Pflanze immer nur Blatt, mit dem künftigen Keime so unzertrennlich vereint, daß man eins ohne das andere nicht denken darf" (HA 11:375).

25 "Wie sie sich nun unter einen Begriff sammeln lassen, so wurde mir nach und nach klar und klärer, daß die Anschauung noch auf eine höhere Weise belebt werden könnte: eine Forderung, die mir damals unter der sinnlichen Form einer übersinnlichen Urpflanze vorschwebte. Ich ging allen Gestalten, wie sie mir vorkamen, in ihren Veränderungen nach" (FA 24:748).

26 "[W]ie ich, auf eine kindliche Weise, den Begriff der Plfanzenmetamorphose gefaßt" (FA 24:413).

27 "Ideal, als das letzte Erkennbare, real als erkannt, symbolisch, weil es alle Fälle begreift, identisch mit allen Fällen" (HA12:366; "Maxims and Reflections" 15).

28 "Durch die Erfahrungen, welche wir an dieser Metamorphose zu machen Gelegenheit haben, werden wir dasjenige enthüllen können, was uns die regelmäßige verheimlicht, deutlich sehen, was wir dort nur schließen dürfen; und auf diese Weise steht es zu hoffen, daß wir unsere Absicht am sichersten erreichen" (HA 13:65).

29 "Denn wie in jenem Fall, die Natur vorwärts zu dem großen Zwecke hineilt, tritt sie hier um eine oder einige Stufen rückwärts" (HA 13:65).

30 "Warum sollte das Mittel nicht durch eine uns unbekannte Ursache Doppelbilder hervorbringen können?" (LA 1/3:158). Iceland spar, or cropped pieces of calcite, are not the only minerals displaying birefringence, but the difference of indices of refraction for the ordinary and extraordinary rays are large (about forty times larger than in ice). Goethe systematically explored domains of optical phenomena in which Newton's theory could be criticized. Birefringence in the seventeenth century was used by C. Huygens to propose a non-corpuscular theory of light.

31 "Ein besonderes Stück aber dieses Minerals besitze ich, welches ganz vorzügliche Eigenschaften hat. Legt man nämlich das Auge unmittelbar auf den Doppelspat und entfernt sich von dem Grundbilde, so treten gleich ... zwei Seitenbilder rechts und links hervor, welche, nach verschiedener Richtung des Auges und des durchsichtigen Rhomben, bald einfach ... bald doppelt ... erscheinen" (*Doppelbilder des rhombischen Kalkspats*, 1813).

32 "Dieses nunmehr genugsam entwickelte farbige Phänomen lassen wir denn nicht als ein ursprüngliches gelten, sondern wir haben es auf ein früheres und einfacheres zurückgeführt und solches aus dem Urphänomen des Lichtes und der Finsternis, durch die Trübe vermittelt, in Verbindung mit der Lehre von den sekundären Bilderns abgeleitet" (FL-DT §247). "[W]ir [haben] die erstgedachten ziemlich einfachen Phänomena aus dem Vorhergehenden abzuleiten oder, wenn man will zu erklären" (FL-DT §218).

33 "Alles, was wir an Materialien zur Geschichte, was wir Geschichtliches einzeln ausgearbeitet zugleich überliefern, wird nur der Kommentar zu dem Vorgesagten sein" (FL-HAT 94).

34 "Jene früheren Geographen, welche die Karte von Afrika verfertigten, waren gewohnt, dahin, wo Berge, Flüsse, Städte fehlten, allenfalls einen Elefanten, Löwen oder sonst ein Ungeheuer der Wüste zu zeichnen, ohne daß sie deshalb wären getadelt worden. Man wird uns daher wohl auch nicht verargen, wenn wir in die große Lücke, wo uns die erfreuliche, lebendige, fortschreitende Wissenschaft verläßt, einige Betrachtungen einschieben, auf die wir uns künftig wieder beziehen können" (FL-HT 94).

35 "Der Konflikt des Individuums mit der unmittelbaren Erfahrung und der mittelbaren Überlieferung, ist eigentlich die Geschichte der Wissenschaften" (FL-HT 87).

36 "Indem wir nun von Überlieferung sprechen, sind wir unmittelbar aufgefordert, zugleich von Autorität zu reden. Denn genau betrachtet, so ist jede Autorität eine Art Überlieferung" (FL-HT 92).

37 "Die Schriften Bacons zeugen von großer Ruhe und Besonnenheit. Er fühlte sehr tief den Kampf, den er mit der Natur und mit der Überlieferung zu bestehen hat" (FL-HT 101).

38 "Man bedenkt niemals genug, daß eine Sprache eigentlich nur symbolisch, nur bildlich sei und die Gegenstände niemals unmittelbar, sondern nur im Widerscheine ausdecke" (FL-DT § 751).

39 "Wir mußten annehmen: Es walte in der Vegetation eine allgemeine Spiraltendenz, wodurch, in Verbindung mit dem vertikalen Streben, aller Bau, jede Bildung der Pflanzen, nach dem Gesetze der Metamorphose, vollbracht wird" (FA 24:794).

40 Bacon's natural development in a retrogressive period displays the two forces. The individual and collective dimensions are joined: "ist die Geschichte der Wissenschaften mit der Geschichte der Philosophie innigst verbunden, aber eben so auch mit der Geschichte des Lebens und des Charakters der Individuen, so wie der Wölker" (FL-HT 68).

41 "Die Naturwissenschaften haben sich bewundernswürdig erweitert, aber keinesweges in einem stetigen Gange, auch nicht einmal stufenweise, sondern durch Auf- und Absteigen, durch Vor- und Rückwärtswandeln in grader Linie oder in der Spirale, wobei sich denn von selbst versteht, daß man in jeder Epoche über seine Vorgänger weit erhaben zu sein glaubte" (FL-HT 94).

42 "Szepter einer aufgedrungenen Autorität" (FL-HT 94).

43 "Ein großer Teil dessen, was man gewöhnlich Aberglauben nennt, ist aus einer falschen Anwendung der Mathematik entstanden, deswegen ja auch der Name eines Mathematikers mit dem eines Wahnkünstlers und Astrologen gleich galt. Man erinnere sich der Signatur der Dinge, der Chiromantie, der Punktierkunst, selbst des Höllenzwangs; alle dieses Unwesen nimmt seinen wüsten Schein von der klarsten aller Wissenschaften, seine Verworrenheit von der exaktesten" (FL-HT 102).

44 "Ein Entscheidendes Aperçu ist wie eine inokulierte Krankheit anzusehen: man wird sie nie los, bis sie durchgekämpft ist" (HA 14:263).

WORKS CITED

Amrine, Frederick. *Goethe in the History of Science. Studies in Modern German Literature.* Vol. 29–30. New York: Peter Lang, 1996. Print.

Amrine, Frederick. "Goethean Intuitions." *Goethe Yearbook* 18 (2011): 35–50. Print.

Amrine, Frederick. "The Metamorphosis of the Scientist." *Goethe Yearbook* 5 (1990): 187–212. Print.

Arber, Agnes. "Prefatory Note to the Fragment Afterwards Known as Die Natur." *Chronica Botanica* 10 2 (Goethe's Botany) (1946): 119–22. Print.

Garani, Myrto. *Empedocles Redivivus: Poetry and Analogy in Lucretius.* Studies in Classics. New York: Routledge, 2007. Print.

Goethe, Johann Wolfgang von. *Goethe's Botanical Writings.* Ed. B. Müller. Woodbridge, CT: Ox Bow P, 1989. Print.

Goethe, Johann Wolfgang von. *Goethe's Collected Works.* 12 vols. New York: Suhrkamp, 1983–9. Print.

Goethe, Johann Wolfgang von. *Goethes Werk.* 14 vols. Hamburger Ausgabe: DTV, 1994. Print.

Goethe, Johann Wolfgang von. *Poems of the West and East.* Trans. John Whaley. Berne: Peter Lang, 1998. Print.

Goethe, Johann Wolfgang von. *Sämtliche Werke.* 40 vols. Frankfurt: Deutscher Klassiker, 1987. Print.

Goethe, Johann Wolfgang von. *Die Schrifte zur Naturwissenschaft.* Weimar: Leopoldina, 1947– . Print.

Hegge, Hjalmar. "Theory of Science in the Light of Goethe's Science of Nature." *Goethe and the Sciences: A Reappraisal.* Ed. Frederick Amrine, Francis J. Zucker, and Harvey Wheeler. Vol. BSPS 97. Dordrecht/Boston: D. Reidel, 1987. Print.

Huxley, Thomas Henry. "Goethe: Aphorisms on Nature." *Nature* 1.1 (1869): 9–11. Print.

Lindroth, Sten. "The Two Faces of Linnaeus." *Linnaeus, the Man and His Work.* Ed. Tore Frängsmyr. Canton, Mass.: Science History Publications, 1994. 1–62. Print.

Marcum, James A. "The Nature of Light and Color: Goethe's 'Der Versuch Als Vermittler' Versus Newton's Experimentum Crucis." *Perspectives on Science* 17 (2009): 457–81. Print.

Portmann, Adolf. "Goethe and the Concept of Metamorphosis." *Goethe and the Sciences: A Reappraisal.* Ed. Frederick Amrine, Francis J. Zucker, and Harvey Wheeler. Boston: Reidel Publishing Co., 1987. 133–45. Print.

Ribe, Neil, and Friedrich Steinle. "Exploratory Experimentation: Goethe, Land, and Colour Theory." *Physics Today* (July 2002): 43–7. Print.

Schaffer, Simon. "Glass Works: Newton's Prisms and the Uses of Experiment." *The Uses of Experiment: Studies in the Natural Sciences.* Ed. David Gooding, Trevor Pinch, and Simon Schaffer. Cambridge: Cambridge UP, 1989. 67–104. Print.

Sepper, Dennis L. *Goethe Contra Newton: Polemics and the Project for a New Science of Color*. Cambridge: Cambridge UP, 1988. Print.

Westphal, Jonathan. *Colour: Some Philosophical Problems from Wittgenstein*. Aristotelian Society Series. Oxford: Oxford UP, 1987. Print.

Zemplén, Gábor Á. "Auxiliary Images – Appropriations of Goethe's Theory of Colours." *Variantology 2: On Deep Time Relations of Arts, Sciences and Technologies*. Ed. Siegfried Zielinski and David Link. Köln: Verlag der Buchhandlung Walther König, 2006. 169–202. Print.

Chapter Seven

Vertiginous Life: Goethe, Bones, and Italy

ANDREW PIPER

Dann ist Philosophie immer etwas Verrücktes.

– Martin Heidegger

Turns

What does *revolution* look like? This was one of the more urgent questions hanging over Europe after 1789, not just in the world of politics, but in those of literature and natural history. In Italy, Goethe thought he had found an answer. On 22 April 1790, while walking along the Lido outside of Venice, Goethe found a sheep's skull, and, upon inspection, realized that the skull was nothing more than the fusion of vertebrae (six in total). For Goethe, and for many others to follow, the vertebra, or *Wirbelbein*, was thought to be the basis of all animal form. At the beginning of life was the turn.

In the early nineteenth century, bones, like books, were big business. As fossils, they made possible ideas like species extinction, "global revolutions," in Georges Cuvier's words, among animal forms.[1] For others, like Lorenz Oken, Étienne Geoffroy Saint-Hilaire, and, later, Richard Owen, bones like the furcula or wishbone in the ostrich were signs of a fundamental formal continuity in nature, that nature left vestiges of prior forms in newer ones in a more evolutionary sense.[2] Of the thousands of bones marshalled in Romantic debates about the meaning of calcified forms – and it was the sheer plurality of bones that was surely one of the key factors in the scientific ambiguity that surrounded them – the vertebra was, perhaps not surprisingly, most often at the centre of such debates. The vertebra came to stand as a particularly *Romantic* bone, one that embodied the tensions between competing notions of change in the early nineteenth century, between the ruptures of revolution on the one hand and the continuities of evolution on the other. The vertebra became a quintessential figure through which to imagine the form of change.

Since Aristotle, the vertebra was the primary classificatory instrument of the animal kingdom between vertebrates and invertebrates. Until the early nineteenth

Figure 7.1 A geometrical reconstruction of the ideal vertebra, from Carl Gustav Carus, *Von den Ur-Theilen des Knochen und Schalengerüstes,* (1828), Fig. XV, Plate II.

century, in other words, it marked the ultimate species dividing line. To imagine the vertebral origins of the skull as Goethe and others were doing was to shift dramatically the vertebra's core identity. The vertebra no longer stood for a fundamental distinction within nature, but represented the formal mechanism through which categorical continuity could be conceptualized. The vertebra's relationship to skeletal structure was no longer *pars pro toto*, as one bone among many that could stand for a heterogenous whole; rather, the vertebra *was* the whole, the inner core of skeletal diversity itself. As Lorenz Oken would argue in his opening lecture at the University of Jena, setting off a decades-long dispute with Goethe over the priority of its discovery, not to mention years of pan-European wrangling over its validity as a theory:

> A bubble calcifies and it becomes a vertebra. A bubble lengthens itself into a tube, is then divided and calcifies. It becomes a spine. The spine creates side channels, these calcify and become the rib cage. This skeleton repeats itself at both poles and becomes head and pelvis. The skeleton is nothing more than a developed, branching, repetitive vertebra; a vertebra is the preformed seed of the skeleton. Man is just a vertebra. (5)

By the 1820s, Geoffroy Saint-Hilaire would go so far as to argue that the vertebrate/invertebrate distinction, which Cuvier had refined into four categories or *embranchements*, was no longer tenable.[3] Not just man, but the entire animal kingdom was now a vertebra. The German anatomist, artist, and gynecologist Carl Gustav Carus would subsequently attempt to work out the complex geometric laws of how the spherically derived vertebra could transform itself into the linear shapes of spines, ribs, and femurs – in other words, how circles could become lines, or perhaps even more abstractly put, how closures could become openings (Figure 7.1).

Retro-spection

In locating his discovery of the vertebral structure of the skull in Italy, Goethe was not only intervening into osteological debates about the nature of man; he was also making a major intervention into literary historical debates about the genre of autobiography and the nature of "life" that it claimed to represent. *Italian Journey* (1816/17) would emerge as "Section Two" of Goethe's larger eleven-volume landmark autobiographical project, *From My Life* (*Aus meinem Leben*). As a work, *Italian Journey* was deeply invested in the problem of representing radical change. Not only was Italy referred to as a place of personal "rebirth," thereby combining the neoclassical Grand Tour with the Christian poetics of the Pauline conversion (*Italienische Reise* 15.1:158),[4] but as a book, *Italian Journey* also marked out the dual problems of narrative and bibliographic rupture. *Italian Journey* was initiated in lieu of the unfinished fourth volume of Section One (known as *Poetry and Truth* [*Dichtung und Wahrheit*]); the division of the autobiographical "section" or *Abtheilung* thus shifted the narrated time of the autobiography by ten years, even as it slowed down the narrative time so that a single year occupied multiple volumes. It marked an engagement, in other words, with the problem of heterochrony, the different scales of time that surrounded autobiography, and increasingly, theories of organic life more generally. But Goethe's Italian project was also an engagement with what it meant to think about bibliographic change, not in terms of the corporal "section," but in how one moved from handwriting to print, how to think about the relationship between the massive trove of manuscriptural remnants, both drawings and notes, that belonged to Goethe's Italian experience and that would form the basis of the eventual printed volume. The deeply vorticular identity of the book – the turning that belonged to the recto/verso identity of the book's page – was a key medial backdrop to thinking about the revolutions of writing.

Italy and *Italian Journey* thus served as spaces to think about the correspondence between personal and bibliographic change, between the forms that life can take and the forms of its representation. Section One of the autobiography had largely been concerned with the problem of *aspiration* – what it meant to think of life as a vertically oriented form of striving beyond oneself always only in order to become oneself – in short, to be like a plant. The central icon of such entelechial organicism would, in characteristic Romantic fashion, be figured most prominently in the shape of the gothic spire, the stone that desired to become a plant. For Goethe, it was Strasbourg Cathedral that would stand as the most decisive architectonic proxy for this notion of aspirational life. By Section Two of the autobiography, however, questions of vertical continuity and the Germano-Frankish seam of Strasbourg to which they belonged would give way to a variety of torsional figures that were grounded above all else in the geographic space of Italy. In the calcified, time-worn figures of the fossil, but also the ruin, the torso, and even the anatomical sketch, Goethe would inquire into the problem of the *rotatory* nature of life, the idea of *conversion* or the complete turn, that had stood at the basis of life writing since Augustine. The turn from the aspirational to the rotational was in some sense also a turn away from the divinatory identity of the spire – its reaching up toward God – and toward a far more terrestrially oriented theory of life, one that points down and around. It marked an epochal rewriting, or conversion if you will, of the idea of conversion.[5]

Along with a number of other thinkers from the period, Goethe was asking what it meant to conceptualize life as both radical change *and* as that which rotated on its own axis, as a form of incessant recurrence, drawing together the double valence of the notion of "revolution" available at the time.[6] The question for so many Romantics was how the opposing notions of revolution and evolution could be integrated with one another. On the one hand, it meant moving beyond an eighteenth-century notion of evolution in Blumenbach's terms as *bloße Entwickung*, as *mere* development,[7] that which unfolded in a predetermined sense from an origin. On the other hand, it meant rejecting the idea of revolution as nothing more than a radical or complete break, as an end*point*. Instead, what was required was a theory of development that could incorporate knowledge of radical discontinuity, one that could bring forth novelty in the world and that was premised in some sense on an idea of the unknown. It required that revolution be inscribed in, not opposed to, an idea of evolution. In the broadest possible terms, it was an attempt to think through the possibility of representing *fracture*, in a personal, political, natural, aesthetic, and even bibliographic sense.[8]

What I want to explore in this essay is the way two forms of knowledge – the autobiographical and the natural scientific – converge in Goethe's work to generate not only a new theory of life, but also a new theory of its observability, one that has important implications for how we think about this relationship between the categories of revolution and evolution in a Romantic context, between radical change and categorical continuity. Discussions of the relationship between Goethe's autobiography and the natural sciences almost always focus on the "autobiographical" nature of science for Goethe, the way knowledge of the self is important to knowledge of the world.[9] Seldom do scholars engage with how the autobiographical texts themselves initiate new forms of scientific inquiry, and the way they often do so *differently*. My aim in this essay, and the larger project from which it is drawn, is to undo the categorical distinctions between the "autobiographical" and the "scientific" and see instead how they work together to produce a variety of new modes of understanding "life" during the Romantic period. I am interested in showing how autobiographical texts can and should be read as natural scientific treatises in their own right, to engage with the particular epistemological work that they perform as one component of the broader problem of "writing life" at the turn of the nineteenth century.

Goethe's work in particular is significant in this regard because of the way it established a persistent working toward a new kind of looking that could accommodate new objects of natural knowledge, or, in more circular fashion, the way new objects of knowledge also produced new ways of looking at the world. Indeed, it would be precisely this aspect of circularity, revolution in a rotational sense, that would be a core component of the visual paradigm that Goethe was working out in his reflections on Italian space and osteological form.[10] Autobiography was essential to this project, and not only because it involved a self-reflexivity that had become central to Goethe's understanding of scientific knowledge. On an even more local level, it was significant because of the way it operated under the principle of *Rücksicht* (retrospection, but also, more literally, "back-looking"), a particular type of looking that was integrally related to a particular anatomical object. There was an osteological concern, a care for the skeletal, if you will (*Rücksicht* can also mean considerateness), that was encoded in the project of life writing. The back, and the vertebra at its centre, emerged as objects capable of engendering this new faculty of biological as well as autobiographical insight.

Few objects have been more canonical for autobiography than the skull. The skull is the ultimate sign of *memento mori*, the conjunction of memory and death under which autobiographies are most often written (Olney). And yet in conjoining the skull and vertebra, Goethe was significantly altering the skull's metaphorical valence and the principles of "retrospection" for which it stood. It was a shift in meaning (*rücken* in German) that corresponded to a new anatomical understanding of the human body, one that now rested on the centrality of the spine, the behind, and the back (*der Rücken*). No longer a simple peak, point, or crown (the skull's verticality was always essential to its cultural meaning), the vertebral skull was understood as a summation of turns, a space of synthetic spirality. In this, it combined the two etymological strands of vertex/vortex that lay behind the vertebra as a site of both verticality and vorticularity. Retrospection was no longer coded as a simple linear process, but was informed by a sense of *circumspection* as well, a looking *around*. The skull, that most mimetic of all bones, was endowed in Goethe with a genealogical abstraction that challenged its otherwise performative simplicity. No longer a cavernous outside, a shell to humanity's psychological core, it emerged as an ambiguation of this very inside/outside dichotomy. The vertebral head was a helix.[11]

Knowledge of life, and the vertebral nature of Italian life in particular, thus depended for Goethe upon a profound sense of medial and ocular disorientation (*verrücken*). There was an inherently vertiginous quality to looking onto vertebral forms. As the stakes of such disorientation suggested – *verrückt* is the word for going mad – vertebral knowledge always verged on a fundamental errancy. As Goethe remarked on the impossible finality of life's representation, "One cannot make it round and finished" (*Morphologie* 24:504). The Goethean contribution to the science of life that emerged from *Italian Journey*, I want to suggest, was marked not just by the way it stands as one of the most profound inquiries into the continuity of difference that would be so fundamental to Darwinian evolution (as well as modern autobiography), the way revolution is reinscribed into a concept of evolution. Rather, Goethe's contribution lies most pronouncedly in the way revolution is reinscribed into how we come to *know* about life, the way error is inextricably bound up with knowledge of life.[12] Not the archetype, but *misprision* becomes the fundamental feature of theorizing life after Goethe, an incompletion that belonged to knowledge of such vertebral forms, what it meant to think about *conversion* in a deep sense. Italy and the suite of fractured remnants that one could find there become the geo-bibliographical sites through which these reflections on the torsional nature of life and the tropological nature of its representation – life's essential errancy – could be most fully articulated.

Ruins

The vertigo of Italian life is introduced in *Italian Journey* in programmatic fashion during Goethe's first recorded encounter with a classical ruin, the amphitheatre in Verona (Figure 7.2):

> Verona, 16. September
>
> The amphitheater is the first meaningful monument of antiquity that I have seen – and so well preserved! As I entered, but even more so when I ascended and wandered around the outer edge, it seemed strange to see something so grand and yet actually to see nothing. Nor does it wish to be seen empty, but rather full of people, as was the case recently with festivals for Joseph I and Pius VI. The emperor, who is accustomed to seeing masses of people, was rumored to be astounded by the sight. Still, only in its earliest existence was it able to produce its entire effect, since the Folk was then still more of a Folk than it is today. Such an amphitheatre is actually made to impose the Folk upon itself, to see itself at its best. (*Italienische Reise* 15.1:44)

The ovular amphitheatre is, not surprisingly, coded by Goethe as a space of self-reflection, one where the spectacle is not the performance, but the public itself. As Sigrid Weigel has argued in her reading of this passage, such collective self-reflexivity becomes the condition of a political *Bildung*, the improvement of the "body politic." We move in that final sentence, with no more justification than the caesura of the comma, from the verb "impose [imponieren]" to the phrase "to see itself at its best [das Volk mit sich selbst zum besten zu haben]," in which imposition becomes the condition of self-improvement. Through the architectural

Figure 7.2 Arena di Verona. Courtesy of the Klassik Stiftung Weimar, Goethe-Nationalmuseum 70–2010–0526.

space of the amphitheatre, the Folk is gradually brought to realize a sense of self-possession, of *having* itself at its best (das Volk mit sich selbst zum besten *zu haben*). What Goethe refers to at one moment as "the many-headed, multisensory, oscillating animal erring here and there" is here unified through the amphitheatre "into a noble body," and, one presumes, given the vertical metaphorics at work, an upright body. The ovular amphitheatre is a perfect example of what Peter Sloterdijk would call an anthropotechnology, a cultural technology of "humanization."

What interests me about this brief scene of speculative history, created by Goethe, of how society came to be is the way underneath, or rather, at the very centre of this unity produced through self-reflexivity lies an abyss – literally, a nothingness: "As I entered, but even more so when I ascended and wandered around the outer edge, it seemed strange to see something so grand and yet actually to see nothing [*nichts* zu sehen]" (*Italienische Reise* 15.1:44; emphasis added).

The problem, or perhaps we could call it the truth, that the self-reflexive amphitheatre as ruin introduces is the way it points to a nothing at its centre. In distinction to the entelechial spire of the incomplete gothic cathedral – upwardly striving and still in the making – the ovular *ruin* serves as a material portal into an emerging notion of a centreless subjectivity in Goethe's work.

The word Goethe uses for this negative space will be *Krater* – from the Latin *crater,* meaning a vessel or bowl, but also found in a natural scientific context for the geological opening of a volcano. There is a circularity established to the distinction between nature and culture as the word for a cultural vessel is used to denote the natural phenomenon of an eruptive shape, which is in turn used to understand the larger cultural formation of the amphitheatre. Perhaps even more significant is the way the *Krater* gestures toward the notion of the Latin *mundus* or pit that was thought to reside at the heart of all Etruscan societies and that served as a gateway to the world of the dead (as well as a place to store food). The nothing at the heart of culture was death, the ruin the instrument (gateway, opening, aperçu) of its recognition.

To understand Goethe's framing of the ruin in these terms is to see the extent to which he rewrites much of his contemporaries' understanding of ruins, as well as a great deal of the subsequent scholarship on the matter. By the close of the eighteenth century, the ruin had emerged as a central aesthetic object informing the picturesque view. As Sophie Thomas has written, the ruin set one in relation to time (see also Dubin). It didn't so much represent a particular moment in history; instead, it represented *relationality* itself, a point that could be gleaned from theoretical treatises such as William Gilpin's three essays on the picturesque or Constantin Volney's *Les ruines* that triumphed the "*juste équilibre*" (xiv) which ruins afforded the soul of the viewer. As Matthias Schöning has argued, the ruin was where Romantic individuals came to feel time in two particular forms, what he calls "time giving time" and "time consuming time." In this, the ruin can alternately represent a sense of decline (time consuming time) or a sense of duration, of something that has outlived itself (time giving time). For Goethe, by contrast, the emptiness of the ruin was precisely what *disrupted* any sense of spatio-temporal relationality or scale: "Now when one sees it empty one has no sense of scale [*kein Maßstab*, literally a measuring stick], one does not know if it is big or small" (*Italienische Reise* 15.1:45).[13] Far from setting the viewer within a comfortable relationship to either time or space, the ruin for Goethe cancelled any sense of measure or equilibrium. It lacked sense.[14]

Crucial to the logic of Goethe's encounter with the ruin is the fact that it is only when one ascends to the upper and outer margin, or *Rand,* of the amphitheatre that one finally becomes aware of the nothing in the middle. For Carl Philip Moritz, whom Goethe had befriended in Italy, being on the outside of an enclosed structure had been one of the fundamental tropes through which his autobiographical project was expressed in *Anton Reiser* (1785). The autobiographer was not only an outsider to his or her own life, but this very outsideness was the condition of possibility of lending that life meaning, of seeing it as a totality. As Moritz writes, "Only in that moment when he cast a glance at human life in its entirety did he first learn to differentiate those monumental aspects of life [das Große im Leben] from their detail" (277). For Goethe, on the other hand, being on the outside of the arc of life looking in (and down) was not to sense the monumentality of life, life understood as a totality. It was instead to feel a vertiginous sense of personal and sensory experience, one in which meaning turned into a "seeing nothing." Verticality and marginality are the conditions of the knowledge of nothingness at the heart of the practice of self-reflection.

Again and again as Goethe travels through Italian space, visual experience is coded by a vocabulary of disorientation, endlessness, a crisis of scale, and even madness. There is a basic negativity encoded in the Italian encounter with objects. Upon seeing Tintoretto's *Paradise*, for example, Goethe remarks: "to admire and enjoy all of this, one would have to possess the piece oneself and have it before one's eyes for one's entire life. The work passes over into endlessness" (*Italienische Reise* 15.1:51). Upon seeing a cloister designed by Palladio, he remarks, "one should have to spend years observing such a work" (15.1:77), and when arriving in Bologna and seeing paintings by Caracci, Guido Reni, and Dominichi, he writes, "Within this heaven new stars appear once again, which I cannot account for and which lead my mind astray [die mich irre machen]" (15.1:112). Upon seeing further works by these artists, he continues, "A significant obstacle to pure observation and immediate insight are the great number of meaningless objects in the images before which one goes mad [toll wird] in trying to honor and venerate them" (15.1:113). In Florence, Goethe writes of a conversation he has: "The good man could not admittedly know that I was silent and reflective because of the way old and new objects so confused my mind [mir den Kopf verwirrte]" (15.1:122). In Rome, he writes, "I finally saw the two colossuses! Neither eye nor spirit are sufficient to grasp them" (15.1:136); and then on the Coliseum: "[W]hen one looks at it everything else appears small; it is so large that one cannot preserve the image in one's soul" (15.1:145); and the Sistine Chapel: "How is one, small as one is and accustomed to smallness, supposed to compare oneself to this noble, monstrous, and well-formed body? And even if one would appropriately like to stand back a bit [es einigermaßen zurecht rücken möchte], then a monstrous hoarde presses in against you from all sides, meets you at every turn, as each demands for itself an offering of attention. How is one supposed to extract oneself from there?" (15.1:157). And finally, upon arriving in Naples, he recites a famous Italian saying: "Veda Napoli e poi muore [See Naples and die]!" (15.1:205). As the proliferating vocabulary of monstrousness, madness, and surplus revealed, Italy was where one went to lose perspective, and ultimately one's self. Italy was a form of death in life.

In order to understand the significance of these statements, it is important to note the extent to which the reception of *Italian Journey* has been overwhelmingly shaped by an understanding of it as a work that argues for and defines a new mode of *objective* visuality in Goethe's life and work.[15] As Goethe himself would write at the opening of the published edition, "I am undertaking this marvelous voyage not to deceive myself, but to get to know things" (15.1:49). Such attention to objecthood was to be the means of stabilizing a new sense of an empirically grounded self. As the critic Horst Althaus has written, "In Goethe's Roman seeing a human totality is reestablished" (145).[16] But when we attend more closely to the *way* Goethe structures his encounters with objects in Italy – and above else, aesthetic objects – we can see the extent to which such subjective recuperation is repeatedly called into question through a recurrent disorientation of visual experience, the way Italian objects do not re-establish a human totality, but rather do quite the opposite. Italian objects seem to undo us. They are spaces of conversion.

Torsos

Rome was in many ways the centre of this perceptual and personal disruption. As Goethe writes at one point, one needed "a thousand pens in Rome, what good was a single quill?" (*Italienische Reise* 15.1:140). Rome stood for a crisis of representation, and it would be one of the city's key aesthetic monuments, the Apollo Belvedere, that would mark for Goethe, as for so many others, a kind of aesthetic epicentre to these visual undulations, indeed, the aesthetic as the site of the unraveling of sight. In words that were a clear reappraisal of Winckelmann's impact on eighteenth-century aesthetics, Goethe writes, "In St. Peter's I learned to grasp how Art as well as Nature can efface all sense of scale. And in this way the Apollo Belvedere shoved me outside of reality [aus der Wirklichkeit hinausgerückt]" (*Italienische Reise* 15.1:144). In place of the still grandeur of the classical object for Winckelmann, the classical work of art for Goethe cancels a sense of scale, it "shoves" us outside of reality.

Goethe had been at work rewriting Winckelmann's position within art-historical debates as early as his Laocoön essay in the 1790s, but even more emphatically in his biography, *Winckelmann and his Century* (1805), which is often read as a key precursor to Goethe's subsequent autobiographical undertaking. It is in this text that we see, first, Rome elevated to an "anarchic" space (*Winckelmann* 19:190), as that which defies visual order. But it is also here that Winckelmann's judgments are both historicized and relativized (his writings are referred to at one point as a "closed book" [19:188]). Indeed, what they show is not some rule or sculptural canon, but life itself. As Goethe writes, "His works are a representation of life, are life itself" (19:200). No other art form at the turn of the century was more strictly tied to the question of canonicity than that of sculpture. In the form of Polykleitos's (lost) "canon," or set of rules, the history of post-classical sculpture was deeply informed by the question of aesthetic precedence (see Leftwich). For Goethe, on the other hand, the value of the sculptural, implicit in its three-dimensionality, was the way it exploded the limits of rule, the way it situated us within a space of circumspection, undoing in the process the fixity of our point of view ("shoving" us out of reality, in his words) that belonged to the other planar visual arts of drawing or painting.[17] Sculptural encounter was, paradoxically, a departure from "reality" and an entry into "life." As Goethe's choice of verb subtly indicated (*rücken*, for shove), the power of the sculptural resided in its central anatomical feature that allowed for the upright figure to turn on its own axis, its torsional core. Unlike a wider Romantic concern with the *stasis* of the sculptural object – the mutual death that transpired between viewer and viewed when looking at the frozen work of art[18] – for Goethe, sculptural encounter entailed an immeasurable movement. Not rule, the straight line of canonicity and precedence, but the rotatory nature of life was the outcome of our encounter with sculptural form.

In order to think further about the meaning of the sculptural torso for Goethe, I want to focus on a particular drawing he made of an anatomical study of the human back that dates to the winter of 1788 during his second stay in Rome and that editors believe was based on the statue of the Hercules Farnese (Figure 7.3).[19] I want to do so because I think it can help illustrate the meaning of the sculptural torso for Goethe not only as site of visual disorientation, but as a space through which one could think about *media* conversion as well, the transformation not of form, but of the materiality of representation. Dwelling on a different medium is a way of dwelling on the problem of medial shifts (*rücken*) that were so essential to the larger project of representing Italy for Goethe. In distinction to the whole Apollonian body or the fluid assembly of the Laocoönian family, the Herculean back, I want to suggest, has something to tell us about the aesthetics of passage or *Übergang* so important to the vertiginous nature of Italian life and the problems of its representation.

Historically, the body of Hercules was significant because of the way it stood for a problem of scale, much like the Verona amphitheatre with which Goethe had begun his travels. The so-called "pillars of Hercules" referred to the geological formations on either side of the Strait of Gibraltar, and during one of his twelve tasks, Hercules would assume Atlas's position as the individual responsible for holding the world aloft for a day. The mortal who could elevate the world, but was also of the world – how large was he (or it)? It was precisely this paradoxical world-relation that Hercules personified, of being both in and without, that perfectly accorded with the conversional paradigm that Goethe was working out in *Italian Journey*.[20] Writing an autobiography was, in this sense, quite literally a Herculean task. But just as important, and here Goethe's particular rendering of the Herculean back reveals its deeper significance, is the way this sense of disorientation for which Hercules stood

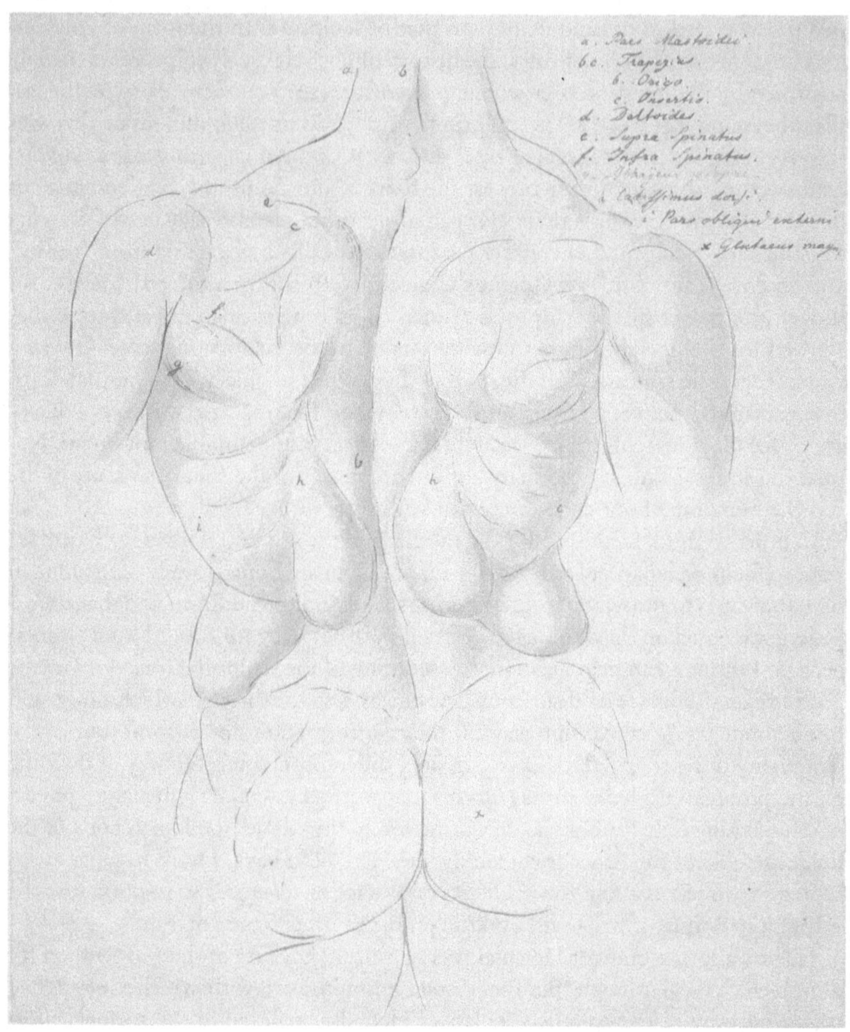

Figure 7.3 J.W. Goethe, "Anatomiestudie." Courtesy of the Klassik Stiftung Weimar, Goethe-Nationalmuseum GGz1782.

was compensated for by the way the back or spine provided an important visual reorientation, an understanding of conversion not in the circular sense, but in a more transformational sense. On one level, the drawing is a profound representation of the rotatory problems I've been discussing so far. As it shifts between anatomical and artistic registers, it combines a scientific attention to the different muscle groups (whose key stands in the upper-right corner) with a voluptuous attention to form, embodied most pronouncedly – and indeed jarringly – by the two female torsos that are imbedded within the male back (look on either side of the spine just beneath the shoulder blades). The Herculean back is thus an emphatic example not only of Goethe's claim about the need to combine art and science for a fuller knowledge of the world,[21] it also stands for the aim of seeing difference more generally, the perspectival *shifts* that are encoded in the act of retro-spection, of back-looking.

But on another level, what seems most striking about the back-view on display here is not the duality that it encodes within itself (whether sexual or disciplinary), but rather the leading-off or *ableiten* to which the eye is invited.[22] Unlike the mirror-images of the female torsos (reflecting each other, reflecting masculinity back to itself), we should consider the serpentine line that is one of the drawing's most important features and that runs down its middle, drawing together the anatomical spaces of the spine, buttocks, and inner thigh. The serpentine line was, of course, the ultimate sign of beauty in an eighteenth-century context, but unlike one of its most popular subsequent Romantic incarnations in the arabesque – in the flattened play between text and image, between decoration and object – in the context of the anatomical reproduction its meaning is integrally related to its three-dimensionality, to its demarcation of perspectival space. Not only is it visible only in relief, or as relief (the spine in the muscular back does not protrude, but is visible as an indentation or intrusion), it also conjoins the vertical wave-form that runs top to bottom with the horizontal turns of the rotating torso. The serpentine line here more properly approximates, or rather turns into, the spiral line. Curvature is the entry into space, into another dimension, but one that could only be grasped in relief, as negation. The serpentine line as spiral line marks a vanishing point, an act of *ableiten* or abduction in the intellectual sense (Piper, "Vanishing Points"). Unlike the disorientation (*verrücken*) of moving among visual registers (art/science, male/female), the spiral line of dorsality provides for a reorientation beyond the demarcated planar space of the page.

The spiral line would go on to become one of the most important figures of thought for Goethe's ideas about the nature of nature, an ideal figure to capture a sense of both continuity and change. Its particular meaning within the Herculean back could be seen in an art-historical tradition in which Hercules was often depicted bearing a column (rather than a globe) on his back, prints of which Goethe held in his collection (Figure 7.4).

Figure 7.4 Marcantonio Raimondi, *Hercules*. Courtesy of the Klassik Stiftung Weimar, Goethe-Nationalmuseum IK 3463/93.

Rather than stand for a figure of either exile or containment, much like his agon Atlas – the back that could hold the world, that was exterior to it and thus the condition of its spherical nature – Hercules's back stood instead for a sign of a vertically oriented rotationality (the textilic fold, which was also the condition of his death prior to divination, his death in life so to speak, was its soft-tissue equivalent). The geo-architectural column was thus the extension of, rather than the opposition to, the spinal column upon which the "world" (but also "mind") rested. Goethe's insight into the vertebral nature of the skull can be read on one level as emerging from, or at least confirmed by, a visual tradition that was grounded in the column-bearing Hercules. But on another level, we could say that Hercules's columnar identity – in the sense of the physical transitions envisioned among those biological, architectural, and geological columns – provided the viewer with knowledge of the *medial* transitions that so heavily marked out Goethe's Italian experience and that were integral to knowledge of the vertebral nature of Italian life. The Herculean back, and the spiral line at its core, invited one to conceive of the representational multidimensionality and attendant vanishing points necessary for an understanding of life manifested in a sense of the *Wirbel* or elevational turn.

Bones

I have so far been arguing for the variety of ways in which Italian objects were coded by a visual sense of disorientation, the way "life" in Italy, whether natural or cultural, was figured as initiating a visual problem, one fundamentally bound up with the dual concerns of verticality and torsionality. It was through such rotational Italian things, I am suggesting, that Goethe would begin to conceptualize an observational as well as representational paradigm that underpinned both his scientific and autobiographical projects, his notion of "writing life" in a larger sense. I want to conclude by turning to the way such cultural remnants set the stage for a new natural scientific, as well as poetic, paradigm to emerge in his work, one that depended principally on his engagement with bones. It is here that Goethe will work out most clearly a poetics of revolution as a science of evolution.

In the second issue of the first volume of his natural scientific periodical, *Zur Morphologie*, published in 1820, three years after the second volume of *Italian Journey*, Goethe would mention his theory of the vertebral basis of the skull, a conviction that he had, he tells us, "for many years" (24:506). Three years later, in the same periodical, he would publish his famous anecdote that located the discovery in Italy in 1790 during his second Italian journey (24:597–8). It was in many ways a companion piece to his later description of his insight about the *Urpflanze* in the botanical garden in Padua, which he published as part of his bilingual edition of *The Metamorphosis of Plants* in 1831 just prior to his death

(24:747). The lacunae surrounding Goethe's scientific "insights" that are absent from the autobiographical documents are gradually filled in over the course of the 1820s within the natural scientific writings.

In the extraordinary temporal delay between discovery and communication that characterized his work, we can see Goethe making an argument about the necessary relationship between inspection and retrospection, that all insights were in some sense coded by the composition of time. Indeed, in placing his discovery of the vertebral nature of the skull within the *second* visit to Italy – whether true or not – he was coding the category of "insight," which he would variously name in the same essay *Aperçu*, *Gewahrwerden*, *Auffassen*, *Vorstellen*, *Begriff*, and *Idee* (insight, revelation, apprehension, imagination, concept, and idea), with a sense of return and repetition, with a fundamental aspect of secondariness. But in locating the discovery in the serial "*Heft*" or pamphlet, and not in the autobiographical corpus, Goethe was adding yet another dimension of bibliographic circumspection as well, one that required a reader to look bibliographically elsewhere to apprehend the truth of this apprehension, much in the same way as the manuscriptural remnants were understood to be both elaborations on and preconditions for the Italian autobiography.

This bibliographic, as well as circumspective, basis of Goethe's project – that is to say, the circumspective nature of the bibliographic for Goethe – was underscored even further in the initial morphological essay through his use of a particular analogy to make sense of his osteological insight, which he likened to decoding fifteenth-century manuscripts. At times, Goethe tells us, they contract through the use of abbreviations what they would otherwise have made graspable, and at times they spell out in far too expansive terms what could have otherwise been contracted, producing the effect of an "unbearable boredom" (*Morphologie* 24:505). Scriptural incompletion *and* the long duration of articulation become the dual conditions of osteological insight – but also error. The bibliographical analogy is a sign of the limits of meaning within a significatory record that is at once incomplete and impossibly overseeable. As Goethe would write of these manuscripts, "They reveal what they had hidden and hide what they just revealed" (24:505).

The temporal delay in Goethe's work between apprehension and communication, between the swiftness of insight and the long duration of its unfolding, was itself a communicative reflection of the very object about which it intended to speak, the bibliographical correlate to the osteological item. Such delay had, of course, also been the grounds upon which Goethe's dispute with Lorenz Oken over the priority of the discovery was and continues to be waged.[23] Oken, we remember, announced his theory of the vertebral nature of the skull during his inaugural lecture at the University of Jena in 1807. Thirteen years later, Goethe would announce his discovery of the same phenomenon and date it seventeen years *before* Oken. Whether this was true or just a cheap way of winning priority over Oken, I want to suggest that in either case Goethe was making an important argument about the mediality, but also circularity, of vertebral knowledge. Back-looking, in both senses of the word – as a form of retrospection and looking at vertebrae – implied a looking around, a looking back at oneself in the process, a recursiveness that in no way allowed for something like "priority" to emerge, the very bedrock of science studies. Goethe's attention to the vertigo of biological knowledge was a means of challenging the value of priority as an analytical category, the way firstness was of questionable use when it came to thinking about scientific insight. Priority not only cancelled the inherent temporality of observation, the discrepancy between its experience and its communication, it also overlooked the temporality that inhered in the natural object being observed – that it, too, existed in time, and that its meaning was a function of time. But priority also overlooked the importance of aesthetic knowledge to scientific insight; it created an epistemological paradigm that could not account for the entirety of how we come to understand things. In privileging the scientific, we overlook the value of conversion to knowledge of the world.

It would, accordingly, be the poetic device of chiasmus that gradually emerged in Goethe's late writing as the preferred, or perhaps only, adequate rhetorical vehicle to capture this nature of understanding life. Again and again in his late work, we find increasingly chiastic structures, such as tautology, but also the trope of inversion, served as the new grounds of insight, further underscoring the centrality of poetic knowledge for scientific understanding. As he writes in a poem that concludes the first printed announcement of his discovery of the vertebral nature of the skull in 1820:

Klein das Große, groß das Kleine,
Alles nach der eignen Art.
Immer wechselnd, fest sich haltend,
Nah und fern und fern und nah;
So gestaltend, umgestaltend. –
Zum Erstaunen bin ich da.

[Small the large, large the small,
Everything in its own way.
Always changing, ever remaining,
Near and far and far and near;
Thus forming, deforming. –
To wonder am I there.]

Like his later, and more famous, poem on Schiller's skull ("Wie sie das Feste läßt zu Geist verrinnen, / Wie sie das Geisterzeugte fest bewahre [The way it allows the firm to trickle into spirit, / The way it firmly preserves spiritual creation]"), there is a circularity, a roundedness, to Goethe's rhetorical attempts to articulate the mode of insight appropriate to the conceptualization of skeletal objects. In a move of beautifully minimalistic proportions, we see how the precondition of that final act of "*Erstaunen* [astonishment or wonder]" takes the form of the double gerund, "*gestaltend, umgestaltend* [forming, reforming]," whose repetition is divided by nothing more than the grammatical sign of the comma and the semantic prefix of the "*um*" or "around." The roundedness of the poetic object, like that of the vertebra or skull, is divided by an aroundness, by an act of circumspection and circumnavigation, deliciously captured in the incomplete curvature of the comma that separates the recurring word for (re)formation. The mid-point or division of the poetic line, like that of the natural object about which it speaks, is coded as an outside or exterior, but also as a fragment or part; it is at once both kernel and shell. Form, *Gestalt*, is always also *Umgestalt*, deformation, or perhaps better, circumformation.[24] In thinking about bones, we can see Goethe working out a science, as well as a poetics, of revolution.

It is important to remember that Goethe was not the only one making such insights about the circular nature of biological insight. His work ran deeply parallel to that of Geoffroy Saint-Hilaire, who would write, upon inspecting a dissected lobster in 1822, that an invertebrate was nothing more than an inverted vertebrate:

> What was my surprise ... in seeing an arrangement that placed before my eyes all of the organic systems of the lobster in the order in which they are arranged in mammals? ... And yet in what relations do these neural and vascular systems find themselves in regards to the case that contains them? In an inverse state, relative to the idea that we construct for ourselves for the words *back* and *stomach* [dos et ventre] ... Look at a crayfish turned on its back, and the entire order that I have pointed out is that of its various systems, just as it is also that of the same systems in the higher vertebrates. ("Considérations" 113)

Geoffroy's work was significant, not only because of its larger cultural influence during the period and for long after. It was important because of the way its origins lay in a remarkably similar cultural environment to that of Goethe's Italy. Geoffroy's first article on the topic of osteological continuity had been published while he was in Egypt accompanying Napoleon's team of savants, where he spoke of the wishbone as an osteological "vestige" (Geoffroy, "Observations"). When we look at the massive, multi-year and multi-authored documentation that would emerge from that voyage, *Description de l'Egypte* (1809–22), we see page after page of monumental ruins, imaginative reconstructions of tropical landscapes, and encounters with scales of great dimension, both spatially and temporally. Whether it was the massive remnants of Ozymandias or the partially consumed columns of the enormous Grand Temple of Edfou, the torso and the ancient ruin would function for Geoffroy as crucial motors of biological and, in particular, osteological insight. They would serve as generative objects into the knowledge of the torsional nature of life and the shifts (*rücken*) that accompanied this breakdown of both scale and perspective.

Conclusion

If gothic cathedrals served as important backdrops to reflect on the monstrous nature of life during the Romantic period – like the hybrid Quasimodo, resident of Notre-Dame – bones, ruins, and torsos were very good things with which to think about the fractures of time. Like ruins, bones persist, but they also break. As Geoffroy recognized, they are the most "vestigial" of all organic forms. They endure well past the ontogenetic boundaries of an individual life, and thus allow us, as observers, to ask after the shape and meaning of continuum – if there is indeed a continuum to life. Bones persist as a reminder, but also a question. In many ways, they are like books. They make possible thought between not just generations or even strata, but categories themselves.

In its combinatory potential, the vertebra was a special kind of bone. As a building block, it served as a key cumulative object, a sign of the archonic structure of nature. But in its openness and protrusions, or, in more technical terms, its foramina and processes (Figure 7.5), the vertebra allowed for the entwining of heterogenous elements with one another, the interrelation of the skeletal with the soft tissue of nerves, vessels, muscles and ligaments, circulatory structures out of which osteological elements themselves emerged and remained in vital contact. As Carl Gustav Carus had argued in his osteological treatise, what made bones unique, when compared with horns or shells, for example, was the way they persisted in a relational exchange (*Wechselwirkung*) with the soft tissue around them. Their solidification was never absolute, referred to by Carus as a "relative death," a death within life that was still part of vitality. The vertebra, as the original bone for Carus, was the most developed sign of this notion of entanglement, of the way verticularity conjoined difference. It made possible the generation of heterogeneity, and complicated issues of identity and differentiation. As Geoffroy argued, if you folded a vertebrate back on itself, you created an invertebrate. The vertebra was an embodiment of the vertigo of categorical distinction.

In placing the vertebra at the centre of the skull, and in placing a rotational force at the centre of life, Goethe's Italianism served as an occasion to explore this vertiginous quality of life, which at its self-reflexive core resided a permanent sense

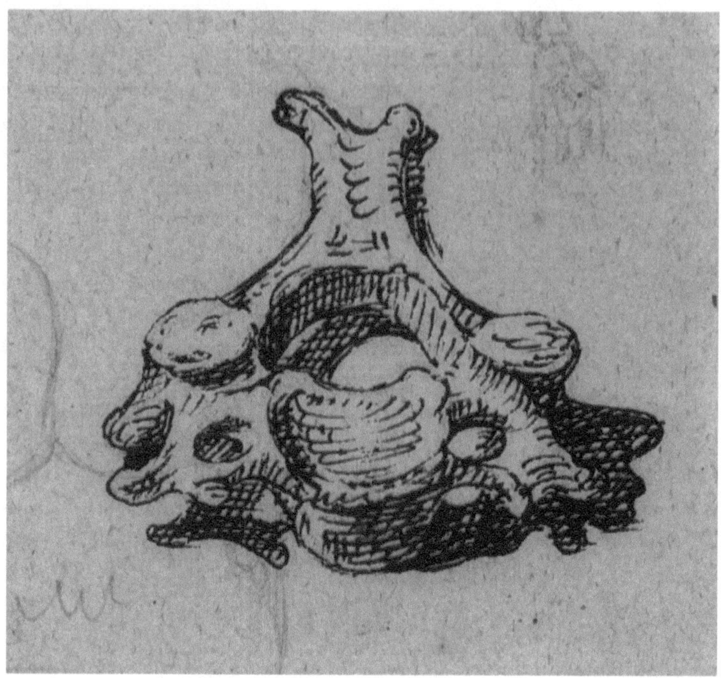

Figure 7.5 J.W. Goethe, Vertebra. Courtesy of the Klassik Stiftung Weimar, Goethe-Nationalmuseum 80–2010–0409.

of disorientation. To think about life, and to write life, one had to account for the retrospection, the *Rücksicht*, of autobiography, but also the displacement, the circumspection and *verrücken*, of self-reflection. Rebirth, the condition of discontinuity, was predicated on revolution, the condition of circularity and recurrence. Where Augustine's conversion had taken place in the intimate space between the grasped book and the domestic garden, Goethe's conversion was a function of distance – from home, from the objects around him, and from the massive trove of notebooks, papers, and sketches he had accumulated during his journey. There was an aroundness, a circularity, to such conversional experience, but also a lack of closure. Unlike the totality and the singularity of Augustine's conversion, there was an absence of finality, a repetitiveness, but also an openness of meaning in Goethe that was not ultimately confinable to the autobiographical narrative or its bibliographical container. Fracture was never entirely complete.

The final image with which Goethe would conclude the third and last volume of his Italian autobiography more than ten years after he had begun was that of the Roman Coliseum. It was a significant replaying of his initial encounter with the Verona amphitheatre, only now on a grander scale, one last reflection on the structural condition of self-reflection. But instead of ascending to the upper and outer edge of this rotational form and looking down, as he had in Verona, he looks through a gated entrance into the emptiness of the middle. It is at this moment, he tells us, where he draws together "ein unübersehbares Summa Summarum [an incalculable summary]" of his stay in Italy (*Italianische Reise* 15.1:596). The end of his conversion narrative is coded by a fundamental inaccessibility, gated and unoverseeable. Goethe will conclude by invoking Ovid's third elegy, written after his own banishment from Rome, a "*Rückerinnerung*" or back-memory of his melancholic departure. The important point of the scene is not its nostalgia, but the way the return home is framed as a form of banishment, a death in life. The circular form of the Coliseum and the poetic roundedness of the Ovidian elegy combine to become the conditions of a personal sense of disorientation and errancy, one that leaves Goethe unable to continue writing: "I could not get his reflection out of my mind. As I repeated the poem, which rose up almost exactly in my memory, it truly disoriented me and hindered me in terms of my own production [mich an eigner Produktion irre werden ließ und hinderte]; even when undertaken later it never materialized" (15.1:596). Conversion is not only the condition of autobiographical narrative; it also marks out its impossible completion. It is here, in the conjunction of the circular and the erroneous, the *irren* or erring that belongs to the vertiginous quality of knowing life, that we can see the most concrete example of the way evolution recapitulates revolution.

NOTES

1 For a discussion of Cuvier and revolution, see Rudwick, *Georges Cuvier, Fossil Bones, and Geological Catastrophes* and *The Meaning of Fossils.* For a discussion of the relationship between geological revolution and poetic form during the Romantic period, see Bewell 237–80, and Heringman.

2 For a thorough discussion of the debates between Cuvier and Geoffroy, see Appel.

3 Geoffroy's assault on Cuvier's distinctions would take several stages, beginning first with the similarity between humans and crustaceans and then, later, going on to that between humans and mollusks, even more rudimentary invertebrate forms. See his "Considérations générales sur la vertèbre" (1822) and then, later, the work that was the result of his famed confrontation with Cuvier, *Principes de philosophie zoologique* (1830), which Goethe reviewed and promoted in his morphological journal.

4 All translations are my own unless otherwise noted. For an inspired reading of Italy as an underworld, and more specifically, as a womb and thus a space of birth/rebirth, see Amrine. For a discussion of Goethe's paradoxical union of a Christian narratological motif of rebirth within the classical aesthetic space of Italy, see Rüdiger, Kiefer, and Requadt.

5 For a reflection on Goethe's theory of conversion in a theological sense, see Hörisch.

6 Such rotatory theories of life were significant not only to the osteological inquiries of Goethe, Oken, and Geoffroy, but also in the field of optics, with the question of the polarity of light, as well as in the field of crystallography and the work of Hegel's colleague, Christian Samuel Weiß in Berlin. For the most extensive discussion of Goethe's contribution to Darwinian evolution, see Richards 407–502, in particular 476–91. For Richards, Goethe's relationship to Schelling in 1800 is one of the decisive influences for Goethe's embrace of a more circular theory of knowledge (463–71). For a discussion of the importance of rotatory forces in Schelling and *Naturphilosophie* more generally, see Rajan.

7 I want to capture here a sense of the way the term "evolution" shifts dramatically from the 1770s as a synonym for preformationism to its later Darwinian context to mean species change, a semantic shift that is both remarkable and yet still largely uncharted. Blumenbach's use of the term appears in Blumenbach 6.

8 For a companion piece that explores a notion of optical *Brechung*, or fracture, as central to Goethe's knowledge of life, and that I argue grows out of *Italian Journey*, see my "Egologies: Goethe, Entoptics, and the Instruments of Writing Life."

9 For previous discussions of Goethe's "autobiographical science," see Koranyi, Richards 325–511, von Mücke, and Kuhn.

10 As Goethe would write in his essay on the discovery of the sheep's skull in Italy, "Man only knows himself insofar as he knows the world, which he can only grasp in himself and him in it. Every new object, well observed, opens up a new organ within us" (*Morphologie* 24:596). New objects do not just give rise to new ways of looking for Goethe; they also give birth to entirely new *organs* of envisioning. In this, there is an emphatically evolutionary aspect to scientific observation itself, the way it can contribute to the emergence of new organs.

11 As Goethe would write in his scientific periodical, *Zur Morphologie*, "Natur hat weder Kern/Noch Schale,/ Alles ist sie mit einemmale [Nature has neither kernel/ Nor shell,/ She is everything at once]" (*Morphologie* 24:523). The syntactic location of the shell in the centre (or "kernel") of the three-line sentence performs the very semantic inversion that is the poem's theoretical argument. See also Goethe's earlier helical drawing of the structure of the plant's leaf that he made while in Italy and that was accompanied by his famous statement, "Alles ist Blat [All is leaf]" (Goethe- und Schiller-Archiv, 26:LV, 13,1, Bl. 167).

12 As Nietzsche would write in this post-Goethean vein in *The Birth of Tragedy*, "All life rests on appearance, art, deception, optics, the necessity of perspective and error" (*Werke* 1:12).

13 In this, Goethe's encounter echoes Henry Fuseli's famed image of the artist overwhelmed by the grandeur, as well as the monstrosity, of classical remnants (in his case a giant hand and foot). For a discussion of the way the ruin signalled a problem of scale for early nineteenth-century viewers, see Dillon and Siegel. It is important to note the extent to which these arguments run counter to one of the most canonical pieces of scholarship on the ruin, Georg Simmel's "The Ruin," in which he speaks of the ruin as a sign of a cultural *Gleichgewicht* or equilibrium – the ruin's sinking into the earth serves as a sign of balancing the cultural "upheaval" that belongs to monuments. Equilibrium is precisely what is undone for Goethe, as well as Fuseli, by the ruined encounter.

14 As Friedmar Apel has suggested (141–9), the classical artefact resists a contextualizing gaze, the production of sense or *Sinn* in the double sense of the word as making meaning and constructing sensory perception. Apel will go on to suggest, however, that this idea of *Sinn* does eventually emerge in *Italian Journey* through the perception of landscape at the close of volume 2, thereby ensuring the program of self-creation that he sees as the point of the autobiography. As I will argue, the aim of *Italian Journey* for me is a didactic one of learning to dwell in negativity or non-sense, not to resolve it. The project of recuperation belongs to a later part of the autobiography and depends on new scientific instruments and new thinking about the medium of the printed book as technologies of reflective socialization. For a further elaboration, see Piper, "Egologies."

15 See Jane Brown's marvelous essay on the birth of an objective style in Goethe's post-Italian writing that is an outcome of this new visual objectivity. Norbert Puszkar has argued that the *Scheideblick* or "parting view" is the most marked feature of *Italian Journey*, where Goethe takes leave of his own aspirations to being a visual artist. For an emphasis on the ugliness of Italian life, and, with it, a sense of death, see Pfotenhauer.

16 Such totality is often associated with a sense of a revived pre-modernism in the Romantic period, a return to classical origins as a medicinal cure to modernity. For a reading of Italy as a pre-modern space, see Rohde 326.

17 Discussions of Goethe and sculpture often revolve around questions of "distance," the negotiation of the object's reception between a sensory and intellectual response. My interest is the way sculpture undoes this sense of distance precisely through its circular receptivity. For a discussion of sculpture and distance in Goethe, see Kaiser.

18 For an excellent discussion of this idea of mutual death, see Haley 193–218.

19 It is important to note the way this study is also connected with another sculptural object located in the Cortile del Belvedere, the famed seated Belvedere Torso, thought at the time to be of Hercules, and whose back Goethe remarked in his notes was "exceptionally beautiful," and which he would expressly compare to the Hercules Farnese (Goethe, *Werke* I.32: 448).

20 It would also fit with the criss-crossing of Christian and classical motifs that suffused the work, and that were most visible in the figure of Hercules bearing a column as a prelude to Christ's carrying the cross.

21 "Since no totality can be brought together in either knowledge or reflection – because the former lacks interiority and the latter exteriority – it is necessary to conceive of science as art if we are to expect any kind of totality" (*Zur Farbenlehre* 23.1:605).

22 In the second announcement about his discovery of the vertebral nature of the skull, Goethe would write, "I have found that my entire practice rests on abduction [Ableiten]" (*Morphologie* 24:597). I have translated *Ableiten*, following C.S. Peirce, as "abduction," literally a leading-off, and not more colloquially as deduction, which does not correspond to Goethe's scientific method.

23 For the repeated attempts at trying to make sense of this problem of priority, see Zaunick, Sievers, Zittel and Richards 491–502.

24 As another indication of the significance of Goethe's epistemological project for subsequent developments in the life sciences, see the work of the microbiologist Ludwig Fleck, who would use remarkably similar language to describe a new notion of scientific facticity in the 1920s: "Knowing is neither passive contemplation, nor acquisition of a singularly possible insight into a final given. It is an active, vital relational event, a deformation and being deformed [ein Umformen und Umgeformtwerden], in short creation" (426). Discussed in Rheinberger 28.

WORKS CITED

Althaus, Horst. "Goethes Römische Sehen." *Ästhetik, Ökonomie und Gesellschaft.* München: Francke, 1971. 142–62. Print.

Amrine, Frederick. "Goethe's Italian Discoveries as a Natural Scientist (The Scientist in the Underworld)." *Goethe in Italy, 1786–1986.* Ed. Gerhart Hoffmeister. Amsterdam: Rodopi, 1988. 55–76. Print.

Apel, Friedmar Apel. "Aufmerksamkeit ist Leben: Goethes 'Italienische Reise' als Sehprojekt." *Literature ohne Kompromisse.* Ed. Sabine Kyora. Bielefeld: Aesthesis, 2004. 141–9. Print.

Appel, Toby A. *The Cuvier-Geoffroy Debate: French Biology in the Decades Before Darwin.* Oxford: Oxford UP, 1987. Print.

Bewell, Alan. *Wordsworth and the Enlightenment: Nature, Man, and Society in the Experimental Poetry.* New Haven: Yale UP, 1989. Print.

Blumenbach, Johann Friedrich. *Über den Bildungstrieb.* Göttingen: Dieterich, 1789. Print.

Brown, Jane. "The Renaissance of Goethe's Poetic Genius in Italy." *Goethe in Italy, 1786–1986.* Ed. Gerhart Hoffmeister. Amsterdam: Rodopi, 1988. 77–93. Print.

Carus, Carl Gustav. *Von den Ur-Theilen des Knochen- und Schalengerüstes.* Leipzig: Fleischer, 1828. Print.

Dillon, Brian. "Fragments from a History of Ruin," *Cabinet Magazine* 20 (2005/06). http://www.cabinetmagazine.org/issues/20/dillon.php. Web. 12 December 2012.

Dubin, Nina. *Futures and Ruins: Eighteenth-Century Paris and the Art of Hubert Robert.* Los Angeles: Getty Research Institute, 2010. Print.

Fleck, Ludwig. "Zur Krise der 'Wirklichkeit.'" *Die Naturwissenschaften* 17 (1929): 425–30. Print.

Geoffroy Saint-Hilaire, Étienne. "Considérations générales sur la vertèbre." *Mémoires du Muséum d'Histoire Naturelle* 9 (1822): 89–119. Print.

Geoffroy Saint-Hilaire, Étienne. "Observations sur l'aile de l'autruche." *La décade égyptienne* 1 (1798): 46–51. Print.

Geoffroy Saint-Hilaire, Étienne. *Principes de philosophie zoologique.* Paris: Pichon et Didier, 1830. Print.

Gilpin, William. "Three Essays: On Picturesque Beauty; On Picturesque Travel; and On Sketching Landscape." *Aesthetics and the Picturesque, 1795–1840.* Vol. 1. Ed. Gavin Budge. Bristol: Thoemmes P, 2001. Print.

Goethe, Johann Wolfgang von. *Die Italienische Reise. Sämtliche Werke.* Vols. 15.1 and 15.2. Ed. Christoph Michel und Hans-Georg Dewitz. Frankfurt/Main: Deutscher Klassiker Verlag, 1993. Print.

Goethe, Johann Wolfgang von. *Schriften zur Morphologie. Sämtliche Werke* Vol. 24. Ed. Dorothea Kuhn. Frankfurt/Main: Deutscher Klassiker Verlag, 1987. Print.

Goethe, Johann Wolfgang von. *Werke.* 144 vols. Weimar: Böhlau, 1999. Print.

Goethe, Johann Wolfgang von. *Winckelmann und sein Jahrhundert. Sämtliche Werke.* Vol. 19. Ed. Friedmar Apel. Frankfurt/Main: Deutscher Klassiker Verlag, 1998. Print.

Goethe, Johann Wolfgang von. *Zur Farbenlehre. Sämtliche Werke.* Vol. 23.1. Ed. Manfred Wenzel. Frankfurt/Main: Deutscher Klassiker Verlag, 1991. Print.

Haley, Bruce. *Living Forms: Romantics and the Monumental Figure.* Albany, NY: SUNY P, 2003. Print.

Heringman, Noah. *Romantic Rocks, Aesthetic Geology.* Ithaca, NY: Cornell UP, 2004. Print.

Hörisch, Jochen. "Religiöse Abrüstung. Goethes Konversionstheologie." *Spuren, Signaturen, Spiegelungen. Zur Goethe-Rezeption in Europa.* Ed. Bernhard Beutler and Anke Bosse. Köln: Weimar, 2000. 31–44. Print.

Kaiser, Gerhard. "Idee oder Körper: Zu Schillers und Goethes Rezeptionsweise antiker Plastik." *Goethe. Nähe durch Abstand.* Weimar: VDM, 2000. 147–62. Print.

Kiefer, Klaus. *Wiedergeburt und neues Leben: Aspekte der Strukturwandel in Goethes Italienische Reise.* Bonn: Bouvier, 1978. Print.

Koranyi, Stephan. *Autobiographik und Wissenschaft im Denken Goethes.* Bonn: Bouvier, 1984. Print.

Kuhn, Bernhard. "Goethe's Autobiographical Science." *Autobiography and Natural Science in the Age of Romanticism: Rousseau, Goethe, Thoreau*. Burlington, VT: Ashgate, 2009: 63–96. Print.

Leftwich, Gregory. *Ancient Conceptions of the Body and the Canon of Polykleitos*. Ann Arbor: UMI, 1999. Print.

Moritz, Carl Philipp. *Anton Reiser: Ein Psychologischer Roman*. Stuttgart: Reclam, 2006. Print.

Müller-Sievers, Helmut. "Skullduggery: Goethe and Oken, Natural Philosophy and Freedom of the Press." *Modern Language Quarterly* 59.2 (1998): 231–59. Print.

Nietzsche, Friedrich. *Werke. Kritische Studienausgabe*. Vol. 1. Ed. Giorgio Colli und Mazzino Montinari. München: Deutscher Taschenbuch Verlag, 1988. Print.

Oken, Lorenz. *Über die Bedeutung der Schädelknochen. Ein Programm beim Antritt der Professur an der Gesammt-Universität zu Jena*. Jena, 1807. Print.

Olney, James. *Memory and Narrative: The Weave of Life-Writing*. Chicago: U of Chicago P, 1998. Print.

Pfotenhauer, Helmut. "Der schöne Tod. Über einige Schatten in Goethes Italienbild." *Um 1800. Konfigurationen der Literatur, Kunstliteratur und Ästhetik*. Tübingen: Niemeyer, 1991. 113–36. Print.

Piper, Andrew. "Egologies: Goethe, Entoptics, and the Instruments of Writing Life." *Goethe's Ghosts: Reading and The Persistence of Literature*. Ed. Simon Richter and Richard Block. Rochester: Camden House, 2013. 17–36. Print.

Piper, Andrew. "Vanishing Points: The Heterotopia of the Romantic Book." *European Romantic Review* 23.3 (2012): 381–91. Print.

Puszkar, Norbert. "'Scheide Blick': Abschluß und Ausschuß in Goethes *Italienischer Reise*." *Études Germaniques* 57 (2002): 649–70. Print.

Rajan, Tilottama. "Excitability: The (Dis)Organization of Knowledge from Schelling's First Outline (1799) to Ages of the World (1815)." *European Romantic Review* 21.3 (2010): 309–25. Print.

Requadt, Paul. *Die Bildersprache der deutschen Italiendichtung*. München: Francke, 1962. 15–101. Print.

Rheinberger, Heins-Jörg. *Epistemologie des Konkreten. Studien zur Geschichte der modernen Biologie*. Frankfurt/Main: Suhrkamp, 2006. Print.

Richards, Robert J. *The Romantic Conception of Life: Science and Philosophy in the Age of Goethe*. Chicago: U of Chicago P, 2002. Print.

Rohde, Carsten. *Spiegeln und Schweben: Goethes autobiographisches Schreiben*. Göttingen: Wallstein, 2006. Print.

Rüdiger, Horst. "Goethes Rom-Erlebnis." *Goethe und Europa*. Berlin: de Gruyter, 1990. 280–99. Print.

Rudwick, Martin J.S. *Georges Cuvier, Fossil Bones, and Geological Catastrophes*. Chicago: U of Chicago P, 1997. Print.

Rudwick, Martin J.S. *The Meaning of Fossils: Episodes in the History of Palaeontology.* London: Macdonald, 1972. Print.
Schöning, Matthias. "Zeit der Ruinen: Tropologische Stichproben zu Modernität und Einheit der Romantik." *Internationales Archiv für Sozialgeschichte der deutschen Literatur* 34.1 (2009): 75–93. Print.
Siegel, Jonah. *Desire and Excess: The Nineteenth Century Culture of Art.* Princeton: Princeton UP, 2000. Print.
Simmel, Georg. "The Ruin," *A Collection of Essays.* Ed. Kurt H. Wolff. Columbus, OH: Ohio State UP, 1959. 259–66. Print.
Sloterdijk, Peter. *Regeln für den Menschenpark.* Frankfurt/Main: Suhrkamp, 1999. Print.
Thomas, Sophie. "Assembling History: Fragments and Ruins." *European Romantic Review* 14.2 (2003): 177–86. Print.
Volney, C.F. *Les ruines, ou Méditation sur les révolutions des empires.* Paris: Desenne, 1791.
Von Mücke, Dorothea E. "Goethe's Metamorphosis: Changing Forms in Nature, the Life Sciences, and Authorship." *Representations* 95 (Summer 2006): 27–53. Print.
Weigel, Sigrid. "'Irrendes Tier' und 'edler Körper': der Gründungsmythos der Kunst als Urszene des zweigestaltigen Volks in Goethes *Italienische Reise.*" *Die andre Stimme: das Fremde in der Kultur der Moderne.* Ed. Alexander Honold and Manuel Köppen. Köln: Böhlau, 1999. 21–7. Print.
Zaunick, Rudolph. "Oken, Carus, Goethe. Zur Geschichte des Gedankens der Wirbel-Metamorphose." *Historische Studien und Skizzen zur Natur- und Heilswissenschaft.* Berlin, 1930. 118–29. Print.
Zittel, Manfred. "Oken und Goethe." *Lorenz Oken. Ein politischer Naturphilosoph.* Ed. Breidbach et al. Weimar: Böhlau, 2001. 149–82. Print.

Chapter Eight

Taking Chances

THERESA M. KELLEY

This essay surveys an argument about history, futurity, and narrative shapes in Romantic writing by way of some working principles and specific instances that together articulate the inquiry implied by my title. Briefly put, I understood "taking chances" as something like a slogan for a way of understanding possibility that returns, with different emphases across Romantic writing. This essay accordingly maps a larger terrain, but does so by showing how that terrain might also be grasped more locally, in writing drawn from different strands of Romantic thought.

My working argument is that the notion of progress, and the hope of it, that Romantics carried on from late Enlightenment writers, took a very different direction after the Reign of Terror. Reinhart Koselleck invites notice of this folding of the question of the future into the history of the Revolution in the German title of the work whose English translation is *Futures Past: Vergangene Zukunft,* literally the lost or past future (xi). Reading this title along side Koselleck's claim that the Revolution as Diderot foresaw it was both inevitably on its way and unplottable (23–4), it is easy enough to recognize how emphatically the Terror pushed the future of the Revolution to the wayside before it arrived.

In this essay, I read the impact of this revolutionary "new time" on Romantic writing about whether and how life and matter exist and develop. In the fold where these two strands and arguments meet, chance and accident complicate and even derail the narrative of progress and forward evolution that some Romantic writers, and many of their readers, have claimed for the era. For precisely this reason, Romantic writing offers a powerfully interrogative challenge to its desire for progress and organic development as one version of a progress narrative. In what follows, I make three related claims: that contingency wounds the hope for a rationalized future; that Romanticism's inability to recover from that wound creates a differently structured sense of narrative possibility; and that this chastened (and enlivened) view of narrative futurity allows Romanticism to awaken to its material life and temporality. The first part of this argument concerns Romantic works in which progress and futurity take chances or are taken up by chance and accident, together with the strange and arresting fortunes of the word "accident" itself. The second part considers how Romantic writing about nature understands the work of taking chances.

By now, the sceptical reader will have already constructed a long list of Romantic prophecies, together with an equally impressive list of critical books about the genre as a very nearly homegrown product of Romanticism. Let Ian Balfour's recent *The Rhetoric of Romantic Prophecy* be the standard-bearer for this critical tradition. Yet if there is evidently no dearth of Romantic works that imagine progressing toward a better future, what I find compelling is how frequently such imagining does not arrive at that end. Sometimes getting there goes badly, or the future is, in a critical human sense, no future, as it is in Mary Shelley's *Last Man* (1826) or even *Frankenstein* (1818), in which a genealogy that might end up somewhere in the future comes to a dead end for the monster and Frankenstein, putting paid to this creator's plans to establish a new species and thus command evolutionary possibility before it happens. Even the emphatically prophetic argument of Percy Shelley's *Prometheus Unbound* (1820) embeds its announcement of the destined or sought future in Demogorgon's dry reminder that some watchwords must be preserved in case radical revolution needs to happen all over again: "These are the spells by which to reassume / An empire o'er the disentangled Doom" (4.569–70). Or hope for a future may turn inside out, inaugurating a series of counterfactuals that must be put aside, as the speaker of Wordsworth's 1805 *The Prelude* does when he lists the various topics his narrative might take up but will not. Here, too, gestural renunciation has happenstance at its back. Just before, the speaker admits that his sense of being inspired by Nature's Aeolian visitations has been "defrauded" (Wordsworth, *The Prelude* 1.105). In Mary Shelley's *Valperga*, the introduction of two invented female characters, Euthanasia and Beatrice, are also counterfactuals. Their challenge to Castruccio's realpolitik manoeuvres to gain control of central Italy are, in the end, unsuccessful, yet their presence in the novel nonetheless invites a sense of possibility, however unrealized. In Shelley's *The Triumph of Life* (left unfinished in 1822), "Life" goes on in ways that exceed the grasp of the speaker and Rousseau, the guide who loses his way and hangs on too long. In this, both resemble the imperial types who conduct their "triumph" over the vanquished, for none of them knows what possibilities for life lie ahead. In distinctive ways, these narratives include moments of accident, interruption, or regress that complicate Romantic thinking and writing about the future.

These "impediments" to seeing or finding a narrative path, as the speaker of *The Prelude* identifies them (4.141), speak to and for narrative stops and starts that recur in Romantic writing with something like the power of an obsession. It is as though the need to mark the liabilities of future talk became something like the shadow structure of its narrative. The difficulty of getting narratively to the future, despite the oracular project of Romantic prophecy, invites notice of, among other features, a reflective turn away from ideal notions of progress in Romanticism.[1] It is as if the looping folds of Hegel's narrative of history are finally waylaid, despite the call of Spirit to keep going, or at least to keep the ideal of historical consciousness in view. As such, the impediments that lie in narrative wait for attempts to get there temper the Romantic disposition toward the heroic, itself a symptom, if not in fact a genetic trait, of Romanticism.

One reiterated location for the first of these is Wordsworth's notice of what the speaker of the two-part *Prelude* (1799) calls "numerous accidents in flood and field" (1.280) that halt or deflect the purportedly simple and determined narrative bounds that the speaker of the 1805 poem claims. That some of the most prominent of these – the boy of Winander, the drowned man, and the gibbet scene in Book XI of the expanded 1805 poem – end or begin with a death, has been widely noted. In each of them, narrative desire goes unsatisfied: the boy of Winander does not continue those calls to nature; the boy who discovered the drowned body returned to the surface cannot avoid this discovery, although he can aestheticize it; and (not a death but another end to desire) the boy who hoped to hoard enough money to buy all volumes of the Arabian tales fails to do so. These episodes belong to a series of moments in *The Prelude* in which impediments make it impossible to go forward, among them the opening of Book 5 (1805), in which the speaker laments being unable to save knowledge from an impending deluge or cataclysm, as well as the speaker's notice that his "harp" was "soon defrauded" (1.105), the declaration that sets in motion the series of narrative stoppages or failures that I have been tracking.

It is, of course, fundamental that death is among them, along with not possessing a book whose tales save the life of the teller (and recount many deaths) and go on and on. For what is at issue in this conjunction is the relation between life and going on as narrative events that are repeatedly joined in the poem. It is a remarkable paradox of the expanded *Prelude* that many of these episodes cluster in Book 5, the book that Wordsworth stopped and started repeatedly in the throes of deciding how the early efforts to write an autobiographical poem would need to expand to take in the French Revolution and its aftermath, the event horizon that prompted, the speaker insists, a near-definitive collapse. Book 5 is also the book that takes aim at "Sages, who in their prescience would controul / All accidents, and to the very road / Which they have fashioned would confine us down / Like engines" (5.380–4). That the 1805 poem does go on, even as it harbours death, loss, and accident within its rhetorical shape and texture, suggests how we might understand Romantic narrative in its relation to the bumpy, lurching sense of history that succeeds as the story of the Revolution becomes, I surmise, the story of how to go forward and thus how to narrate toward the future.

For Ross Hamilton, the long, largely European investigation of Aristotle's distinction between accident and substance takes a decisive turn when writers begin to suggest that the accidental qualities of one's life constitute its meaning, or substance – whether spiritually so, as in the case of the accident that prompts Augustine's conversion to Christianity, or imaginatively, as in Wordsworth's alignment of the accidental and the unexpected with sublimity and thus with a poetic version of Kant's argument that what is sublime is not the natural world but the mind that recognizes what cannot be contained by finite nature (Hamilton 202–7). What sense of history or temporality might, I ask, be embedded in this view of the modern self, an account that echoes Charles Taylor's understanding of how we became modern, but then slides away from the collected, near-hegemonic certitude of the self that emerges in Taylor's vision of Western modernity? What understanding of futurity is available to the modern self? What traps, or hidden encumbrances, might we find there?

If, as Aristotle argued, substance defines essence, its accidental qualities are those aspects of a substance that can change in ways that do not alter its essence. Socrates's substance, Hamilton notes, is the fact that he is a philosopher, whereas his height, hair colour, his complexion, his dwelling place, his activities are all accidents that could change without compromising his substance (Aristotle 6.2.1026b28–32; cited in Hamilton 12). And yet it has long been difficult to fix exactly what absolutely belongs to something, and thus constitutes its substance. For if we determine (again, this is Hamilton's example) that the oak or maple of a table is accidental in the sense that one could have a table made of almost anything that is structurally solid, we find ourselves positing something strangely unspecified in form and matter: "tableness," the character of which is required for a table to exist (Hamilton 13). So understood, substance is, as it were, without phenomenal substance: good for Plato, but not good for Aristotle, and perplexing for those who insist on some sense of form and matter as necessary for substance – and as clearly attractive, insofar as this line of inquiry allows us to posit something behind or beneath appearances that in some sense constitutes essence without being fully apprehended as material substance. What substance opens up is a space, or aporia, that is available for occupying, either by Platonic essences or by accidents.

As Hamilton observes, Aristotle made it fairly easy to fold accident into substance by defining *accident* as both that which is not essential to a substance and as an unexpected event (Hamilton 4). This double meaning informs what Michael Witmore has called the "culture of accidents" in early modern England, where accident signifies both what is mutable and thus not substance, as well as the cultural force of unexpected events as the way Elizabethans understood the task of being alive. As accident becomes inextricable from the early modern sense of events and plots, Witmore argues, temporality is articulated by accidents, not by causes with inevitable or recognizable effects (2–4). The early modern development of a mathematics of probability inflects subsequent claims about chance in at least two directions.

For David Hume, it is not "chance" that prompts billiard balls to knock together in the assorted ways they do (or do not, this too being a matter of expertise), but a complex interaction of causes and effects that we cannot trace and so dismiss (or elevate) as chance (56–9). Yet our modern sense of how to talk about historical change, as Michel Foucault and then Alain Badiou differently recognize, is in some sense bound up with the accidental, the unexpected. For Foucault as for Hume, chance (whether veridical or chimerical) is bound up with, marked by, the inability to track causes and effects adequately (Foucault xii). If the accidental is embedded in what Badiou argues we can say about being and about event, an "eventual site" like the French Revolution is "historical" in, at best, a roughly provisional sense, a stitching together "at the edge of the void." As do early modern mathematicians, Badiou manages this difficulty by imagining the event as a set of mathematical values, held together by the work of thought (177). Noting how surprisingly allergic the eighteenth-century realist novel is to probabilistic, rationalized plots, Jesse Molesworth argues that novelistic chance re-enchants an Enlightenment more chary of probability than it knew (12–14). If chance is or looks like magic, its magic abides, Demogorgon-like, in the deep temporality of narrative, holding there oddities that might be taken up, or not. Across this philosophical terrain, which Hamilton visits in different ways and for different purposes, chance keeps getting folded in, as though it were shadow boxing with, and perhaps overcoming, what we take to be substantial, real, and necessary to being.

As such, chance marks a way to think about history at the moment when the French Revolution challenged the possibility that history would continue as it had long been. Edmund Burke's view of the Revolution in 1789–90 insists instead on its discontinuity, its lack of coherence. That is to say, for Burke there is no "set," Badiou's highly provisional, mathematicized term for what might be bundled up to indicate an event – just an incoherent, fundamentally lawless series of acts or events that lack the warrant of tradition and habit (Burke 39–43). Whatever the moment of their retrospectives on the Revolution, and whatever the political lens they used for looking back, Romantic writers may have mostly experienced the Revolution as Burke did, not as a set of events with a legible, coherent event horizon, but instead as a modern instantiation of the unexpected turns of a plot that was more hauntingly centrifugal than centripetal, more like tragedy than comedy – and, as such, marked by the difficulty of going forward and a narrative suspicion of progress, as though it were an engine whose cogs and movements were inherently untrustworthy.

Consider, for example, how Charlotte Smith's *Beachy Head* (1807) rocks from an opening address that invokes a vast geological and national history to repeated denials of this historical and epic ambition that keep shifting the lens of the poem between the sea seen "afar" and a super-local particularity (217–47). John Keats's *Hyperion* (1820) poems step away differently from the progressive mythological narrative each purports to begin, failing to take the story forward to Apollo's conquest of the Titan gods. Entranced instead by the strange spectacle of the Titans wondering over their lost or departed divinity, neither version wants to tell the story that lies ahead, available in any history of mythology, that would arrive at the dominion of Apollo, Keats's early and most favoured figure of the poet. What is compelling about these poems is less that they are fragments than that they halt the narrative teleology that they so evidently invoke.

Thinking with Romantic writers about these questions urges recognition of Romanticism as a force field rather than a compliant history in which progress and development go on. This is not to say that writers, scientific and poetic, did not at some level entertain hope for a progress narrative of history, for they certainly did, but rather it is to say that such a hope is marred, or differentiated, or reworked as doubt and difference. If the subject of history turns, as Heidegger argued, not on facticities, but possibilities, then to think about the future is also to think about history. Koselleck's reflections in *Futures Past* on these Heideggerian positions remains drawn to the new time (*Neuzeit*) that propels the Revolution and modernity forward (xi, 138–140). I emphasize the counter-impulse to this argument in Koselleck's notice that the macro histories of great movements and events cannot account for the micro level of events that continually crosscut and disturb macro analysis (106–14). Those micro disturbances offer one speculative analogy for how the materiality of nature disturbs the plot of progressive development that Samuel Coleridge and many of his Romantic contemporaries called organic form.

In the midst of efforts to classify the natural world, the heavens, and everybody from everywhere, late eighteenth-century and Romantic writers identified or stumbled over evidence that lines of difference fissured this project at nearly every turn of the taxonomic high road, and at times with about as much regularity as chaotic, or at least unexpected, contingent intrusions allow. Around 1800, as scientific inquiry began to offer a much grainier view of the organic, the call to imagine organic life as a single, coherent, and developing form had to compete with a more differentiated map of matter and formal possibility. As Andrew Piper notes, emergent biological thought about life forms insisted that their origin was fluid and their interrelations equally so. Wet, as Piper puts it, became dry, or at least chemically constituted out of two elements when Lavoisier discovered the chemical composition of water in 1775. As the key conceptual point of convergence for matter and life, Romantic-era chemistry rendered unstable the barrier that earlier systematic thought had put up between matter as such, and life forms. As compounds were identified in terms of Galvanic, organic, then electrochemical, thermoelectric and electromagnetic discoveries and processes, it became likely that all forms of life, both anorganic and organic, had some unspecified set of internal relations (Piper 8; Golinski 203–18).

Although Goethe's prescient management of this recognition in the *Metamorphosis of Plants* (1790) understands plant development as an organic process that culminates in the flower, his account of the successive "openings" (*anastomoses*) by which a plant transforms its parts and functions acknowledges that at every one of these "openings," something different might emerge, including an "irregular metamorphosis" that would in turn reveal something before or otherwise hidden in the inner workings of botanical development (31–2). "Regular metamorphosis" is here companioned by its developmental other, which articulates a potential development that is typically left aside, not taken up. Even here, in the essay that declares Goethe's view of metamorphosis, going forward involves successive moments or events when going forward might roll elsewhere, where each opening is a switching point that could produce different motions, both retrograde and going forward.

This possibility hovers in Romantic thinking about deep time, occurring even in geological writing that seems, at least superficially, less tuned to the discoveries and hypotheses that announce, with Georges Cuvier, the bursting of time (Rudwick 586). James Hutton's 1785 *Theory of the Earth* imagines a history of the planet that is in one direction anchored by the conviction that the processes that shaped the earth have always been the same, hence "uniform," even when those processes involved volcanic action, the favourite of the Wernerian or "catastrophic" theory of the earth. Yet Hutton embeds these uniform processes in a timeline that is strangely split between a proximate teleology (the earth's inhabitants and its possibilities available for human consumption) and an untethered reach of unimaginable time. If for Hutton the processes of the earth's formation are relatively stable, reiterated, unchanging in character, they nonetheless exist in a time that is wholly unanchored, having "no vestige of a beginning, – no prospect of an end" (128). Hutton's departure from the highly fixed, numerologically exact timelines of earlier theories of the earth invokes a remarkable dance in which particular instantiations of long-range geological processes hover in space, ongoing yet unlocatable along a specified continuum, lost in time yet vestigially still around.

This more particularized, unanchored focus on life and events works against claims for the seamless whole of organic form, and against a more subtle understanding of fluctuation in form and narrative as tending, irregularly and haltingly, toward an as-yet unreached totality. At such moments, the promise of organic form looks much grainier, much less seamless and commanding, as life forms invite migratory thought to other kingdoms, and thus away from the holistic device or impresa of a plant, tree, person, or rock.

As a motif or tendency, with some unsettled staying power in Romantic writing across what we now regard as different disciplines or knowledge regimes, the difficulty of writing the future or writing to the future should, for these reasons, compel our interest. For how Romantic writers deal with their instantiation of a world and history in which the unexpected, the uncharted, keeps turning up requires a declination toward resourcefulness that is at the very least salutary, because it attempts to confront, or at least recognize, a world and events that cannot be fitted into anybody's theory of history as the march of progress or mind. Minds and writers, it turns out, do not proceed with this degree of cohesion.

Consider, for example, those individuals or particulars that require taxonomic generalization to the level of species and beyond. Reading a wobbly through line from Charles Darwin to contemporary gene-centred research, Elizabeth Grosz notes the "impossibility of any one-to-one mapping of genes to phenotypical characteristics ... even with an accurate map of the genome." Hence, she adds, "the need to generalize about individuals" seeks to overwhelm, statistically, the fact of individual difference" (41–2). In the late eighteenth century, identifying individual specimens that could be abstracted into a species type became the work of many hours, until the difficulty of that work, and the difficulty that had by then accrued to the question of what a species is, prompted Charles Darwin to suggest that it was whatever his learned colleagues supposed a species to be (47).

For Darwin, the narrative irregularity implied by monstrous forms, be they double flowers or some unlikely excrescence at some stage of development, would rarely be sustained in natural selection. For as they crossed in the first and second generations with "the ordinary form ... their abnormal character would almost inevitably be lost" (Darwin 6; Grosz 5). Taking chances, as nature does, contriving little machineries of pollination that seem to be inventive and creative, but which Darwin wants nonetheless to be the work of chance – in these half-contradictory instances, chance keeps cropping up, its effects normalized, even as the array of possible, chance turnings in evolutionary development contribute to the array of diversity that Darwin's *On the Origin of Species* (1859) pursues.

Darwin's theory of natural selection, in which chance both is and is not in play, looks askance and backward to the way Erasmus Darwin embedded the appearance of chance resemblance in the logic and form of "Loves of the Plants," Part II of *The Botanic Garden* (1789).[2] Presented as a verse explanation of Linnaeus's plant systematics, the poem repeatedly undermines a taxonomic narrative whose sense of hierarchy and order is as tight as Darwin's is loose, in ways that the exchanges between the poet and the bookseller in interludes between cantos of the poem pretend to ignore. For although both speakers suggest that loose poetic analogies, such as those that drive the verse of the poem, are of a different order from the more strict, scientific or philosophic analogies that Erasmus Darwin's notes offer, this distinction is, practically and imaginatively speaking, bogus. What the poem and its scientific notes offer instead is an exuberant tangle of life forms and meandering connectivity. Here, the analogical engine Gillian Beer recognizes in her essay in this volume is less a precision instrument than a persistent catching up of possibilities whose very strangeness speaks for a potentiality that risks being too much, thereby avoiding the same dull round of too little.

Consider, for example the note in "Loves of the Plants," Canto 1 for the plant "Callitriche," a pond plant that recalls Ovid's Narcissus, who gazes into the pond as his plant "tresses" float there, looking at his own image. The two virgins who share him are smitten with his starry eyes and radiant hair. The note reads:

> Callitriche. L. 45. Fine-Hair, Stargrass. One male and two females inhabit each flower. The upper leaves grow in form of a star, whence it is called Stellaria Aquatica by Ray and others; its stems and leaves float far on the water, and are often so matted together, as they bear a person walking on them. The male sometimes lives in a separate flower. (4)

It is the short distance from the person who walks on the mat created by the plant's stems and leaves to the "male" that sometimes lives in a separate flower that interests me here. Walking on a matted plant that floats on water edges toward the unbelievable; it is as though Darwin's poet has backed into a fantastical, fictional world. The segue from the person who walks to the male that lives in another flower is odder still: read glancingly, one could for a moment imagine a real male, not simply a Linnaean stamen, living on its own, separate for whatever reason from those competing virgins.

I risk this highly figured reading of Darwin's scientific note to specify via its formal excess how fluid the line between science and figure is in this poem, a point not inadvertently aided by the poem's frequent consideration of water plants. Darwin's persistent notice of similarities that cross the kingdoms of nature in the notes to the poem indicates relationships that exceed the separation of kingdoms that Linnaeus's taxonomic project had urged. Darwin's swinging analogies between the "respiration" of plants and animals, among others, instead imagine a network of functional resemblances that travel across the kingdoms of nature. The speaker of "Loves of the Plants" suggests that its project is Ovid in reverse: to restore to plants "some of their original animality, after having remained prisoners for so long in their respective vegetable kingdoms" (x).

At times, the poem pursues plant-to-animal analogies as though these imply an evolutionary telos, as in the long note to Curcuma or Turmeric, which begins by noting that this flower structure includes one male and one female and four "imperfect males" – that is, stamens without anthers on them, which, Darwin inevitably explains, Linnaeus called eunuchs. The second paragraph of the note detours from plants to insects that have two wings "analogous to the rudiments of stamens above described." These "little knobs," as he calls them, appear to indicate hind wings, albeit underdeveloped. From here Darwin moves to "the existence of teats on the breasts of male animals ... which are generally replete with a thin kind of milk at their nativity" (Darwin 7–8). The closing speculation of the note seems badly stitched to the preceding analogies. "Perhaps," writes Darwin, "all the productions of nature are in their progress to greater perfection." (8). If nature's productions are in such a progress, it is difficult to understand how the allusive path that goes from stubby stamens to milky male teats by way of rudimentary hind wings on insects signals a perfecting evolutionary process. For these shifts are oddly lateral, sidewinding across kingdoms instead of gathering individuals and species to assemble them into settled plant genera. Instead, what the note offers is a romp among odd similarities across kingdoms or at least phyla that doesn't really add up to a telos or progress but does offer occasions for imagining that plants live as well as love.

The ease with which Erasmus Darwin reverses the expected direction of plant-animal analogy can be arresting. The note for *Anthoxanthum* or Vernal grass declares that it is "viviparous" because it "bears sometimes roots or bulbs instead of seeds, which ... drop off and strike root into the ground." A second paragraph finds an analogous instance of the "double production" by root and seed in the animal kingdom: "the same species of Aphis is viviparous in summer, and oviparous in autumn" (Darwin 11). Although the note includes an intervening example, a plant named *Polygonum viviparum* or viviparous bistort, the claim that plant reproduction may include "live birth" puts the figured use of this term very much in the lead, and bizarrely so. Somehow "viviparous" gets captured as a figure before its usual, non-figurative application gets into the picture. Carnivorous plants create a field day for Darwin. A "wonderful contrivance" in the *Dionaea muscipula* keeps insects out: "the leaves are armed with long teeth, like the antennae of insects," leaves so "irritable, that when an insect creeps upon them, they fold up, and crush or pierce it to death." Another plant, the *Arum muscivorum*, has a flower that smells like carrion, but other carnivorous plants, the note winds down, "give an agreeable odour in the night" (16).

Bad smells, good smells, leaves like teeth that crush or pierce, or, in other notes, plants "furnished with arms for their protection" (18), which is to say, as Darwin does go on to say, prickles, spines or thorns: these are plants that make good on the project of the "Loves of the Plants," that is, returning an original animality to plants, as note after note does. Viscous fluids that prevent small insects from eating leaves are like "the ear-wax in animals" (25), and the common name for the fern *Polypodium barometz*, the Tartarian Lamb, launches a note that works the figural relay from plant to lamb and back again. This fern has a root that is "thick, and everywhere covered with the most soft and dense wool, intensely yellow." It resembles a lamb standing on four legs. Its "down" is used in India to stop hemorrhages, where it is called "golden moss." The next paragraph lays down more tracks: "The thick downy clothing of some vegetables seems designed to protect them from the injuries of cold, like the wool of animals." The next stop on the analogy train is the fat or oil of whales, then, rounding back, snow as a protective covering for "vegetables" (30–1). A long note on the *Mimosa* or sensitive plant ends "see note on vegetable respiration in Part I," referring to one of the substantial essay notes that Darwin produces at the end of Part 1 of *The Botanic Garden.* Or a discussion of air vessels or bladders in a seaweed that moves to the air bladders of fish and finally to the unfortunate Mr. Day, drowned in a diving-ship of his own construction because he failed to attend to evidence (from plants, from fish) of the effects of water pressure on air chambers. Along the way, Darwin considers air-vessels in animal placenta and whether such structures are respiratory rather than nutritive in function (42).

The wild, careening ride of so many of the notes that refer to each other in the *Botanic Garden* mark resemblances across the poem, both within cantos and across them. Here, Darwin seeks out functional analogies without making a sustained claim for a progressive, evolutionary *telos* akin to that proposed in *Zoonomia* (1794) or the *Temple of Nature* (1803). Instead, the narrative momentum of *Botanic Garden* conveys a duration that is marked by chance interruptions. For precisely because a reader never knows when one analogy in the poem or its notes will prompt another that may or may not be hinged by functional analogy (remember poor Mr. Day), each nip and turn of the verse and note shapes a narrative moment that chance rather than organic design might inhabit. A reader could end up anywhere, then bounce off to somewhere else. Like the increments of an infinitesimal calculus, each moment has a force of its own, and its impulse may be irruptive rather than sequential.

If this reading of Darwin's *Botanic Garden* finds there only scant support for the evolutionary claims he develops in *Zoonomia* and elsewhere, it may nonetheless make available an aspect of evolutionary possibility that more tightly wound organic models foreclose or occlude. At any moment in development, even in that of a single plant, that project may take an unexpected detour of the kind that Goethe briefly mentions in the *Metamorphosis of Plants*. Such detours mark where something untoward or unexpected may enter as a matter of importance. Indeed, contemporary notice of the role of contingency in evolutionary processes insists further that matter, environment, and external events are co-evolutionary, not inert elements that DNA simply inhabit (Oyama, Griffiths, and Gray 6).

Grosz's reflections on time and evolutionary futurity, and recent discussions of the plasticity of evolutionary change both inside and outside the mind, extend lines of inquiry that emerged in Romantic scientific thought around the late 1780s. As a real, "regulative force of life as we know it," time is for Grosz "a precondition of matter's emergence, and the force that, surprisingly, without predictability, rends life from its more unstable interactions" (247). The questions that she asks belong to an overtone series that echoes and exfoliates Romantic questions:

> Out of what raw materials and using what processes did the simplest forms of life emerge? What were the non-organic ingredients of the prebiotic soup out of which elementary life appeared? What is the point of conversion from chemical to biological components? How closely tied are biological life forms to the particular chemistry of those forms of life with which we are familiar? We, and all creatures on earth, are carbon-based life forms. The question contemporary a-life [*sic*] scientists ask is: Is life essentially tied to those accidental, carbon-based life forms we know today? Can there be a silicon-based, or, say, a nitrogen-based life form? What, for example, are those open-ended computer programs that exhibit reproductive, regulative, and emergent properties, similar, at least in some respects, to other forms of life? (22–3)

This insistence that time, specific moments of time, matter to matter is worth lingering with. Such moments of unpredicted change or development constitute a resistance that is latent in, as I would put it, form, matter, and even organization.

The Romantic-era philosopher whose thinking about nature, change, and life that this essay has long had in its rearview mirror is G.W.F. Hegel. In her essay for this volume, Tilottama Rajan argues that both Hegel and Friedrich Schelling recognize an underlying resistance in nature with which idealist philosophy must deal. Schelling writes about variations in being that constitute Nature's "apprenticeship in learning to make man," variations that Schelling calls "misbegotten attempts" that fail to achieve absolute or ideal being (22). For Rajan, Hegel makes this resistance the ongoing vehicle of evolutionary striving, with the absolute as its goal. For this reason, she observes, Hegel's *Lectures on Aesthetics* end not with Classical but with Romantic art, wherein the form and content that were well-matched in the Classical phase are again separate, "this time because of a deficiency in matter that repeats and reverses the problems of the Symbolic" (23–4). Hegel's preference for Romantic art exhibits what Jean-Luc Nancy characterizes as the "restlessness of the negative" (5), a phrase that Rajan understands as the fundamental impulse of Hegel's philosophy because it impels his pursuit of a perfectibility that aims at the ideal even as it falls short.

What Hegel's pursuit of a negativity that continually discloses more possibilities ahead puts aside or, more precisely, subsumes, is contingency, the slippage of accident and contingent materiality that articulates, even as it disturbs, Romantic history and narrative form. Here, a return to Grosz's account of evolution helps to specify the difference between Hegel's view of the role of chance, contingency, and accident and one that would understand these as constitutive of narrative possibility. For contingency and accident are at once what Hegel seeks to fold into the work of spirit in the *Phenomenology* and what resists the very logic of that folding in. The space that evolution leaves open for a latent potentiality is not, I think, admissible into the Hegelian project that Rajan, Nancy, Rebecca Comay, and Theodor Adorno have considered as the engine, but also the staying power, of the Hegelian dialectic.

Put another way: in Hegel's *Phenomenology*, the work of contingency, chance, caprice, or accident must be lifted up into the work of the ideal Spirit such that the very terms we use to define the unexpected are taken up into the virtual, anticipated life of the Spirit. Whatever the deepening and widening advantages of this supersession, it can never make an unchallenged place for what Grosz calls "the untimely, the dislocated, that which precedes, surpasses, and moves beyond man, that which goes beyond the human and unhinges progress and continuity, displacing the known and the present for a future that does not yet exist" (98). If it is not clear whether Grosz would also claim here that the residues left unincorporated into evolutionary selection might one day be taken up elsewhere, she clearly insists that the innumerable chances at work in evolutionary history are what make futurity happen, whatever that futurity is. I have argued here that Romantic narrative makes a place for what is (yet) untimely, but also for accidents as material and figurative witnesses that no history or narrative can proceed in full knowledge of futurity.

I return and conclude with Mary Shelley's *Valperga* (1823) to characterize why its historical project appears also to shadow a Romantic future bound to history and the implosive present of Romantic Italy around 1820. By inventing two women characters that inhabit this fictional retelling of Machiavelli's story of the prince Castruccio, Shelley creates a fictional version of something like chance, or, to be more precisely adherent to Hume's argument, a Humean billiard ball thrown into the juggernaut history of Castruccio's rise and fall that might, from a certain perspective, look like chance. Euthanasia, who seeks to change the course of that history, shifting it away from war and conquest and toward something like an enlightened peace, takes her chances with and then against Castruccio. She fails abysmally, and the other character, Beatrice, gets rolled under the cart of despotic authority. Yet the novel's unfulfilled imagining of a future for history that unrolls from the fourteenth-century Italian setting of the novel to Shelley's Romantic present both indicates a future that never arrives, in which rational and beneficent human communities might thrive without pillage, and posits a potential future (Kelley 625–52). On the off chance that some such interruption in the historical plottedness of the same, dull round of conquest and repression just might happen, *Valperga* suggests how some other outcome might or could have been different and, further, that such a difference would require an event that interrupts, just then, a historical engine and narrative that intend to go elsewhere.

As each of these characters moves across the narrative and thus across the historical moment, they generate narrative possibilities that remain imaginatively in play, even after these possibilities are foreclosed by and within the historical events of Castruccio's life and times. Put another way, the narrative threads that Euthanasia and Beatrice spin out belong to a sheaf of causalities that may not be discernible in that historical moment, but remain available to futurity. Whereas early modern efforts to tame chance rely on probability to make a numbers game out of what might otherwise appear to resist such rationalization, the presence of something like chance, contingency, and accident in Romantic temporality, both lived and invented, speaks to a different regard for interruptions and possible swerves that do not carry forward or achieve an imagined telos or end. What they do convey instead is an invitation to take up the unexpected, to recognize in its very latency some inkling of how possibility, chance, and even counterfactuals, as the work of world and plot, might be opportunities for moving ahead.

NOTES

1 For a philosophical analysis of those moments when Romantic writing turns from a declared program, see Terada, *Looking Away.*

2 Parenthetical citations of Darwin's "Loves of the Plants" refer to the pagination internal to that part of the poem, which became Part II in the 1791 edition I use here.

WORKS CITED

Adorno, Theodor. *Hegel: Three Studies.* Trans. Shierry Weber Nicholsen. Cambridge, Mass.: MIT P, 1999. Print.

Aristotle. *Metaphysics. Complete Works.* 2 vols. Ed. Jonathan Barnes. Princeton: Princeton UP, 1995. Print.

Badiou, Alain. *Being and Event.* Trans. Oliver Feltham. New York: Continuum, 2005. Print.

Balfour, Ian. *The Rhetoric of Romantic Prophecy.* Palo Alto: Stanford UP, 2002. Print.

Beer, Gillian. *Darwin's Plots.* 2nd ed. Cambridge: Cambridge UP, 2000. Print.

Burke, Edmund. *Reflections on the Revolution in France.* Ed. L.G. Mitchell. Oxford: Oxford UP, 1993. Print.

Comay, Rebecca. *Mourning Sickness: Hegel and the French Revolution.* Palo Alto: Stanford UP, 2011. Print.

Darwin, Charles. *On the Origin of Species.* Ed. Gillian Beer. Oxford: Oxford UP, 1996. Print.

Darwin, Erasmus. "Loves of the Plants." Part 2 of *The Botanic Garden.* London: J. Johnson, 1791. Print.

Foucault, Michel. *The Order of Things.* New York: Vintage, 1970. Print.

Goethe, J.W. von. *Goethe's Botanical Writings.* Trans. Bertha Mueller. Woodbridge, CT: Oxbow P, 1952. Print.

Golinski, Jan. *Making Natural Knowledge: Constructivism and the History of Science.* Chicago: U of Chicago P, 1998. Print.

Grosz, Elizabeth. *The Nick of Time: Politics, Evolution, and the Untimely.* Durham, NC: Duke UP, 2004. Print.

Hacking, Ian. *The Taming of Chance.* Cambridge: Cambridge UP, 1990. Print.

Hamilton, Ross. *Accident: A Philosophical and Literary History.* Chicago: U of Chicago P, 2007. Print.

Hegel, G.W.F. *Phenomenology of Spirit.* Trans. A.V. Miller. Oxford: Oxford UP, 1977. Print.

Hume, David. *Enquiries concerning Human Understanding and concerning the Principles of Morals.* 3rd ed. Ed. L.A. Selby-Bigge and P.H. Nidditch. Oxford: Clarendon P, 1975. Print.

Hutton, James. *System of the Earth.* 1787; rpt. New York: Hafner P, 1973. Print.

Keats, John. *Complete Poems.* Ed. Jack Stillinger. Cambridge, Mass.: Harvard UP, 1978. Print.

Kelley, Theresa M. "Romantic Temporality, Contingency and Mary Shelley." *ELH* 75: 3 (Fall 2008): 625–52. Print.

Koselleck, Reinhart. *Futures Past: On the Semantics of Historical Time.* Trans. Keith Tribe. New York: Columbia UP, 2004. Print.

Molesworth, Jesse. *Chance and the Eighteenth-Century Novel: Realism, Probability, Magic.* Cambridge: Cambridge UP, 2010. Print.

Nancy, Jean-Luc. *Hegel: The Restlessness of the Negative.* Trans. Jason Smith and Steven Miller. Minneapolis: U of Minnesota P, 2002. Print.

Oyama, Susan, Paul E. Griffiths, and Russell D. Gray. "Introduction: What is Developmental Systems Theory?" *Cycles of Contingency: Developmental Systems and Evolution.* Cambridge, Mass.: MIT P, 2001. 1–11. Print.

Piper, Andrew. "Writing Life: Autobiography, the Life Sciences, and the Book of Life ant the End of the Eighteenth Century." Presented at the Ninth Annual Bloomington Eighteenth-Century Studies Workshop, Indiana University, 12–14 May 2010. Print.

Rudwick, Martin. *Bursting the Limit of Time.* Chicago: U of Chicago P, 2005. Print.

Shelley, Mary. *Valperga.* Ed. Stuart Curran. New York: Oxford UP, 1997. Print.

Shelley, Percy. *Shelley's Poetry and Prose.* 2nd ed. Ed. Donald H. Reiman and Neil Fraistat. New York: Norton, 2002. Print.

Smith, Charlotte. *The Poems of Charlotte Smith.* Ed. Stuart Curran. New York: Oxford UP, 1993. Print.

Taylor, Charles. *Sources of the Self: The Making of Modern Identity.* Cambridge, Mass.: Harvard UP, 1989. Print.

Terada, Rei. *Looking Away.* Cambridge, Mass.: Harvard UP, 2009. Print.

Witmore, Michael. *The Culture of Accidents: Unexpected Knowledges in Early Modern England.* Palo Alto: Stanford UP, 2001. Print.

Wordsworth, William. *The Prelude, 1799, 1805, 1850.* Ed. Jonathan Wordsworth, M.H. Abrams, and Stephen Gill. New York: Norton, 1979. Print.

PART FOUR

Evolutionary Idealisms

Chapter Nine

Did Goethe and Schelling Endorse Species Evolution?

ROBERT J. RICHARDS

Charles Darwin (1809–82) was quite sensitive to the charge that his theory of species transmutation was not original but had been anticipated by earlier authors, most famously Jean Baptiste de Lamarck (1744–1829) and his own grandfather, Erasmus Darwin (1731–1802). The younger Darwin believed, however, his own originality lay in the device he used to explain the change of species over time and in the kind of evidence he brought to bear to demonstrate such change. He was thus ready to concede and recognize predecessors, especially those who caused only modest ripples in the intellectual stream. In the historical introduction that he included in the third edition of *On the Origin of Species* (1861; first edition, 1859), he acknowledged Johann Wolfgang von Goethe (1749–1832) as "an extreme partisan" of the transmutation view. He had been encouraged to embrace Goethe as a fellow transmutationist by Isidore Geoffroy St Hilaire (1805–61) and Ernst Haeckel (1834–1919).[1]

Scholars today think that Darwin's recognition of Goethe was a mistake. Manfred Wenszel, for instance, simply says: "An evolutionism ... establishing an historical transformation in the world of biological phenomena over generations lay far beyond Goethe's horizon" (784). George Wells, who has considered the question at great length, concludes: Goethe "was unable to accept the possibility of large-scale evolution" (45–6). A comparable assumption prevails about the *Naturphilosoph* Friedrich Joseph Schelling (1775–1854). Most scholars deny that Schelling held anything like a theory of species evolution in the manner of Charles Darwin – that is, a conception of a gradual change of species in the empirical world over long periods of time. Dietrich von Engelhardt, in commenting on an enticing passage from Schelling's 1799 *Erster Entwurf eines Systems der Naturphilosophie* (*First Sketch of a System of Nature Philosophy*), declares, "Schelling is no forerunner of Darwin." Schelling, according to Engelhardt, advanced no real theory of descent, rather only "a metaphysical ordering of plants and animals" (312–13).

The assessments of these contemporary scholars are, however, opposed to judgments by earlier thinkers. In the 1860s, when Haeckel, while at Jena, began championing Darwin's evolutionary ideas, his friend Kuno Fischer (1824–1907) – the great neo-Kantian historian of philosophy – offered his colleague a mild rebuke. In his two-volume history of Schelling's thought, Fischer claimed: "Schelling was the first to enunciate with complete clarity and from a philosophical standpoint the principle of organic development that is fundamental to the Darwinism of today" (2:448). And, as I mentioned, Haeckel himself convinced his English friend of Goethe's priority in holding the transformational hypothesis. I believe Fischer and Haeckel were entirely correct. I will argue that Goethe and his young protégé, Schelling, mutually reinforced each other's theories of species evolution.

Prehistory of the Evolutionary Conception

The reflexive denial to individuals like Goethe and Schelling of any tincture of the notion that empirical species might change over time gains strength from the assumption that species evolution was not a conceptual option prior to the nineteenth century. But this assumption is simply incorrect. Speculations about species change far antedate Lamarck and Darwin. Aristotle, for instance, entertained the idea, in *De generatione animalium*, that men and quadrupeds might originally have been spontaneously generated from something like insect grubs, with a later development into recognizable form, a metamorphosis comparable to the caterpillar into the butterfly. He seems not to have believed this, but did think it a conceptual possibility (361).

In the mid-eighteenth century, Charles Bonnet (1720–93) proposed that originally, God created a plenitude of germs, each encapsulating a miniature organism that in turn carried germs containing yet more homunculi and their germs, enough to reach the Second Coming. He thought that within each line of germs one species might have given rise to another species according to a "natural evolution of organized beings [d'Evolution naturelle des Etres Organisés]" (1:250). When Bonnet used the term "evolution," he adapted it from its use in embryology, wherein it was synonymous with preformationism: that is, the conceit that the embryo was already an articulated organism that simply had to unroll (that is, evolve) during gestation. This term, which had its original provenance in embryology, was used by Bonnet, then, to refer to species unfolding.[2] Schelling would adopt the same term, but with him it would shed something of its preformational character.

Johann Friedrich Blumenbach (1752–1840), in the 1780s, advanced the idea that a special force, a *nisus formatives – ein Bildungstrieb* – caused the embryo to develop in an epigenetic fashion, that is, become articulated out of a homogeneous mass. Like Bonnet, he thought the earth was salted, as it were, with germs that under the aegis of the *Bildungstrieb* would unfold new species to replace the old that were wiped out by the catastrophes of which fossils of extinct organisms gave evidence.[3]

Though Immanuel Kant (1724–1804) condemned the notion of species evolution when it was suggested by his one-time student Johann Gottfried Herder (1744–1803), he had a change of heart after reading Blumenbach, a real natural philosopher in his estimation.[4] In his *Kritik der Urteilskraft* (*Critique of the Power of Judgment*, 1790), Kant allowed as a conceptual possibility that species might be transmuted through a mechanical expansion or contraction of a basic organization. He did, however, reject the further idea that living organization might arise out of the inanimate, the unorganized, in a kind of spontaneous generation. Nonetheless, he tolerated the conception of a change of species over time and described that notion as a "daring adventure of reason." He ultimately refused to participate in this daring adventure, since he thought the evidence of such species transmutation to be wanting.[5] Species change was obviously a live conceptual option in the last part of the eighteenth century.

Kant would be challenged by both Schelling and Goethe on two counts. Both would allow "mother earth" to generate organic life because, as Schelling would maintain, the earth and its chemical processes were already organic, not dead matter, not mechanical; and second, as Goethe would show, fossils, as well as the metamorphosis of plants and insects, provided ample evidence of species transformation.

Finally, Darwin should be mentioned, not Charles but his grandfather Erasmus, who in the late eighteenth century advanced an evolutionary theory according to which God created the first living filament, after which natural processes – mostly in the form of the inheritance of acquired characters – gave birth to all the animal and plant species populating the world. Darwin's book *Zoonomia; or the Laws of Organic Life*, the first volume of which appeared in 1794, was immediately translated into German and read by both Goethe and Schelling.[6] These *Naturphilosophen* were thus quite familiar with transformational ideas coming from England and the continent.

Goethe's Early Morphological Theories

By the mid-nineteenth century, Hermann von Helmholtz (1821–94) recognized Goethe as having founded the dominant theoretical framework in biology during the earlier part of the century. He credited his great predecessor with establishing a science of morphology that became vital to evolutionary conceptions in the second part of the century.[7] And Helmholtz's judgment was entirely correct. For

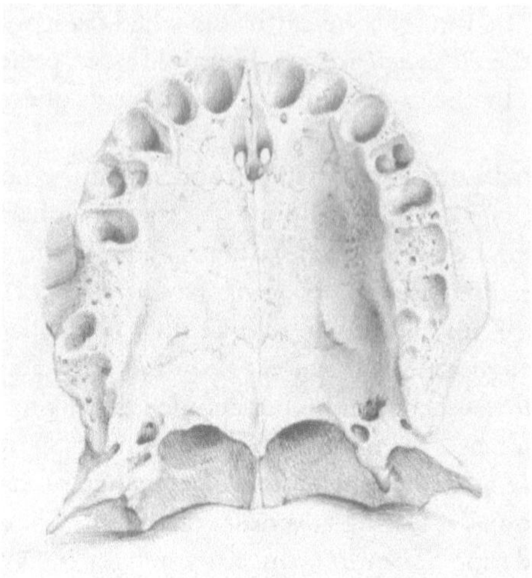

Figure 9.1 Wilhelm Waitz's illustration prepared for Goethe's essay on the *Zwischenkiefer*. It shows the faint suture of the intermaxillary bone in the human upper jaw (top quarter), similar to that found in apes and other animals. From Goethe's *Über den Zwischenkiefer des Menschen und der Thiere* (*Leopoldina*, 15.1, 1831). Courtesy of the University of Chicago Library.

Goethe, morphology was the doctrine of plant and animal forms, especially in their dynamic properties. As he put it in the late 1790s, "The doctrine of forms is the doctrine of transformation. The doctrine of metamorphosis is the key to all signs of nature" ("Morphologie" 4.2:188).

In the previous decade, Goethe had begun to develop ideas about the dynamic character of animal forms, especially in his discussions with Herder, who was composing his *Ideen zur Philosophie der Geschichte der Menschheit* (*Ideas towards a Philosophy of the History of Mankind*) in the early 1780s. Their mutual interest revolved around ideas of the unity of nature and the transformations within that unity. In 1784, Goethe discovered the os intermaxillaris, or *Zwischenkiefer*, in the human fetus.[8] It is barely visible in the adult skull (Figure 9.1).

Most anatomists thought that this bone in the upper jaw was characteristic only of animals. For Goethe and Herder, the discovery meant that the human vertebrate form displayed a unity with other vertebrates. At this time, both individuals began to speculate on the development of the universe and its various creatures. As Goethe recalled of his discussions with his friend, "Our daily conversation concerned the origins of the water-covered earth and its organic creatures, which have developed [entwickelt] on it from very ancient times" ("Der Inhalt" 12:16). This was the kind of speculation that Kant initially regarded as uncontrolled fantasy, and so chided his former student in a stunning dismissal of Herder's *Ideen*. Charlotte von Stein (1742–1827), Goethe's intimate friend, wrote to the court Administrator Karl Knebel (1744–1834) in May 1784 to describe the extent of the pair's transformational ideas:

> Herder's new work makes it probable that we were first plants and animals. What nature will further stamp out of us will remain well unknown. Goethe now ponders [grübelt] thoughtfully these things, and anything that first has passed through his imagination becomes extremely interesting. (Düntzer 1:120)

Two years later, in 1786, Goethe began conducting experiments on spontaneous generation, watching as microscopic animalculae seemed to arise out of bits of plants soaked in water and sealed in containers.[9] Many theories of evolution in the eighteenth and nineteenth centuries postulated some sort of spontaneous generation, of the sort that Kant wished to condemn.

In 1790, Goethe saw his *Metamorphose der Pflanzen* through the presses. That tract argued that the various parts of a plant – the stem, petals, stamen, and other organs – could be understood as transformations of a fundamental, underlying structure, which he denominated the "ideal leaf." This underlying structure, only graspable by the mind's eye, contained, as it were, all the potential of its many instantiations in different parts of the plant and in different plant species. This plenum conception would distinguish Goethe's archetypal concept from the alternative, developed particularly in Britain, where the archetype was treated as a minimal structure, not a plenum structure. Also in 1790, Goethe undertook an intensive study of the new critique by Kant, the *Kritik der Urteilskraft*. His reading of Kant, from whom he had originally kept his distance, helped consolidate a set of ideas about aesthetics and teleology that he had been turning over in his imagination for some time.[10]

Through the 1790s, Goethe composed five essays, mostly uncompleted, on the morphology of animals.[11] In these essays he attempted the kind of developmental analysis he had conducted on the morphology of plants. He formulated a theory of the animal archetype, or *Urtypus*, that would be comparable to the plant archetype. These essays maintained that just as the plant archetype served as the pattern exhibited by all plants, so the animal archetype was the pattern by which all animals – at least the vertebrates – could be comprehended in unity.

He conceived this archetype as a common osteological pattern of bones; so, for example, the fox and the sea-lion have skeletal features that display an underlying pattern that has been altered in respect of their different environments – the sort of unifying pattern that Kant imagined could be mechanically deformed or altered to correspond now to this vertebrate frame, now to that. Further, Goethe adopted the idea of Blumenbach's *Bildungstrieb* to maintain that a creative force was exhibited in embryological development, so that all vertebrates, for instance, displayed the common form. Like Lamarck and the young Darwin, he also recognized another, external force, which adapted animals to their environments. Thus, while the fox and the sea-lion exhibited the same general body structure, that structure had been altered to adapt it to their particular circumstances – the land, on the one hand, and the sea, on the other (Goethe, "Versuch" 4.2:182).

In the Third Critique, Kant maintained that the organization of living creatures had to be understood according to archetypal ideas; plants and animals could not simply be regarded as mechanisms, but had to be understood as if they originated from an ideal plan. He further argued that archetypal ideas suggested an intentional will that causally imposed organizational structure on living creatures, providing the characteristic design that particular species exhibited. Of course, for Kant, this assumption of an archetype and its intentional, creative force was regulative, guiding our human understanding in the quest for more scientifically appropriate mechanical conceptions. Goethe, however, still in thrall to a latent Spinozism in the 1790s, regarded archetypal ideas as constitutive: that is, they were *adequate* ideas having creative potency really residing in nature. They existed in nature as a dynamic force, which Goethe regarded as instances of Blumenbach's *Bildungstrieb*. Thus Goethe amalgamated ideas from several of his contemporaries, and gave them his own particular enticing twist. One who was so enticed was the British morphologist Richard Owen (1804–92).

In British biology, Owen – the most influential biologist of the first half of the nineteenth century, and one thoroughly immersed in German biological ideas – would recognize the two forces that Goethe discerned. He postulated one force producing homologous relations among animal organisms, and another, a teleological force, adapting those homologous creatures to particular environments. Owen showed, for instance, that the vertebrate limb displayed both archetypal unity across a variety of species and teleological adaptations characteristic of particular species. So the same topological arrangement of bones could be found in the forelimb of a mole, the paddle of the porpoise, and the wing of a bat, though each has been adapted for use in its particular circumstance, digging in the ground or flying through the air (4–14). More generally, Owen contended that the vertebrate skeleton displayed a common plan or archetype that was specified in different vertebrate species according to their environment (Figure 9.2).

Owen based his own theory of the archetype on that of Goethe's protégé Carl Gustav Carus (1789–1869). In his *Von den Ur-theilen des Knochen- und Schalengerüstes* (*On the Fundamental Parts of the Bones and the Hard Structures*, 1828),

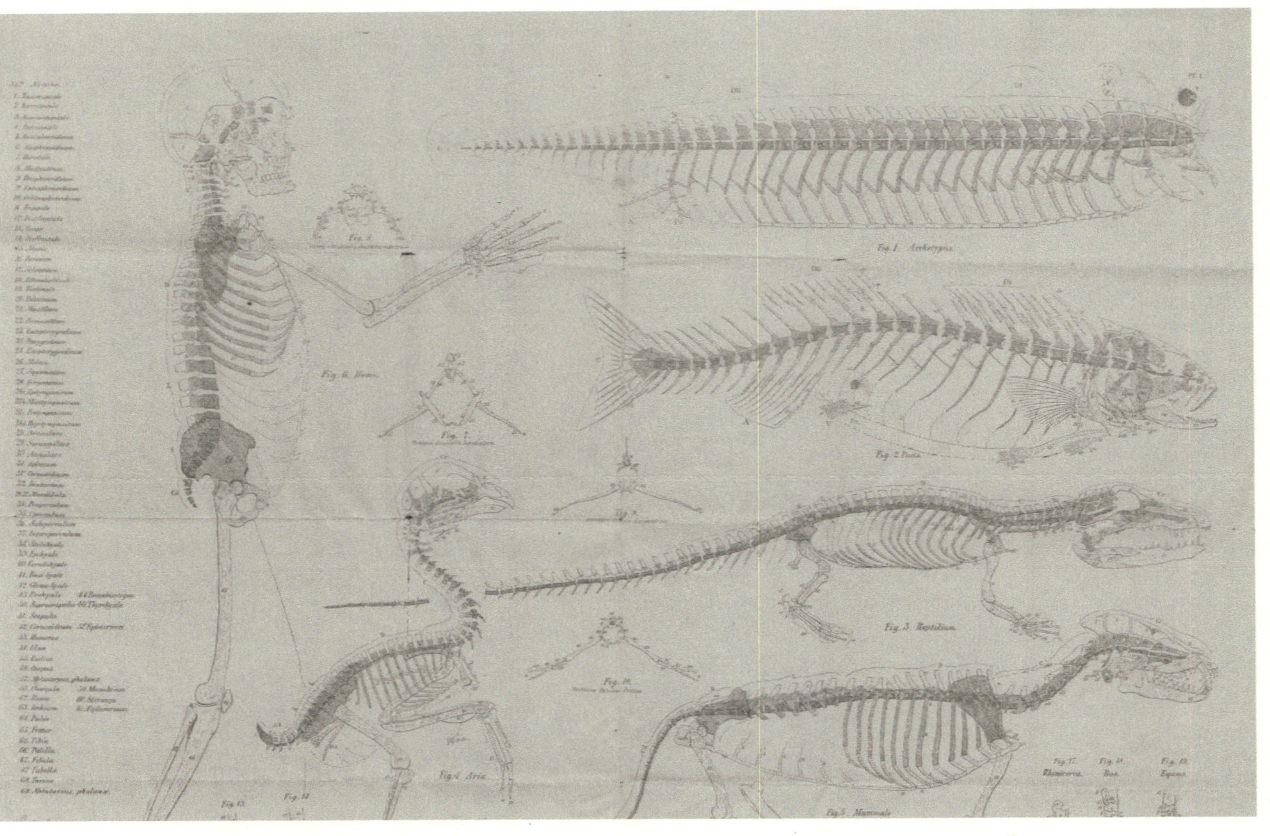

Figure 9.2 Plate from Richard Owen's *On the Nature of Limbs* (1849), showing the vertebrate archetype at the upper right, the homologous pattern for the skeleton of all vertebrate species. From the author's collection.

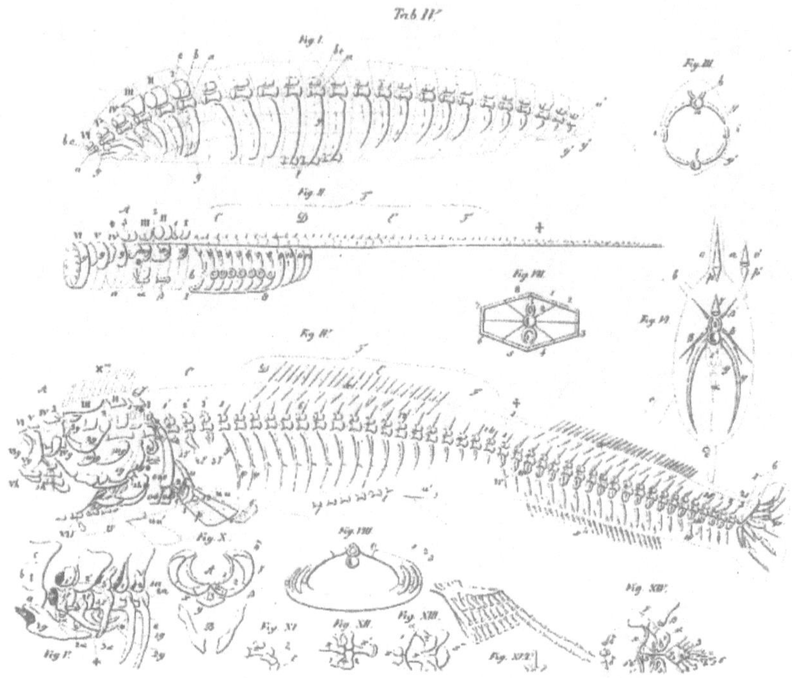

Figure 9.3 Illustrations of the vertebrate archetype (upper left) and of the ideal vertebra (upper right), from Carl Gustav Carus, *Von den Ur-Theilen des Knochen und Schalengerüstes* (1828). From the author's collection.

Carus synthesized ideas from several sources to construct the Goethean ideal type of the vertebrate skeleton (Figure 9.3).

Carus emphasized the "idea of parallelism between the development of the higher animal forms – yes, even man himself – and the development of the particular classes and species in the animal kingdom" (vii). Owen, not always forthcoming about his sources, utilized Carus's ideas for his own theory of the archetype, as is obvious from notes he took on his predecessor's book and a comparison of the illustrations from the books in question.[12] In his little book *On the Nature of Limbs* (1849), Owen drew proto-evolutionary conclusions from his application of archetype theory, even though evolution – at least in its Lamarckian version – was highly suspect in Britain. Only after Darwin published did Owen begin to make bolder claims of priority for his implicit theory of the descent of species. This is simply one vein of Goethe's thought that led to the evolutionary hypothesis, but it was a telling one.

Goethe's Aesthetics

From his morphological ideas, the ever-synthesizing Goethe drew implications for understanding the other end of the Kantian Critique, namely aesthetics. Kant had defined artistic genius as

> the talent (natural gift) that gives the rule to art. Since the talent, as an inborn productive ability of the artist, itself belongs to nature, we can also express it thus: genius is the inborn mental trait (*ingenium*) through which nature gives the rule to art. (*Kritik der Urteilkraft* 5:300 [A34–5, B34–5])

The artist of genius executes a work of fine art through an aesthetic feeling generated by unconscious considerations of rules of the beautiful with which the artist of genius is endowed. These rules, according to Kant's conception, remain embedded, as it were, in the artist's nature, guiding the artist's hand, not through conscious, rational consideration, but only through aesthetic feeling.

Prior to reading Kant's Third Critique, Goethe came to a quite similar view, one of the reasons he found the new Kantian conjunction of art and science so congenial. Goethe maintained that the artist of genius created his products by comprehending archetypal ideas, the same *adequate ideas* (in his Spinozistic terms) as the biologist; and the artist executed the art-object by exhibiting the same creative force as nature herself displayed. As he put it in a letter to von Stein during his Italian Journey in 1787,

> These great works of art are comparable to the great works of nature; they have been created by men according to true and natural laws. Everything arbitrary, imaginary collapses. Here is necessity, here is God. (*Italienische Reise* 15:478)[13]

A similar comparison between artistic and natural production occurred in an essay of 1789 that Goethe jotted in his travel diary (and published shortly after his return from Italy that year). In "Einfache Nachahmung der Natur, Manier, Styl" ("Simple Imitation of Nature, Manner, Style"), he distinguished artists of modest ability, who faithfully copied from the surface of nature, from those who also expressed a deeper part of themselves, which he called "style"; and he distinguished both of these from gifted artists who became more deeply aware of what lay behind nature's productions. The artist of great talent would be able to combine all of these modes of artistic expression. He or she would be able to utilize the laws – or archetypal ideas – that nature herself deployed in her creations to execute a work of art that was not simple imitation, but deeply expressive of nature's own principles.

Prior to 1800, Goethe had thus developed a conception of morphology as postulating dynamic forces resident in nature. These forces, he contended, explained patterns to be found in animal and plant organisms. He merged this conception with his aesthetic ideas, suggesting that the artist of genius employed the same power in artistic creation as nature did in her organic creations. In this respect, the artist was nature – a characterization, incidentally, used by Schiller to describe Goethe's particular kind of naive genius. For Goethe, this meant that artistic representation could reveal the deep laws of nature, or, as he epigrammatically put it, "The beautiful is a manifestation of secret laws of nature, which without its appearance would have remained forever hidden" (*Maximen und Reflexionen* [no. 1344] 942).

These Goethean notions would be quite favourable to the conception of species transformation; but aside from von Stein's mention that Goethe was speculating that we were once fish and then animals, there is only circumstantial evidence that prior to 1800 he endorsed species evolution. I believe that he came to hold firmly such a theory, at least in a manifest way, as a result of his interactions with an individual in whom he took a paternal interest, Friedrich Wilhelm Joseph Schelling.

Schelling's Biological Theories

The editor of the *Philosophisches Journal*, Friedrich Immanuel Niethammer (1766–1848), the idealist philosopher Johann Gottlieb Fichte (1762–1814), and the theologian Heinrich Paulus (1761–1851) conspired to bring the twenty-three year old Schelling to Jena, while his father attempted to secure him for Tübingen, where he had finished his university studies.[14] Schelling's glittering reputation as a philosophical Wunderkind had impelled the three co-conspirators to seek his appointment at Jena. Initially Goethe was opposed, being greatly suspicious of anyone spouting Fichte's kind of idealism; but even this empirically grounded spirit was won over by Schelling himself, who, during a party thrown by Friedrich Schiller (1759–1805), quite generously recognized the older man's artistic and scientific acumen, especially in respect to essays on optics that Goethe had recently published. In May 1798, Goethe wrote Christian Gottlob Voigt (1743–1819), chief administrator for the Duchy of Saxony-Weimar-Eisenach, that the young philosopher had "a very clear, energetic, and according to the latest fashion, a well-organized head on his shoulders." Moreover, he gave "no hint of being a sansculotte," unlike Fichte (*Goethes*

Briefe 2:349). With Goethe's endorsement, Schelling became extraordinarius professor in 1798 at Jena.

Though yet suspicious of Fichtean idealism, Goethe began reading Schelling's *Weltseele* (*World Soul*) in late June of 1798. He later remarked in his diary that he saw Schelling's *Weltseele* "incorporated into the eternal metamorphosis of the external world" (*Tag* 58). In the preface to *Weltseele*, Schelling made a claim that would later catch the eye of Kuno Fischer (1827–1904), a claim concerning the transmutation of species. The passage referred to Kant's assertion in the Third Critique that organic life could not be derived from the inorganic according to any natural laws. Schelling countered that it was "vintage delusion" to hold that "organization and life cannot be explained from natural principles." He further proclaimed:

> One would at least take one step toward [such] explanation if one could show that the stages of all organic beings have been formed through a gradual development of one and the same organization. – That our experience has not taught us of any formation of nature, has not shown us any transition from one form or kind into another (although the metamorphosis of many insects ... could be introduced as an analogous phenomenon) – this is no demonstration against the possibility. For a defender of the idea of development could answer that the alteration to which the organic as well as the inorganic nature was subjected ... occurred over a much longer time than our small periods could provide measure. (*Weltseele* 416–17)

Schelling thus would chance that "daring adventure of reason" from which Kant himself shied.

Schelling, though, agreed with Kant that the organic could not be derived from the inorganic by some kind of spontaneous generation. He differed from Kant in holding that mother earth herself was organic, so that perfectly natural principles of development could produce organic life out of the *apparently* inorganic. He suggested in the above passage that the evolution Kant rejected because of lack of empirical evidence might yet occur if we considered that the transition took place over a very long span of time. During the period when Schelling was writing, scholars had already stretched the earth's history back farther than any biblical chronology would indicate. Already in the mid-eighteenth century Georges-Louis Leclerc, Comte de Buffon (1707–88) had estimated the world to be at least ninety thousand years old – not, of course, within the range of our contemporary estimates, but far beyond the age calculated for Adam and his brood.

During the winter term 1798–9, Schelling and Goethe met often to discuss the subject that the young philosopher was lecturing on at the time, namely *Naturphilosophie* (*Tagebücher* 12–13, 16 November 1798; III 2.2:222–3). And from mid October 1799 to mid-November, the two companions met almost every day to discuss Schelling's *Erster Entwurf eines Systems der Naturphilosophie*; they spent an intense week discussing his *Einleitung* (*Introduction*) to the *Entwurf*.[15] These works, especially the *Einleitung*, show the impact on Schelling of Goethe's insistence that knowledge claims should be empirically grounded. In these tracts, Schelling claimed that all knowledge came through experience and that empirically acquired laws could be cast into an autonomous system, one that yet reflected an ideal set of deductive considerations. This was a first big step away from his mentor Fichte, and a move toward his objective idealism.

In the *Erster Entwurf*, Schelling did seem, however, to take back what he declared as a possibility in the *Weltseele* concerning a temporal transformation of species. In the *Erster Entwurf*, he asserted:

> Several naturalists seem to have harboured the hope of being able to represent the source of all organization as a successive and gradual development of one and the same original organization. This hope, in our view has vanished. The belief that the different organizations are really formed through a gradual development out of one another is a misunderstanding of an idea that really lies in reason. (2:62–3)

Von Englehardt latched on to this passage in his dismissal of the suggestion that Schelling held anything like a Darwinian thesis. Schelling did reject the Darwinian thesis, but it was that of Erasmus Darwin that he rejected.

What brought a shift in Schelling's attitude was the reading of Erasmus Darwin's *Zoonomia*, which, as I mentioned above, was translated into German almost immediately.[16] Darwin's genealogical theory supposed that all organic features of living creatures had been mechanically derived, during the deep past, from a simple structure bereft of any tincture of more advanced organization. That original living filament had been endowed by God to be sensitive to the external environment and to respond in Lamarckian-like ways. In his lectures at Jena, Schelling frequently derided the kind of flatfooted English empiricism found in John Locke and Erasmus Darwin.[17]

It was, I believe, Darwin's concept of the mechanistic evolution of organisms, in a genealogical fashion, that Schelling rejected, not the fundamental idea of species change in the empirical world. In place of Darwin's conception of the foundations of species evolution, Schelling instead proposed a principle of *dynamische Evolution*, which, as he explained it, postulated a rational archetype that served as the ideal standard for empirical instantiations. This archetype

> would be the absolute, the sexless condition that is neither the individual nor the species, but both together, in which the individual and the species are conjoined. This absolute organization cannot be represented through a particular product, but only through an infinity of particular products, which particulars deviate from the ideal in infinite ways, but in the aggregate are congruent with the ideal. (Schelling, *Erster Entwurf* 2:63–4)

Like Goethe's conception of the archetype, Schelling's was that of a plenum standard, one that included all its differentia. But it was also an ideal that would, nonetheless, be realized in time through the temporal development of a huge variety of types responding to natural forces. Schelling made this clear in a letter to Goethe in January1801, after having spent the Christmas period with his mentor. He wrote:

> The metamorphosis of plants, according to your theory, has proved indispensable to me as the fundamental scheme for the origin of all organic beings. By your work, I have been brought very near to the inner identity of all organized beings among themselves and with the earth, which is their common source. That earth can become plants and animals was indeed already in it through the establishment of the dynamic basic organization, and so the organic never indeed arises, since it was already there. [This was his answer to Kant's objection that organic life cannot arise out of inorganic earth.] In the future we will be able to show the first origin of the more highly organized plants and animals out of the mere dynamically organized earth, just as you were able to show how the more highly organized blooms and sexual parts of plants could come from the initially more lowly organized seed leaves through transformation. (*Briefe* 1:243)

For Goethe, of course, the plant does go through a temporal transformation, from seed-leaves through stem and mature leaves, to flower, and finally the sexual organs – that is, the archetype is gradually realized in the temporal sphere. And it is this notion of *dynamische Evolution* that Schelling adopted after he had abandoned what he came to identify as Erasmus Darwin's mechanistic version of transmutation.

Carus – a man after Goethe's own heart – yet attempted to render visible what his master contended could be perceived only by the mind's eye. Carus was an artist as well as an anatomist, and the need to illustrate the Goethean ideal required a metaphysical shift. He made the archetype a minimalist structure, essentially a string of vertebrae. Indeed, in his abstractive and mathematizing way, Carus reduced even the string of vertebrae to a single vertebra, and this he attempted to understand as comparable to a mathematic construction out of solid spheres (see Fig. 7.1). And this is how the archetype came to British shores, brought over in the work of Richard Owen.[18] It was the Carus-Owen rendering of the archetype that Charles Darwin historicized.

Goethe's Evolutionary Theory

While Goethe helped shift Schelling toward what became his objective idealism, Schelling moved Goethe to a more idealistic position and, I believe, fostered his incipient evolutionary ideas. Evidence of this comes in March 1813 from Johannes Daniel Falk (1768–1826), a satirical writer and casual friend of Goethe. Falk records a conversation he had with the great man, as they began to talk about Schelling, who had left Jena. Goethe indicated a fundamental agreement with his protégé that "it is as clear as day that the whole realm of appearance is an idea and a thought" (*Gespräche* 2:789). In the course of the conversation, Goethe echoed that earlier letter from Schelling that I have just quoted. Falk had abruptly asked Goethe "whether it did not seem likely to him that all the many different animals have arisen from one another through a metamorphosis similar to that by which the butterfly has arisen from the caterpillar." Though Goethe initially demurred, he did say:

> We are now awfully close to the chemistry of the whole thing, yet we all now choose terminology to disguise the transformations that occur in life ... I have given in my Metamorphosis of Plants the law whereby everything in nature is built up (this is through polarity, through generation). According to this law, things move into ever more splendidly and progressively higher syntheses. (*Gespräche* 2:789)

Goethe's remarks about being close to the chemistry of the whole thing both echoes Schelling's *Weltseele* and is consistent with his own earlier experiments on spontaneous generation. The notion that organic structures move to progressively higher syntheses seems just his way of talking about species development.

The clearest evidence of Goethe's commitment to evolutionary transformation comes, however, in his collection *Zur Morphologie* in the 1820s, when he commented on a new work by Christian Pander (1794–1865) and Eduard d'Alton (1772–1840), their *Die vergleichende Osteologie*. Starting in 1818, Pander and d'Alton visited natural history museums throughout Europe to do comparative studies of mammals and birds, including fossil representations. Their first trip took them to Madrid, where a giant prehistoric monster, about the size of a rhinoceros, was on display (Figure 9.4).

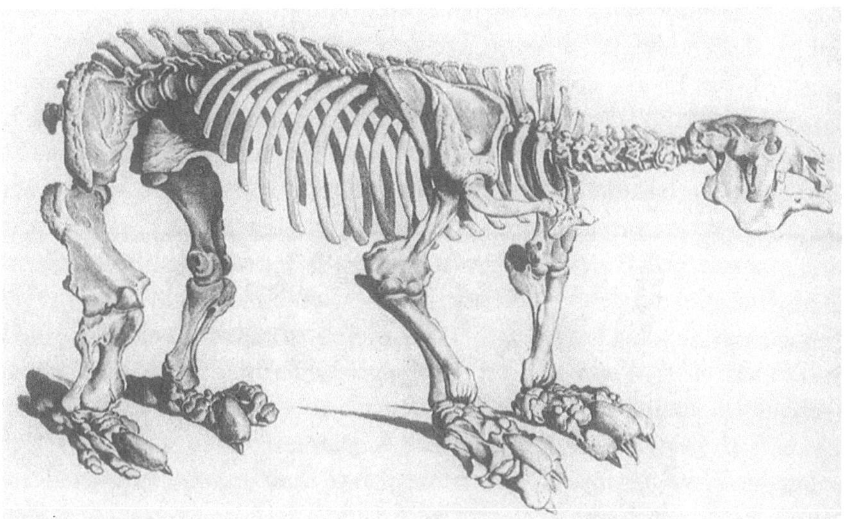

Figure 9.4. Illustration of a Megatherium, from Heinz Christian Pander and Eduard d'Alton's *Das Riesen-Faultier Bradypus Giganteus* (1821). From the author's collection.

It had been dug up in South America. Georges Cuvier (1769–1832) had provided a description of the beast. He called it a Megatherium, that is, "big animal," and remarked on its close affinity to the modern sloth (Figure 9.5). Charles Darwin would dig up another example of the Megatherium during his *Beagle* voyage to South America some fifteen years later.

In the introduction to their work on the Megatherium, Pander and D'Alton asserted that the resemblance between this ancient giant and the modern sloth was the result of a historical process of species transformation, much as Goethe had shown the transformation of the leaf into the various parts of a plant. The authors then generalized their argument:

> The differences in formation of fossil bones in comparison with those of still-living animals are greater the older the rock formations in which they are found (with the fossil remains of the most recent formations quite similar to the structures of living animals). This common observation supports the assumption of an unbroken train of descent [eine ununterbrockenen Folge der Abstammung] as well as of the progressive transformation of animals in relation to different external conditions. The observation that animals during the last millennium have reproduced with specific similarity in no way contradicts the theory of a general metamorphosis; rather such an observation only demonstrates that during this time no significant alteration in the external conditions of development has occurred. (6)

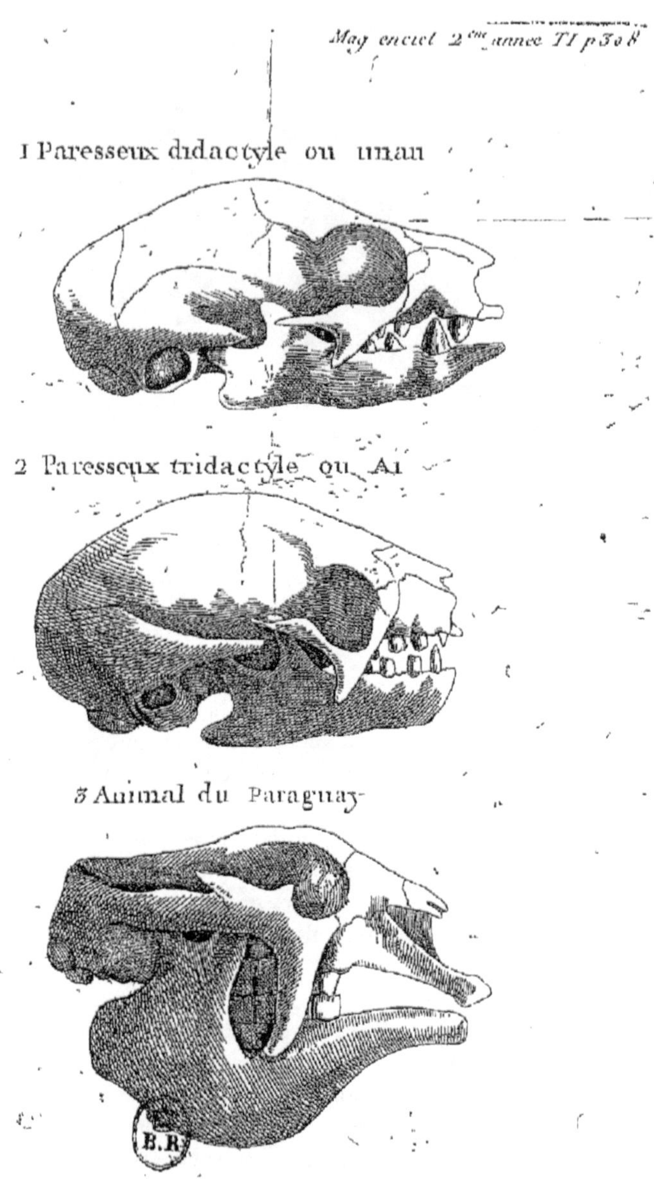

Figure 9.5 Georges Cuvier's comparison of skulls of two species of sloth (the unau and the ai) with that of the megatherium (the Paraguayan animal). From Cuvier's "Notice sur le squelette d'une très-grande espèce de Quadrupède" (1796). *Magazin encyclopédique*, vol. 1, 1796. From the author's collection.

The last sentence of the quotation obviously refers to Cuvier's objection to Lamarck that mummified animals recovered from ancient tombs in Egypt showed no significant deviation from modern forms. Goethe commented on Pander and d'Alton's work in the *Zur Morphologie*. He wrote: "We are in perfect agreement with the authors as concerns the introduction" – and it was, as I mentioned, in the introduction that they expressed their view of species transformation. Goethe continued his comment: "We share with the authors the conviction of a common type, as well as of the advantages of an empirical [sinnig] representation of a sequence of forms; we also believe in the eternal modifiability of all forms in appearance" ("Die Faultiere" 245).

Goethe not only endorsed the authors' evolutionary analyses, he even offered what he called a "poetic" sketch of how the descent of the Megatherium might have occurred. He supposed the giant sloth first existed as a kind of whale that got trapped along a swampy, sandy beach. To bear its great weight on land, it would have had to develop large limbs, which would then be passed to descendants. Subsequent generations would then further adapt to the land, achieving their modern, ungainly structure in the form of the sloth.

When Goethe offered this scenario, Lamarck's evolutionary conceptions were already at least fifteen years in the past. And several more recent German authors – for example, Gottfried Reinhold Treviranus (1776–1837), in 1805, and Friedrich Tiedemann (1781–1861), a student of Schelling, in 1808 – advanced a conception of the transformation of species based on evidence drawn from paleontology and embryology. So it is quite clear that Goethe knew perfectly well what he was endorsing in his comments on Pander and d'Alton.

Conclusion

From our perspective, many loose ends dangle from the transformational theories of Schelling and Goethe. They initially developed their ideas some time before more systematic presentations of evolutionary theory came before the public – those authored by Lamarck in 1800 and Darwin sixty years later. As a result, we cannot expect the kind of conceptual tidiness one finds, for example, in *On the Origin of Species*. And many of our questions of detail have to go unanswered. Yet there can be little doubt, I think, that Schelling and Goethe conceived of transformations in species that were not simply ideal, but that happened in time and through natural forces. Kuno Fischer and Ernst Haeckel were correct: Charles Darwin's theory had its predecessors in the evolutionary conceptions of Schelling and Goethe.

NOTES

1 Darwin mentions in his historical introduction to *Origin* that Geoffroy St Hilaire had recognized Goethe as a transmutationist. See *Origin* 61. Early in their correspondence (Ernst Haeckel to Charles Darwin, 10 August 1864), Haeckel suggested to Darwin that Goethe was one of his predecessors (Darwin, *Correspondence* 12:299). All translations in this essay are my own.

2 I have traced the transition in the usage of the term "evolution" from its provenance in embryology to that in species theory. See Richards, *The Meaning of Evolution* 5–16.

3 Blumenbach part 1, 25. See also my discussion of Blumenbach's notion of species development in *The Romantic Conception of Life* 222–5.

4 Kant reviewed the first two parts of Herder's *Ideen zur Philosophie der Geschichte der Menschheit* (1784–91) in the *Allgemeine Literatur-Zeiting* in 1785–6. See *Rezension* (A17–2, 309–10, 153–6).

5 Kant's construction of the idea of species transformation and his rejection for lack of evidence appear in his *Kritik der Urteilkraft*, in *Werke* 5:538–9 (A363–5, B368–70).

6 Erasmus Darwin, *Zoonomie, oder Gesetze des organischen Lebens,* trans. J.D. Brandis, 3 vols. in 5, Hannover: Gebrüder Hahn, 1795–9. See my discussion of the impact of Erasmus Darwin's work on Schelling and Goethe in *The Romantic Conception of Life* 300–1.

7 See Helmholtz.

8 See Goethe, "Dem Menschen."

9 Goethe's notes on the generation of infusorial animals are in *Sämtliche Werke* 2.2:563. He speculated that various seedlike organisms, if exposed to light, became plants and, if kept in the dark, became animalculae.

10 I discuss Goethe's Kantianism in *The Romantic Conception of Life* 427–30.

11 The essays are: "Versuch über die Gestalt der Tiere" (1790); "Versuch einer allgemeinen Knochenlehre" (1794); "Versuch einer allgemeinen Vergleichungslehre" (1794); "Erster Entwurf einer allgemeinen Einleitung in die vergleichende Anatomie, ausgehend von der Osteologie" (1795); "Vorträge über die drei ersten Kapitel des Entwurfs einer allgemeinen Einleitung in die vergleichende Anatomie" (1796).

12 The notes are kept at the London Museum of Natural History. See also Rupke 121.

13 *Italienische Reise* (6 September 1787). It is unclear whether this entry – in the form of a letter to von Stein and his friends in Weimar – was contemporaneous with the trip or added in 1820, when the book was composed.

14 While a student at the university in Tübingen, Schelling had as roommates Friedrich Hölderlin, whose poetic genius was already in flower, and Georg Friedrich Hegel, who would shortly champion his younger classmate.

15 Goethe read Schelling's *Einleitung* on 23 September and talked with him about it, and then from 2 to 5 October, they read through the work together. See Goethe's *Tagebücher*, in *Goethes Werke*, III 2.2: 261–3.

16 Darwin's *Zoonomia* was translated in multiple parts. While finishing the *Weltseele*, Schelling read part 1, which did not contain Darwin's evolutionary ideas. He read part 2, which does, shortly after finishing the *Weltseele*. See Schelling, *Weltseele* 1:603.

17 Henry Crabb Robinson, an Englishman who studied at Jena, wrote in a letter home that he was amused by Schelling's "contemptuous treatment of our English writers, as last Wednesday I was by his abuse of Darwin and Locke" (1:128).

18 See Rupke's account of the vertebrate archetype (90–140).

WORKS CITED

Aristotle. *De generatione animalium*. Ed. and trans. A.L. Peck. Cambridge, Mass.: Loeb Classical Library, 1953. Print.

Blumenbach, Johann Friedrich. *Beyträge zur Naturgeschichte*. Göttingen: Johann Christian Dieterich, 1790. Print.

Bonnet, Charles. *La Palingénésie philosophique, ou Idées sur l'état passé et sur l'état future des étres vivans*. 2 vols. Geneva: Philibert et Chirol, 1769. Print.

Carus, Carl Gustav. *Von den Ur-theilen des Knochen- und Schalengerüstes*. Leipzig: Fleischer, 1828. Print.

Darwin, Charles. *The Correspondence of Charles Darwin*. Ed. Frederick Burkhardt et al. 18 vols. Cambridge: Cambridge UP, 1985– . Print.

Darwin, Charles. The Origin of Species *by Charles Darwin, a Variorum Text*. Ed. Morse Peckham. Philadelphia: U of Pennsylvania P, 1959. Print.

Darwin, Erasmus. *Zoonomie, oder Gesetze des organischen Lebens*. Trans. J.D. Brandis. 3 vols. in 5. Hannover: Gebrüder Hahn, 1795–9. Print.

Düntzer, Heinrich, ed. *Zur deutschen Literatur und Geschichte: Undedrückte Briefe aus Knebels Nachlass*. 2 vols. Nürnbert: Bauer und Naspe, 1858. Print.

Engelhardt, Dietrich von. "Schellings philosophische Grundlegung der Medizin." *Natur und geschichtlicher Prozess: Studien zur Naturphilosophie F.W.J. Schellings*. Ed. Hans Jörg Sandkühler. Frankfurt a.M.: Suhrkamp Verlag, 1984. 305–25. Print.

Fischer, Kuno. *Friedrich Wilhelm Joseph Schelling*. 2 vols. Vol. 6 of *Geschichte der neuern Philosophie*. Heidelberg: Carl Winter's Universitätsbuchhandlung, 1872. Print.

Goethe, Johann Wolfgang von. "Die Faultiere und die Dickhäutigen." *Zur Morphologie*. Vol. 12. *Sämtliche Werke*. Print.

Goethe, Johann Wolfgang von. "Der Inhalt Bevorwortet." Vol. 12. *Sämtliche Werke*. Print.

Goethe, Johann Wolfgang von. "Dem Menschen wie den Tieren ist ein Zwischenknochen der obern Kinnlade zuzuschreiben." Vol. 2.2. *Sämtliche Werke*. 530–45. Print.

Goethe, Johann Wolfgang von. *Goethes Briefe* (Hamburger Ausgabe). 4th ed. 4 vols. Ed. Karl Mandelkow. Munich: C.H. Beck, 1988. Print.

Goethe, Johann Wolfgang von. *Goethes Gespräche*. 5 vols. Ed. Flodoard von Biedermann and Wofgang Herwig. Munich: Deutscher Taschenbuch Verlag, 1998. Print.

Goethe, Johann Wolfgang von. *Italienische Reise*. Vol. 15. *Sämtliche Werke*. Print.

Goethe, Johann Wolfgang von. *Maximen und Reflexionen*. Vol. 17. *Sämtliche Werke*. Print.

Goethe, Johann Wolfgang von. "Morphologie." Vol. 4.2. *Sämtliche Werke nach Epochen seines Schaffens*. 21 vols. Ed. Karl Richter et al. Munich: Carl Hanser Verlag, 1985–98. 188. Print.

Goethe, Johann Wolfgang von. *Tag- und Jahres-Hefte*. Vol. 14. *Sämtliche Werke*. Print.

Goethe, Johann Wolfgang von. *Tagebücher*. Vol. 3.2. *Goethes Werke* (Weimar Ausgabe). 133 vols. Weimar: Hermann Böhlau, 1888. Print.

Goethe, Johann Wolfgang von. "Versuch einer allgemeinen Vergleichungslehre." Vol. 4.2 *Sämtliche Werke*. Print.

Helmholtz, Hermann von. "Ueber Goethes wissenschaftliche Arbeiten. Ein vortrag gehalten in der deutschen Gesellschaft in Königsberg, 1853." *Populäre wissenschaftliche Vorträge*. Vol. 1 Braunschieg: Friedrich Vieweg und Sohn, 1865. 31–55. Print.

Kant, Immanuel. *Rezension zu Johann Gottfried Herder: Ideen zur Philosophie der Geschichte der Menschheit*. Vol. 6. *Immanuel Kant Werke*. 6 Vols. Ed. Wilhelm Weischedel. Wiesbaden: Insel Verlag, 1957. 781–806. Print.

Kant, Immanuel. *Kritik der Urteilkraft*. Vol. 5. *Immanuel Kant Werke*. Print.

Owen, Richard. *On the Nature of Limbs*. London: John Van Voorst, 1849. Print.

Pander, Christian, and Eduard d'Alton. *Das Riesen-Faultier Bradypus Giganteus*. Bonn: Weber, 1821. Print.

Richards, Robert J. *The Meaning of Evolution: The Morphological Construction and Ideological Reconstruction of Darwin's Theory*. Chicago: U of Chicago P, 1992. Print.

Richards, Robert J. *The Romantic Conception of Life: Science and Philosophy in the Age of Goethe*. Chicago: U of Chicago P, 2002. Print.

Robinson, Henry Crabb. *Diary, Reminiscences, and Correspondence*. 3 vols. Ed. Thomas Sadler. London: Macmillan, 1869. Print.

Rupke, Nicolaas. *Richard Owen: Biology without Darwin*. Rev. ed. Chicago: U of Chicago P, 2009. Print.

Schelling, Friedrich Joseph. *Briefe und Dokumente*. Ed. Horst Fuhrmans. 3 vols. Bond: Bouvier Verlag, 1962– . Print.

Schelling, Friedrich Joseph. *Erster Entwerf*. Vol. 2. *Schellings Werke*. Print.

Schelling, Friedrich Joseph. *Von der Weltseele*. Vol 1. *Schellings Werke*. 3rd ed. 12 vols. Ed. Manfred Schröter. Munich: C.H. Beck, 1927–59. Print.

Wells, George. *Goethe and the Development of Science, 1750–1900*. Alphen aan den Rijn: Sijthoff & Noordhoff, 1978. Print.

Wenzel, Manfred. "Naturwissenschaften." *Goethe Handbuch*. 4 vols. Ed. Bernd Witte et al. Stuttgart: Metzler, 1996–8. 2:781–96. Print.

Chapter Ten

The Vitality of Idealism: Life and Evolution in Schelling's and Hegel's Systems

TILOTTAMA RAJAN

Encyclopedics and the Life Sciences

Shelley's last poem, *The Triumph of Life,* has at its core the question, "Then what is life?," posed by the character Shelley to Rousseau, who has returned from the grave in the vegetable form of what seems to be an "old root" (182, 544). Although my focus will not be Shelley, his existential question also has a scientific resonance. The issue of what constitutes life and organization; what the difference is between plants, animals and man; whether divergent life forms can be arranged in an ascent of man; what role disease, degeneration, and aberration play in this model of ascent that we find in philosophers of nature from Jean-Baptiste Robinet to the early Schelling; as well as the question of where life begins between the organic and inorganic: all these questions traverse and trouble the discourses of the life and earth sciences, biology and botany, physiology, pathology, surgery and medicine, mineralogy and geology.

These sciences, in turn, have metaphoric and analogical effects on a range of areas including history, philosophy, and aesthetics. In this paper, I explore the broader ramifications of the organicist projection in German Idealist and Romantic philosophy. For in this period, there is a profound shift toward understanding forms of thought and culture as themselves in a process of evolution. History becomes the model for thinking other disciplines, such as aesthetics or philosophy, that are no longer approached transcendentally, as in Kant or the early Schelling. Yet history as we conceive it – as development rather than a description not necessarily tied to a sense of time (Rudwick 53) – is itself generated by the philosophy of nature as the place where the "temporalizing of the Chain of Being" (Lovejoy 242) both discloses nature as having a history and submits history to being read in the light of nature. More precisely, history in the thought-formation on which I focus is generated through the evolution of natural history, by a disciplinary species change, into what the Romantics called physiogony or the history of nature. Physiogony, as the British Coleridgean Joseph Henry Green (1791–1863) develops the concept from Kant and Schelling, studies nature as "preface and portion of the history of man" (*Vital* 103). I will return later to this unstably transferential interdiscipline of physiogony. Suffice it to say that the "history of nature," as preface, allegory or mirror-stage of spirit, raises many questions. These include questions both about the figurality of paradigms of development that are naturalized by being projected onto so-called nature, and about the possibilities and dangers that thinking philosophy according to history and nature opens up.

In speaking of the organicist projection, I do not have in mind a series of homologies between the life sciences and other disciplines that would reduce these fields to "the lowest common denominator" of a single type.[1] Rather, I have in mind the use of one area to supplement another with which it cannot be identified, the use of human history to understand an opaque and resistant nature that Hegel describes as "the Idea in the form of otherness," the "negative of itself"; or the contrary recognition that what Schelling calls the "ideal sciences" of philosophy and history must take account of their "real" counterparts as an "alien existence in which Spirit does not find itself" (Hegel, *Philosophy* 3, 13; Schelling, *On University* 75, 115).[2] Kant eschewed as fallacious these disciplinary border-crossings that characterize post-Kantian Idealism (*Metaphysical* 9). But exactly because they are figural, they open a space for questions like those raised by Schelling in his essay *Philosophical Investigations Into the Essence of Human Freedom* (1809): "Does creation have a final purpose at all, and, if this is so, why is it not reached immediately, why does what is perfect not exist right from the beginning?" (66). To think these issues, as Schelling does, through the history of nature, not as a confirmation but as a mirror stage and primal scene of human history, is to recognize the answers given as hypotheses. This is to say, also, that we think something analogically through another discipline precisely to keep it in the realm of speculation, as well as to experiment with possibilities not yet permitted in the original discipline.

Thus, my aim in exploring the impact of the life sciences on philosophy will not be to pick a scientifically correct model such as Darwinian evolution and ask whether thinkers from Robinet to Schelling anticipated it. Such demonstrations do remind us that in this period before the disaggregation of disciplines, philosophers like Hegel, Schelling, and Schopenhauer were well-read in science, and even trained in it: Schelling studied medicine and received awards in the field, Schopenhauer took medicine at Göttingen, and his later book *On The Will in Nature* (1836) is an attempt to give his philosophy a grounding in the contemporary life sciences. But Idealism does not have a single evolutionary model, and the fact that it takes up different models is precisely what allows it to speculate analogically in other areas also in a state of ferment.[3] This is to say that Idealism is just as worth studying when its science is "wrong" as when it is "right." In using the word "evolution" here, I therefore use it in a broad sense. As we see in *The First Outline of a System of the Philosophy of Nature* (1799), Schelling was well aware of debates around preformation, epigenesis, metamorphosis, and recapitulation (38n). But when he writes that "[a]ll evolution presupposes involution" (*Ages* 83), the word does not signify a proto-Darwinian evolution – nor does he mean preformation, which, as Robert Richards points out, was what evolution implied at the time (*Meaning* 1–15).[4] Rather, Schelling means development, but with a question about whether creative evolution and the anthropology it subtends are even possible. My aim, then, in discussing *Naturphilosophie* is to treat this discipline, in Jason Wirth's words, as "a gateway into the original experience of philosophizing." *Naturphilosophie*, as the transference and counter-transference of philosophy and the life sciences into each other, is not, or not just, a "kind of philosophy or a topic within philosophy" (11). Nor is it a kind of naive science that Hegel and Schopenhauer criticized while doing it themselves. Rather, it is a thought experiment, a "way of doing philosophy in accordance with nature," as Wirth says (11), in the process of which the relations between what Foucault calls the empirical and transcendental (319–23) are completely recast. For the empirical, rather than being determined *a priori* by the transcendental, writes back to it, also radically reconfiguring philosophy itself at a transcendental level.

Idealism is a fertile site for exploring the interdisciplinary consequences of the philosophy of nature, because it is a systematic, if perpetually self-questioning, program for thinking disciplines within a larger whole, a project summed up in Hegel's idea of an encyclopaedia of the philosophical sciences, or Schelling's project of introducing "Idealism ... into all the sciences" (*Ideas* 272n). But what does such an encyclopaedia entail, since the very concept of knowledge itself also becomes inflected for Hegel by the question of evolution? For Kant, who gave a course on "The Encyclopaedia of Philosophy as a Whole," the systematic organization of knowledge takes the form of an architectonic, in which "every science [has] its determinate position in the encyclopaedia of the sciences" (*Judgment* 285), and in which this position and the "boundaries" of each individual science are "determined *a priori*" by philosophy (*Pure Reason* 691). Kant thus conceives disciplines as what Bruno Latour in *The Politics of Nature* calls "smooth" rather than tangled objects (22), under the governance of their "*pure*" part (Kant, *Metaphysical* 5). Smooth objects are ones that we grasp separately, whereas tangled objects are reciprocally affected by other objects, and produce risks and possibilities for knowledge (Latour 22–3). To Kant's goal of a smooth system we can oppose Novalis's unfinished *Notes for a Romantic Encyclopaedia*, which gives the name "encyclopaedics" to an interdisciplinary thinking that is not architectonic but in which the parts have effects on the whole. Thus the "encyclopaedization" of a discipline occurs when the parts are not just fitted into a whole but also react upon this whole, both between and within disciplines (76). The result is a tangled system in which the relations of parts and whole are reciprocal and lateral, not hierarchical. Encyclopedics thus transforms individual disciplines from restricted, professionalized "fields," in Bourdieu's sense (97–102), into what Bataille calls general economies capable of reorienting the whole (19–22). The "versability" of disciplines, their contamination by or "translation" into each other, as Antoine Berman calls it (14), exposes disciplines to their "unthought" (Foucault 322–4), through the recourse they often have to other disciplines from which they borrow to understand their objects.

Interestingly, Latour frames his distinction between smooth and tangled objects within the issue of ecology, which for him involves a "crisis" of "objectivity" rather than "nature" *per se* (18): a shift in how we understand material and intellectual objects, including, for my purposes, disciplines. The term "ecology" (though also aimed at ecocriticism's simplifying view of "nature") refers to considering objects in their environment as entangled with other objects. But this epistemological shift can itself be traced back to knowledge-systems of the long Romantic period, which saw a transition from the smooth disciplines of physics and mathematics to the tangled fields of the life and earth sciences as paradigms for how we know. Kant's concept of disciplines as smooth entities precedes the entanglement of "real" and "ideal" sciences that occurs in post-Kantian Idealism (Schelling, *On University* 103-4): the reformulation of ideal sciences like philosophy, history, and aesthetics, by real disciplines such as the life sciences. For Kant, a science may initially be based on "*foreign* principles (*peregrina*)" which it borrows "from another science" (*Judgment* 252). But as David Ferris argues, these borrowings must be domesticated to evolve a science no longer troubled by such foreign matter but committed to "the reproduction," "of itself as a discipline" (1251–2). This internal rationalization of a discipline is what Kant understands by the term "architectonic," by which he means "the art of systems," and which he thinks in terms of the body as a whole, into which all the parts are integrated: a structural and aesthetic, rather than genuinely organic, use of the "body." The architectonic of "all human knowledge," which Kant refers to but does not fully construct, must similarly be a system in which all parts are integrated (*Pure Reason* 691–2). The result would be what Derrida calls a whole "architecture of philosophy," in which "[concepts in] aesthetics, language, logic, history, metaphysics etc. are invisibly interwoven (*Points* 212), but in order to cover over the pattern of closed metaphoric transfers that allows for "manifold cognitions" to be repressed "under one idea" (Kant, *Pure Reason* 691).

The early Schelling theorizes a similar, if more mystical, architectonic in his lectures *On University Studies* (1803), which he describes as an "outline [*Grundriss*]" that "might take the place of a general encyclopaedia of the sciences" (41; 8:481). But he also begins to jeopardize this outline, both by his actual attempts to work it out in areas like speculative physics and medicine, and by the cross-disciplinary mirrors he uses to imagine the interconnections of knowledge. "Mirror" is Schelling's own figure, when he speaks of art, in the *Philosophy of Art* (1803/4), as the "magic and symbolic mirror of philosophy" (8); and Schelling experimented with more than one such mirror, eventually dismissing mathematics on the grounds that Kant's fondness for it produces a paradigm for knowledge that favours a "crystal" over the human body because it never falls ill ("On the Nature" 212–13). The mirror of art in Schelling's early *System of Transcendental Idealism* (1800) and *Philosophy of Art* is not so much an awareness of the role of figure in argument as it is aesthetics, the discipline that Baumgarten defines as the art of thinking beautifully and by analogy.[5] It is in this sense that Odo Marquard can say that *The System* "takes an aesthetic perspective on existence: it determines philosophy primarily as aesthetics" (13). Aesthetics as mirror or supplement to philosophy can transform history, for instance, from a real into an ideal science by "present[ing] real events and histories ... in complete form ... so that they express the highest Ideas" (*On University* 107).

With such idealization in mind, in lectures on *University Studies* Schelling uses physiology as a disciplinary mirror, but filters it through aesthetics (141–2). Thus he conceives the "totality" of knowledge, not as a body without organs, in Deleuze's phrase (44–7), but according to the aesthetically organized figure of an "organic body (*organisches Leib*)" whose life flows from the "central organs" of mathematics and philosophy to "the outermost parts," which include physics and chemistry (*On University* 27, 42; 8:468, 482).[6] A further epistemological analogue for this unification is provided by comparative anatomy: the study of nature in terms of "the unity and inner affinity of all organisms" that "originate in one archetype whose objective aspect" changes but "whose subjective aspect is unchangeable" (142). Disciplines, similarly, are multiple types of a single archetype; or, following a Spinozist model, [they are] modes of one substance, namely "primordial knowing (*Urwissen*)" (42; 8:482). These interlocking models – aesthetics, comparative anatomy, physiology, Spinozist metaphysics – then allow Schelling to organize the rest of his theoretical apparatus: the distinction between the ideal and real sciences as types of the same, whose explosive difference is contained by the parallelism of the eternal and temporal planes (*Ideas* 272); the mapping off of historical and empirical from philosophical or "principled" knowledge (*On University* 82); and the resulting subordination of the empirical to the transcendental in absolute knowledge.

Yet the curious thing is that both Schelling's scientific analogues are potentially problematic. As a figure for the assimilation of parts into a whole, physiology is a troublesome part of Hegel's *Encyclopaedia* (1817), allowing for a part of the body, and by extension the body of knowledge, to function at variance with the whole. Schelling, too, takes up this deviance of the part in the *Freedom* essay, in which he writes that an "individual body part, like the eye, is only possible within the whole of an organism," but also has its own "life" and "freedom" which it "proves through the disease of which it is capable" (18). Yet Schelling does not go in this direction in the lectures *On University Studies*, where he sees comparative anatomy and physiology as "correlative disciplines" that he projects onto each other (141). He thus contains physiological fluidity and its invisible entanglement of systems[7] within the visible smoothness of anatomical structure, making the "external" body of actual knowledge an emanation from the "internal organism of primordial knowledge" (76), and thus reducing the many to the one. Comparative anatomy, with its tropes of consilience, metamorphosis, and recapitulation, is also a highly tangled discipline, as Adrian Desmond's account of the different ways it was politically appropriated tells us. But here, as in Frye's *Anatomy of Criticism,* it is stripped of the more textured debates around transmutationism, evolution, even degeneration, that complicate the static model of anatomy. In short, the Schelling of these early lectures also constructs a smooth system of knowledge whose parts do not conceptually interfere with each other.

Hegel, too, attempts to construct a smooth system that forwards the phenomenology of Spirit: what Derrida criticizes as an "auto-encyclopaedia of absolute Spirit" ("Age" 148). According to Rosenkranz, during the Jena years when many of the elements of Hegel's Encylopaedia were first introduced, Schelling tried to work out the critical foundations of absolute philosophy, while Hegel set to work developing it as a "cycle of sciences" (qtd. in Vater 82). Hegel's system is diachronic rather than synchronic. Ideally, each discipline, though a "sphere" in its own right, is also supposed to be a level or moment in an ascending series. Thus, in *The Philosophy of Nature*, mechanics, physics, and "organics," or the life sciences, are levels in a scale of disciplines that parallels the Chain of Being; together they form the "sphere" of the natural sciences, which is a level supposedly surpassed by the sciences of Spirit. The encyclopaedia or cycle of learning in which consciousness learns how to become Spirit thus becomes an ascent from matter to Spirit through the progression from the real to the ideal sciences. This ladder of disciplines is not unique, though the rungs may be placed differently. Coleridge, who never completed his projected *Encyclopaedia Metropolitana*, may have seen his posthumously published *Theory of Life* (1816) as part of a philosophy of nature that ascends from geology, physics, and mechanics, through zoology and physiology. Brought together, these sciences comprise "the one absolute science of Life," and inaugurate "a new series beyond ... physiology," the ideal series in Schelling's terms, namely philosophy and theology (*Theory* 516, 519n; *Notebooks* 4:4517).

Coleridge also tells us that the *Opus Maximum* was intended as a "compleat ... system of Logic, Natural [Philosophy] and Theology" (*Letters* 4:736), like Hegel's *Encyclopaedia*. In *Mental Dynamics* (1847), Green echoes the title of his earlier *Vital Dynamics* (1840) to effect a transition from nature to Spirit, as a specifically Coleridgean transition from physiological to political "constitution." Using the Hunterian lectures, which were supposed to deal with the always-troublesome life sciences, as a platform to institute the training of a clerisy, Green sketches a disciplinary series proceeding from grammar to "natural history" to what he calls physiogony, through civil history to mathematics and logic and, finally, philosophy (7–19, 41).

But Hegel's system differs from Green's in two ways. First, it is highly tangled, as the levels double as spheres in their own right, and as the ascending structure is complicated by its descent into proliferating microsystems that have to be recontained in an increasingly ramified macrosystem. Thus, "organics" is a "level" in *The Philosophy of Nature*. But as a sphere in its own right, it is further divided into the sciences of the "terrestrial," plant, and animal organisms. Hegel studies the animal (including the human) in terms of physiology rather than anatomy, and physiology in turn contains the sphere of pathology, which cannot clearly be assimilated as a level of normal physiology. The very organization of the *Encyclopaedia* in stages is the symptom of the tremendous labour of the negative that Hegel experiences as he struggles to get difficult material under control. Culminating in illness and death, *The Philosophy of Nature*'s last section on pathology risks derailing the planned transition from nature to spirit. The result is that, though the disciplines are arranged in a progressing series, the parts are dynamically interconnected in the way described by Novalis, forcing us to rethink ideal sciences such as philosophy through such subsystems as medicine and aesthetics (Rajan, "(In)digestible Material").

Second, and most important for our purposes, despite figures of circles that attempt an architectonic containment,[8] Hegel's system is profoundly temporal. Adapting Lovejoy on the temporalizing of the Chain of Being that introduces evolution into nature, we could describe what Hegel does as a temporalizing of the cycle of disciplines. The result is that history and evolution become underlying paradigms for thinking disciplines and their ideas, in terms of their historicity, as still under development. To be sure, Hegel does not favour the word "evolution," either out of a certain conservatism, as J.N. Findlay says, or because he understood the word in an "ideal, not in a real, sense" (xv). Interestingly, he already seems to have associated "evolution," which at the time meant preformation, with some kind of species change. Criticizing the concept of evolution in the sciences, Hegel wants to restrict it to being a metaphor when he writes that it "has been a clumsy idea in the older as well as newer philosophy of nature, to regard the transformation and the transition from one natural form to a higher as an outward and actual production" (qtd. in Höffding 150). But as Rüdiger Bubner says in commenting on Schelling's "On The Nature of Philosophy as Science," an "evolutionary history" was the "model for ... Hegel's *Phenomenology of Spirit*" (163). What Schelling means by "evolution," which in this essay he repeatedly uses in the context of knowledge, is quite complex: "a historical explanation" to begin with, a "living system" that is "not just a sequence of laws," a process in which at different points opposite things can be true, and finally a certain "asystasy" or "inner conflict" as the very core of life (215–16).

Physiogony

In this section, I want to take up one zone of entanglement: the unstably transferential inter-discipline of physiogony. Physiogony, or the history of nature, is a supremely metaphysical conceit: a projection of anthropomorphic models onto nature so as to find in nature a history that confirms certain postulates about human history. This transference, the kind of fallacy Kant criticizes, is invisibly present throughout nineteenth-century models of history. But its institution as a discipline that actually puts itself at risk by disclosing its figural composition is quintessentially Idealist, and is at the very core of Hegel's *Philosophy of Nature* and its intended place in his larger organization of knowledge. For if, as Green says, the history of nature as "preface and portion of the history of man" makes the "knowledge of Nature" a "branch of self-knowledge" (*Vital* 103), the story Nature tells may not be the one we expect. The transference of Spirit onto Nature may well result in the real disciplines of the natural sciences unsettling the ideal disciplines of philosophy, history, and aesthetics.

The curiously Hegelian interpellation of nature into the history of self-consciousness is from Joseph Henry Green's 1828 Hunterian lecture at the Royal College of Surgeons. For it is Green, who had studied in Germany and was well-versed in *Naturphilosophie*, who was a friend and follower of Coleridge, mentor of the leading Victorian biologist Richard Owen (1804–92), Hunterian professor, and then Professor of Surgery at King's, who provides the most succinct account of physiogony. In both his 1827 Hunterian lecture and his 1828 lecture printed in *Vital Dynamics,* Green distinguishes three branches of the study of nature, which are really approaches rather than empirical fields. The first is physiography, or the description of nature's products, what we generally call natural history, rather paradoxically, since it involves an atemporal taxonomy. The second is physiology, or the theory of nature, which is to say the powers behind nature conceived vitally rather than mechanistically: *natura naturans* rather than *natura naturata.* And the last is physiogony, which Green in 1827 imagines as "an agent acting under the analogy of a will and in pursuit of a purpose," but which he describes the following year as part of "self-knowledge" ("Introduction" 307–12; *Vital* 101–3). Green gave the Hunterian lectures from 1824 to 1828 and in 1847, and his task was to provide a narrative to explain John Hunter's cabinet of curiosities, which consisted of various fossil and skeletal remnants: a narrative that is the basis for modern museums of natural history. For Green, as for Coleridge in his *Theory of Life*, which he worked on with Green, physiogony thus becomes anthropology, as the history of nature is subsumed into a temporalized Chain of Being in which nature works her way up from "the *polypi* to the *mammalia*," "labour[ing] in birth with man" (*Vital* 101–6). Yet as we shall see, Green's knowledge of Schelling notwithstanding, Schelling himself is a far more deconstructively speculative thinker.

Green's terms go back semantically to Kant, who, in "The Use of Teleological Principles in Philosophy" (1788), introduces two of the terms: physiography for *Naturbeschreibung*, or the "description of nature"; and Physiogony for *Naturgeschichte*, or what Green, via Schelling, calls the history of nature. Kant's definitions exhibit his usual caution and reserve. *Naturbeschreibung*, more than the account of a mere "empirical traveler," is the "systematic" description of nature exemplified by Linnaeus. It is really what we now know as natural history, the term Kant's translators use for *Naturgeschichte* (197). For natural history, as Foucault points out, was not historical in our sense; instead, it placed the "proliferation of beings occupying the surface of the globe ... [in] the field of a mathesis that would also be a general science of order," making the world codifiable "within a taxonomic area of visibility" (133–7). Kant does not commit himself to the second term, *Naturgeschichte*, glossing the word "history" as both a "narrative" and a "description" of "events in nature [*Naturbegebenheit*]." As a narrative, *Naturgeschichte* "trac[es] back, as far as the analogy permits," the connections between "present-day conditions ... and their causes in earlier times"; it traces things back, but is in no way anticipatory. Moreover, it works only "according to laws of efficient causality." Since giving this narrative a temporal depth by making it one of origins would be a "science for gods," Kant does not really stray from "physics" into "metaphysics" ("On the Use" 195–7), though the very setting up of such boundaries flirts with the possibility of crossing them, especially given Kant's dissatisfaction with being limited to what is known from experience (*Pure Reason* 398), and his delimitation of physics as a science of the mere understanding rather than Reason.

Nevertheless, Kant does not want to go in the metaphysical direction of George Forster, an essay by whom occasioned Kant's response, and who already imagines an "earth in labour" that generates organisms in an "unnoticeable gradation" that can be traced back from man "down the chain of nature" (qtd. by Kant, "On the Use" 214). Indeed, in the slightly earlier *Metaphysical Foundations of Natural Science* (1786), Kant had avoided any sense of deep or future history by defining

Naturgeschichte in the older sense of history, as "a systematic presentation of natural things at various times and places" (4). Yet in the *Opus Postumum* (1804), a kind of universal brouillon, in Novalis's phrase, and thus a work in process, Kant seems to have read Schelling (251, 254), and Schelling's notions of a world-soul and an absolute organism are threaded throughout the text (66–7, 71). Indeed, the *Opus Postumum* resembles nothing more than Schelling's *First Outline*. Its promise of a few sheets making the transition from metaphysics to physics that will complete Kant's system (qtd. in Förster xvi) unravels into a Coleridgean mess of thirteen hundred pages. And similarly, Kant's self-disciplining of his work within the philosophy of science, as it will subsequently be known, also collapses. Physics and metaphysics mutually unground each other, and the analytic and positivistic privileging of physics over metaphysics becomes a finding of "metaphysical foundations of natural science" in "physical foundations" that threaten to break metaphysics open (*Opus* 39).

It is this wavering between physics and metaphysics, understanding and Reason, that allows post-Kantians such as Coleridge to say that Kant could not have meant to be as cautious as he was (*Biographia* 154–5). Here, as in his use of the proto-Romantic term "ideas" or "Idea," Kant, as Philip Sloan disparagingly puts it, "provided the opening ... through which Schelling and his disciples," such as Oken, "could rush with enthusiasm" (27). Commenting on the difference between Kant's actual and potential use of the term *Naturgeschichte*, Schelling notes that Kant's *Naturgeschichte* is not much different from a *Naturbeschreibung*. In this sense, we might add that English translations of Kant's *Naturgeschichte* as natural history are accidentally quite appropriate. In a kind of epigenesis of Kant's term *Naturgeschichte*, Schelling thus tries to give it a "much higher meaning": that of a "*history* of nature itself [*eine Geschichte der Naturselbst*]," in which Nature "gradually brings forth the whole multiplicity of its products through continuous deviations [*Abweichungen*] from a common ideal ... and so realizes the Ideal [*das Ideal*], not indeed in the individual, but in the whole" (*First Outline* 53; 68). Schelling's desynonymization of *Naturgeschichte* and *die Geschichte der Natur*[9] strikingly anticipates Foucault's distinction in *The Order of Things* between natural history and the history of nature. For Foucault, as already noted, natural history has nothing to do with temporality. Although it allows for development, it does so by "traversing [a] preordained table of possible variations" (275) in which time simply unfolds space. By contrast, the history of nature entails a new sense of historicity, especially once it becomes decoupled from Enlightenment discourses of perfectibility. Expanding on Foucault, we could say that the history of nature evolves into an analytic of finitude when geology and biology, rather than physics and even botany (a discourse of plenitude), become synecdoches for natural science. For the geologization of time introduces memory and even trauma into nature. For its part, biology distinguishes animals from plants, which natural history had laid out on the same taxonomical surface. Where the plant, as Foucault says, "held sway on the frontiers of movement and immobility ... the animal maintains its existence on the frontiers of life and death" (277). Natural history thus unfolds within the field of logic, whereas the history of nature, though initially the projection of a form of anthropology, exposes thought to ontology, to life and being, through its human nature.

Yet it would not be correct to credit Kant, except perhaps in the *Opus Postumum*, with an opening that leads to Schelling. For developmental models of nature's history as creative evolution leading to an increasing perfection of man himself precede Kant, and are the subject of his critiques not only of Forster, but also of Herder's *Ideas for a Philosophy of the History of Mankind*, which makes the history of nature a preface and portion of the history of man. These thinkers are constantly criticized by Kant for lack of rigour and for using imagination rather than Reason, thus keeping the study of nature at the level of "a systematic art rather than a science" (*Metaphysical Foundations* 4). Thus Kant, in his review of Herder, comments on Herder's assumption of an evolution from the lowest beings to man, complaining that he (Kant) "does not understand this inference from the analogy of nature, even if he were to concede that continuous gradation of [Nature's] creatures ... For they are *different* beings that occupy the many stages of the ever more perfect organizations" rather than "*the same individual.*" Herder's scheme, whether one describes it as palingenesis or epigenesis, is for Kant an instance of a metaphoric sleight of hand that substitutes "hints" for "determinate concepts," and "force of imagination given wings whether through metaphysics or through feelings" for "observed laws" (131–3). It would be more accurate, then, to say that Kant did not entirely close the door he tried to close on creative evolution, and could be read against the grain, insofar as Kant's authority made it necessary for his successors to find a way through him.

Indeed, the developmental models that Kant found at once seductive and ungrounded go back well before Herder to what Arthur Lovejoy calls the "temporalizing of the Chain of Being" that begins with the *encyclopédiste* Jean-Baptiste Robinet's *de la nature* (1761–8). Robinet introduces the idea of nature as evolving through time, accompanied by the very metaphor of labour that Green will use. Nature, he says, is "never stationary" and is "always at work, always in travail ... fashioning new developments, new generations" (Robinet, qtd. in Lovejoy 275). Insofar as Robinet, as Lovejoy argues (271–80), may have been the first to use the word "prototype," which becomes *Urbild* or *Ideal* in Schelling, Robinet, though a preformationist, theorizes what Schelling in the *First Outline* calls a "dynamic preformation" (37n), in which the "seeds [*semences*]" and the further "germs" they conceal do not all develop together (Robinet [1768] 27). In the "graduated" but still unpredictable and unfinished "sequence of beings" that ensues, nature works with "a single model," such that different beings are "variations" of a generative "prototype," "graduated *ad infinitum*" (Robinet [1761] 2, 12). Robinet even questions the division of the chain of being into four classes (mineral, vegetable, etc.) that persists in nineteenth-century comparative anatomy, where the preservation of these divisions or, alternatively, the construction of links across them analogically constitutes the social spectrum of what Adrian Desmond calls "the politics of evolution," ranging from conservative to Dissenting appropriations of evolution (1–8, 18–21). Thus, for Robinet, the "Scale of Beings" is a "whole infinitely graduated, with no real lines of separation; ... there are only individuals, and no kingdoms or classes or genera or species" (qtd. in Lovejoy 275). Robinet's proto-Romantic theorizing of creative evolution as a ladder of upward mobility is affectively conveyed by the way he follows a pattern of ascent from the simplest to the most complex creatures, rather than a genealogy or pattern of descent that proceeds "from the more complex groups to a termination in the lower orders."[10] That Robinet may not have been a good scientist is not the point. His fictions not only provided an opening for future science, they also marked an important shift in the deployment of nature within a pattern of metaphoric transfers that supports an entire system of aesthetics, history, and the political. This shift allows a structure that was "rigid and static" to become open to change; the "*plenum formarum*," as Lovejoy puts it, also comes to be conceived "not as the inventory of nature" provided by natural history, "but as the program of nature, which is being carried out gradually" (242–4). Paradoxically, this program has the effect of making nature, and perhaps humans, more imperfect than Enlightenment optimism will concede, while also anticipating a future perfection, thus allowing us to see both nature and the history analogically thought through it as being in a process of evolutionary potentiality.

So, if the history of nature goes back to Robinet, and if Kant is cautioning against such metaphysical paradigms, in what does Schelling's theoretical originality consist? Or why is Idealism post-Kantian rather than pre-Kantian? There are two related points to make here, which place Schelling beyond Kant yet distinguish him from British post-Kantians such as Green and Coleridge. First, Schelling's idea that there can be a *history* of nature crosses disciplinary borders that Kant was at pains to maintain. But this does not mean that Schelling made Kant's regulative ideas constitutive (the criticism often made against post-Kantian Idealism); on the contrary, Schelling, unlike Robinet and Herder, is aware that reading nature according to human ends is speculative. The difference between Kant and Schelling is that Kant's aim in emphasizing the dangers of going beyond analogy when one discipline is used to supplement another is critical. Thus, if one brings God into natural science to make "purposiveness explicable," Kant says, and then in a double metaphoric transfer uses this purposiveness "to prove that there is a God, then there is nothing of substance in either of the sciences, and a deceptive fallacy casts each into uncertainty by letting them cross each other's borders" (*Judgment* 253). Schelling, however, is open to the value of analogy in concept-creation, and makes transferences between disciplines constitutive for speculation. But this is not the same as saying that these "Idealizations" ground the Ideas of Reason (in Kant's phrase) in the "archetypes of things-in-themselves" (Sloan 27).[11] To anticipate my further argument, the words "archetype" and "prototype," especially when attached to the word "Idea," are simply ways of giving this Idea a real speculative force, a phenomenological (not noumenal) reality that distinguishes the Idealist Idea from that of Kant, who offers ideas, but then withdraws them as fictions.

In using encyclopedics as a speculative thought-environment rather than a closed system, Schelling also differs from Green. Sloan, in the most detailed reading of Green so far, argues that Green, unlike Coleridge, remained a Kantian, because he maintained the three branches of the study of nature as clearly distinct enterprises, in which the reality-claim of physiogony cannot be established (28, 32–3). Thus Green writes that physiogony sees nature "as an agent acting under the analogy of a will and in pursuit of a purpose, in what sense and whether by a necessary fiction of Science or with some more substantial ground we leave undetermined" ("Introduction" 307–8). But Green, I would argue, also tends to ground the totalizing vision of his physiogony by making God the guarantor of the unfolding archetypes. In his 1827 Hunterian lectures, in which he does allow that physiogony may be a "fiction," Green nevertheless uses the biological term "type" in a distinctly typological way to forward the narrative of physiogony. The "*Types* or characters impressed on animal bodies" form a "visual language" that Green grounds through the trope of the "book" of Nature; "a book not indeed without hiatus and interspaces to be filled up by future discoveries, yet no hiatus of such magnitude or importance as to destroy or even obscure the manifest principles of arrangement that pervades the whole" ("Introduction" 310, 312). But from the 1828 lectures to the late *Mental Dynamics* and the posthumous *Spiritual Philosophy* (1865), which are concerned with the organization of knowledge for the formation of a clerisy, Green grows increasingly metaphysical. Green, that is, jumps from biology into theology via aesthetics, whereas Schelling's *Urbild* undergoes an increasing complication, as the archetype associated with the neo-Platonic Ideas in *Bruno* (125–6), and conceived purely on the level of metaphysics, is made part of the *Ungrund* or *Urgrund* in the *Freedom* essay (62), where it must necessarily be rethought completely through the life sciences as the unconscious of spirit.

Idealism's difference from its British appropriation is summed up by Richards, who points out that in taking over *Naturphilosophie*'s desire to conceive individual organisms and nature itself as teleologically structured, thinkers like Green and

Owen "attributed to God the designing power through which nature came to realize distinctive means-ends patterns" (*Romantic* 518). But to remove divine guarantees and make the purposiveness of nature depend purely on a consilience between aesthetic and teleological judgment also opens up questions about whether nature is indeed teleologically structured. For if one thinks Nature as self-developing or autogenetic, there is no guarantee of where the process will lead; indeed, such a view of nature could also result in aesthetics itself, the substrate of teleology, being radically rethought in accord with nature. This brings me to my *second* point, which is that Schelling does not have a model for the history of nature that is other than conjectural. Between 1797 and 1815, he made several overlapping attempts to think the relation between nature and Spirit, each of which is a speculative foray into the problem. These include a counter-Fichtean *Naturphilosophie* that syncretizes philosophy and nature, Identity Philosophy as an attempt to fit idealism and realism into a single system. They also include the Idealist materialism of the *First Outline,* which begins to understand a certain contention or "asystasy" as inherent within Spirit ("On the Nature" 210) when Spirit is thought as a natural phenomenon, and which thus anticipates the darker *Freedom* essay and *Ages of the World* (1815).

The *First Outline* is the text most complexly engaged with the natural sciences, as well as with physiogony (rather than just physiology, like *Ideas for a Philosophy of Nature*). It is possible on the basis of this text in particular to place Schelling within the debate on evolution, arguing that he postulates a "dynamic evolution" (Richards, *Meaning* 27–9). To expand on this, Schelling's first scientifically precise reference is actually to "dynamic preformation," not evolution. Schelling writes in a long footnote: "I do not yet want to evoke the general principle that no individual preformation, but only *dynamic* preformation exists in organic nature, and that organic formation [*Bildung*] is not evolution, but the epigenesis of individual parts" (37n; 3:61n).[12] Here, Schelling reserves the word "evolution" for an individual preformation, in which the entire development of the individual preexists either in the egg or sperm. He uses the term "dynamic preformation" for a development occurring through the "graduated series" of beings in nature, and he equates this with epigenesis, using the term somewhat incidentally. But in the course of the *First Outline,* Schelling sheds the term "preformation," referring repeatedly to an "evolution of nature" that is not and cannot be completed,[13] a use of the word that is far removed from the closed systems of ovist or spermist preformation – though perhaps consistent with preformation as conceived by Charles Bonnet – and a use of the word that marks an important shift in the meaning of the word "evolution."

As Richards cautions, we should not equate this dynamic evolution with species change (*Meaning* 28–9). Indeed, dynamic evolution may be heterogeny rather than transmutation.[14] But whether or not Schelling anticipates Darwin is irrelevant for my purposes, as he is clear that he is using a kind of metaphor to explore an Idea of Reason in the Kantian sense: "The assumption that different organizations have really formed themselves through gradual development out of each other, is a misunderstanding of an *Idea*, that really lies in *Reason*" (*First Outline* 49; 3:63; translation and emphasis mine). At the level of this Idea, Schelling comes to prefer the word "evolution" because of the continuous historical process it intimates, and the connotation of "*one* production captured at different stages" (39). Schelling does not ally himself with the more technical term epigenesis, probably because, while it may better explain how evolution occurs than does preformation, it does not capture his sense of a larger process at work in nature.

In short, evolution is introduced as a metaphor, though not loosely, since Schelling was a rigorous scientist. Its collateral importance, which is felt not so much in this text as elsewhere in the Idealist corpus, lies in the work it does in areas that can be rethought in the light of nature: history, phenomenology or the history of self-consciousness and its products, and, as I will suggest, aesthetics. But what is this evolution? At first glance, Schelling seems to conceive of it as a relatively seamless process, involving "one organism" or "product" (*First Outline* 149), just as Robinet says that "nature is a single act" ([1761] 2). In other words, "evolution" is Robinet's "progression" ([1761] 5), though "inhibited at various stages" (Schelling, *First Outline* 43), a crucial point to which I will return. In this model, more fully elaborated by Coleridge in *The Theory of Life* and Hegel in the *Philosophy of Nature*, life proceeds from minerals and crystals, through insects and plants by way of aprocess of "vegetation and animalization," up to man (Coleridge, *Theory* 538). This progression is enhanced by the recapitulation of phylo- in ontogenesis, which provides the basis for the *Aufhebung* so central to Idealist phenomenology: once the "original tendencies" of the "formative drive" have been developed, Schelling says, they become inherited without having "to develop all over again in each single individual of the same class" (*First Outline* 46).[15] For Coleridge who, like Schelling, concedes that the formative drive meets resistances on the way, every "grade of ascension" is accompanied by a regression that is, however, recuperated at a new level of integration (*Theory* 548) – a formulation that sounds very much like Hegel's dialectic. Schelling refers to this model of a series of evolutions achieved through recapitulation and sublation as "the graduated series of stages in nature" (*First Outline* 6). And it is the basis for the phenomenologies of spirit that we find in his *Philosophy of Mythology* and for Hegel's accounts of the evolution of aesthetics, philosophy, religion, history itself, and nature: phenomenologies that exist as histories in contemporaneous thought, but that Hegel in particular gives a philosophical cast by making history the history of the "shapes" produced by consciousness in the process of trying to find its way to absolute knowledge.

So the graduated series of stages in nature (or *Stufenfolge*) is the hypothesis, one could say prototype, at the core of the *First Outline,* and, in a way, the Idea that is the very formative drive of Idealism. But is this "deduction" (7–8, 10–11) borne out by nature? For as we have said, nothing is settled here. Despite the title *First Outline of a System of the Philosophy of Nature,* the text contains several competing systems. The "Second Division" alone contains a "First System," "Second System" and "Third Possible System." The whole is supplemented by an outline of the *Outline* at the beginning, and an "Introduction" to the *First Outline* (1799), published separately. In the latter, it seems Schelling wants to fit the philosophy of nature back into the transcendental "system of knowledge," which requires the "real activity to be identical with the conscious or ideal" and which thus tends to "bring back" or "subordinate the real to the ideal" (193–4). But the text's densely textured detail gets in the way of this intention, oddly inserted in an Introduction that is belated, since the philosophy of nature conversely requires Schelling to "explain the real by the ideal" (194). Or, as Hegel also says in frustration at the end of the first stage of his Encyclopaedia, the "ever-increasing wealth of detail" with which "spirit" has to contend in the philosophy of nature makes the latter "refractory towards the unity of the Notion" (*Nature* 444).

Thus, the thirty propositions on the chemical process toward the end of the *First Outline* (172–84) interrupt and retard the straightforward development of Schelling's argument about the graduated stages of nature. These propositions, nomadic fragments or sub-concepts each containing within itself "the principle of its own integrity" as Shelley might say (*Defence* 480), do not so much make the chemical "system" (*First Outline* 110) a key to understanding nature, as they raise the question of how this further system might complicate the biological thinking of nature as an "absolute organism" (54) in the process of evolution. Indeed, throughout Schelling's text, chemistry interferes with the disciplinarity of transcendental biology and zoology dominant in Green's system. At the time, chemistry was a "merely experimental art with no pretension to science," as Schelling says (*On University* 132), following Kant (*Metaphysical* 4–5). As the subordination of chemistry to physics was reversed (*On University* 131), and as it crossed paths with vitalism, chemistry ceased to study tables of compounds and became the study of ungraspably volatile forces, affinities, and mixtures. It thus became a kind of double agent that unsettled the self-certainty of both physical mechanism and organicist teleology, forcing each to think itself from the outside. There is no question that, for Schelling, chemistry is a disturbingly paranormal science: he is reluctant to see life as a "chemical process" because the "chemical system only gives us effects instead of causes" (*First Outline* 110), putting brakes on our ability to know the in-itself. If chemical influences act "externally" on the organism, the organism becomes merely "matter" or "product," and cannot be known from the inside. But if they act internally (109), as a materiality that disturbs spirit's capacity to grasp itself from the inside, the situation is even worse. In this sense, chemistry not only problematizes idealist biology, it is also a kind of symptom. For insofar as chemistry is concerned with mixtures, as Michel Chaouli suggests, it is "the science of all sciences that forever mix and again divide themselves," with the result that a philosophy that takes up chemistry and is inf(l)ected by it can no longer "derive from pure, absolute principles" (208, 212–13).

This is also to say that in the *First Outline,* the model of the graduated stages of nature, developed in idealist biology and zoology, comes under intense pressure from other life sciences: not just chemistry, but also physics and medicine. The very density of scientific and interdisciplinary detail in this text makes its "science" a counter-science through which Schelling must unthink what he hypothesizes. Counter-science is Foucault's term for "a perpetual principle of dissatisfaction" that "flow[s] in the opposite direction" to the established sciences, and "lead[s] them back to their epistemological basis" (373, 379). Schelling's own word for this dissatisfaction that both produces and unsettles systems is *asystasy*. The idea "of contemplating human knowledge within a system," he writes, presupposes "that it does not exist in a system, hence that it is [asystaton] – something whose elements do not coexist, but rather something that is in inner conflict" ("On the Nature" 210).

In concluding this section, let me therefore note two points about the graduated stages of nature as "absolute product" or "absolute organism" that distinguish Schelling's account (*First Outline* 35, 54). First Schelling repeatedly describes the activity of this organism as "inhibited, *retarded*" (5). The graduated stages of nature are not straightforward, but form a negative dialectic in which Nature evolves as "*one* organism inhibited at various stages of development," through a continual series of "deviations from a common ideal" (43, 53). By inhibition, Schelling does not simply mean the delimitation he discussed in the accompanying "Introduction" to the *Outline,* where a bounding line must be imposed on the formless if something is to be produced. Although the word *Hemmung* does sometimes have this meaning of configurative limitation (42, 190), the "retarding force" (190) is more like the tarrying with the negative that Schelling discusses at length in the 1815 *Ages of the World*, albeit in more ontological and even proto-psychoanalytic terms. Here, what is expansive and outpouring is contrasted with "[s]omething inhibiting" that "imposes itself": a "darkening that resists the light," or "obliquity that resists the straight," or an "involution" that resists "evolution" (6, 83). Given this resistance of the Real at the heart of nature, the graduated series of stages in the autogenesis of nature becomes, as David Farrell Krell translates Schelling in his seminal analysis of inhibition, a series of "botched attempts to depict the absolute" (96; cf. *First Outline* 41). And likewise with the autogenesis of Spirit as a natural phenomenon; Spirit, too, in its incomplete form as consciousness, is involved in a series of botched attempts at the Absolute.

But secondly, what happens to the products of these "*misbegotten attempts*" (*First Outline* 35)? As we know, in the theory of recapitulation, lower forms of organization, such as insects and lizards, do not all simply disappear once they have been sublated into the higher, but are "preserved in Nature and maintain the rank of external ... forms" (Green, "Introduction" 310). Indeed, the nightmare of Kafka, Hitchcock, and others is that these forms might actually be more fit to survive. Schelling raises the question of what happens to the inhibited product after its "diremption" (*First Outline* 39) as a philosophical question about the individual entity vs. the universal will. He asks how "this inhibition could be *permanent*," as if perhaps it ought to be – how these "natures which have torn themselves away from universal Nature ... can maintain an individual existence, since all of Nature's activity is directed towards an *absolute* organism." His answer is that "*the individual nature*" cannot "*hold its own against the universal organism*" (53–4). "Nature," as a certain ruthlessness of the Idealist vision of Spirit's evolution, the unmasking of which as a will-to-power forms the symptomatic core of Schopenhauer's *The World as Will and Representation* (1818), assimilates those resistant entities that stand in its way. Earlier, Schelling had noted that "all *permanence* only occurs in Nature as *object*, while the activity of Nature as subject continues irresistibly" (17). If Nature as "subject" is at times the trope for, or personification of, what Schopenhauer calls will, what happens to Nature as object, and the counter-memory that it poses to the narrative of sublation?

The pressure of this question and its implications for consciousness as a "series of evolutions" (Green, "Introduction to the Natural History" 314) are displaced to a later section on disease. Insofar as "deviation ... is intolerable" at the level of the "whole," and insofar as disease is one name for this deviation, this section provides a different "perspective" from that of the narrative of sublation, namely that of the "organic individual," who is a "*limit* to [Nature's] activity, which Nature labours to destroy" (*First Outline* 159, 41). The discussion of disease is curiously "recovered [*nachgeholt*]" as an Appendix to the "third possible system," which is John Brown's theory of "excitability" as the cause of life (158; 220) – third "possible" system, since Schelling is not sure whether Brown's system, as a mediation between the "chemical system" and the "system of vital force" (68), forwards or problematizes the graduated stages of nature. Brown (1735–88) was much taken up in Germany, though with reservations; Schelling, Novalis, and Hegel all wanted to go beyond his empiricism. But was this "empiricism" a deficiency, or did it mark a limit to knowledge, keeping medicine at the level of an "art" (66n), a figural and therefore speculative discipline? If Brown does not finally provide the "third" system, is this simply because of a failure in "execution" (111n), or is the way Idealism tarries with empiricism symptomatic of Idealism's own unthought?

Briefly, Brown classified diseases into sthenic or asthenic, according to whether they arose from an excess or deficiency of excitability. He did not actually use the words "sensibility" and "irritability" sometimes attributed to him. These terms derive from Haller and then Kielmeyer, and become associated with a third term, Blumenbach's force of reproduction (*Bildungstrieb*), the three together making up Brown's "excitability." But what makes Brown unique, as Krell suggests (48–50), is his sense that disease is caused by "the same factors as life" itself (*First Outline* 160). Thus, for Schelling, the "concept of disease" as "the concept of a deviation from a rule ... or proportion" is relative: every "sickness is only a disease in relation to this determinate organism"(159, 159n). Romantic art, for instance, was thought deficient by Goethe and others in terms of classical "health" (Eckermann 154), but might be its own form of aesthetic life. If the normal and pathological are part of a continuum, deviations from the "common ideal" (*First Outline* 53) cannot simply be left behind. Every stage of the evolutionary process must have its own validity, and may indeed make us rethink evolution as a process of division and multiplication in which Nature actually "organizes to infinity," forming spheres in which "other spheres are again formed" (43–4).

Although the Appendix focuses on the individual, in taking up Brown, Schelling goes well beyond thinking the "principles" behind Brown's empiricism (66n), applying the part to the whole according to Novalis's principle of encyclopedics. Brown's *Elements of Medicine* (1780) was a chemistry of the body oriented to treating nervous illness, and is limited to human physiology. Schelling, however, extends Brown's system into the general economy of "physiology," as the term is understood in *Naturphilosophie* as the study of forces operative in Nature as a whole, where it disturbs the ascensionist biological narrative of Romantic physiogony with a more volatile – and empirical, rather than idealistic – chemistry. More specifically, Schelling combines Brown with Haller, Blumenbach, and Kielmeyer, in order to analyse "the synthetic concept of excitability" (160) into "individual systems of *specialized excitability*" that make "organization" an "infinite involution" of "system within system" that puts *evolution* under erasure (127). For Schelling, the detailed study of nature discloses an entanglement of systems produced by the "various *organs* of the same individual" and "the diversity of organisms themselves," along with the diversity of forces they implied (142).

Trying to reunify these systems, Schelling postulates a "gradation of forces" subsumed into "*one* force," so as to posit a "*unity of* FORCE of production throughout ... nature," that results in only "*one* product" (141, 149). This one force is excitability. And yet excitability is far too volatile a force to guarantee a narrative of creative evolution. Excitability is essentially an (im)balance among sensibility, irritability, and reproduction. The "determinate proportion" momentarily established among these forces results in a product, but the constitution of this product upon an imbalance – for instance, irritability in the polyp – results in an attempt to find a balance in another product. Yet once this balance is found, insofar as excitability is the "organic activity" that prevents life from being "exhausted ... in its product" (159–60), this balance seems itself to become an imbalance that requires a further product. In the end, it is hardly possible to disentangle disease and life. And Schelling says as much when he concedes that disease is not "an unnatural state," or, alternatively, that life itself is unnatural, "extorted from Nature ... a state enduring against Nature's will ... a perduring sickness" (160n).

Aesthetics

If life's very vitality is a sickness at odds with evolution, or, alternatively if evolution is actually based on that vitality, what does this mean for the products of consciousness, insofar as the history of nature that Schelling explores in *The First Outline* is part of our self-knowledge? I want to suggest, then, that it is the questions raised by a smooth evolutionary narrative, the resistances of evolution as it were, that are the most important contribution of physiogony as developed by Schelling. These resistances, from the viewpoint of speculative Idealism, cannot just be "convert[ed]" into "willing subjection," as Green says in admitting "the resistance of a contrariant subject" "throughout ... Nature" that troubles his physiogony (*Vital* 54). They radically reorient that key term in Idealism, Idea. For in Schelling's *Freedom* essay, the Idea, which is conceived neo-Platonically in his early *Bruno* and protected within "archetypal" as opposed to "productive" nature (125–6), is relocated to the "anarchy" of the *Ungrund* – a "being *before* all ground" (29, 68) – where it is reconfigured through the life sciences that are the unconscious of Spirit. These resistances also produce new models of mind, in which thinking Spirit according to nature lays the ground for psychoanalysis (Rajan, "Abyss"). And they generate new ways of approaching art.

In conclusion, I want to touch on the last of these areas, the history of art that breaks open the very notion that the principles of aesthetics can be conceived a priori and from a transcendental point of view. In Schelling's early *System,* aesthetics, in Baumgarten's sense of thinking beautifully, underpins a vision of how nature unfolds autotelically in accord with Spirit. But, as we have seen, Robinet also uses the metaphor of nature as artist, in a way that associates art with the unfinished. For Robinet, the series of beings consists of "variations" on an original prototype that, together, comprise Nature's "apprenticeship in learning to make man" through a series of "imperfect sketches" ([1761] 4) that Schelling calls "misbegotten attempts" at the absolute (*First Outline* 35). Although Robinet introduces this metaphor only in passing, his deployment of the figure of nature as artist intimates what will become a paradigm shift from using aesthetics as a way of grounding nature to rethinking art itself in accordance with nature, and with a history that has itself been reconfigured by nature. Here I will turn briefly to Hegel, since if Schelling theorizes evolution, it is Hegel who makes history, and thus the resistances of evolution, the very medium of his thought.

For Hegel, art is supposed to be the "adequate embodiment of the Idea" (*Aesthetics* 1:77), a word ubiquitous in Romanticism, used by Kant, but singularized by Hegel so as to give it a certain drive, and also curiously unreferred, since Hegel's Idea is not the Idea "of" anything. As a concept in Logic, the Idea is "reason identical to itself." But in Hegel's auto-encyclopaedia of disciplines, or the apprenticeship of consciousness in learning to become Spirit, Logic is only the opening proposition. Logic is "the science of the idea in and for itself." But Logic is followed by a phenomenology that has two divisions: the Philosophy of Nature, or "the idea in its otherness"; and the Philosophy of Spirit, or "the science of the idea as it returns to itself from its otherness" (*Encyclopaedia* 54). This schema seems to promise an evolution from nature to Spirit, in which nature provides the phylogenetic preface to the ontogenesis of Spirit. But in practice, the Philosophy of Spirit contains specialized systems of evolution that are not held together by the guarantee of recapitulation that allows one form of consciousness to build on another, so that the Idea must keep going through the same struggle to become identical to itself, only to begin again in a new discipline. The prototype for this process has already been provided by *The Phenomenology of Spirit* (1807): a narrative of the "Calvary" of Spirit that proceeds through several "existential shape[s]" of a consciousness that is never fully raised into Spirit (492–3).

These disciplines in which the Idea keeps beginning again include the philosophy of nature, its subsystems of animal physiology and pathology, the philosophy of history, and the history of philosophy itself, all of which develop rhizomatically metamorphic connections that unsettle their arrangement in a scale or ladder. One of these disciplines is aesthetics, expanded outside the envelope of the three-volume *Encyclopedia* as a specialized subsystem in which art must pass through a long history in which it fails adequately to embody the Idea, and indeed finds its *raison d'être* in this failure. Briefly, Hegel posits three shapes of art – Symbolic, Classical and Romantic – that involve different relations between "inwardness" and its "externalization," or the "idea" and its "embodiment." Or one could speak of three species or genera of art, since the word *Gattung* does double service in both aesthetics and biology. In the earliest or Symbolic phase, represented by the oriental, art fails to achieve identity with itself because of a deficiency in self-consciousness that results in the Idea still being "indeterminate." This problem is overcome in the Classical phase, as art becomes "the adequate embodiment of the Idea" in plastic form. But in the Romantic phase, form and content are once again separated, this time because of a deficiency in matter that repeats and reverses the problems of the symbolic, "even if in a higher way," since external forms have now become insufficient to present an idea that is now fully developed (*Aesthetics* 1:77–81). Theoretically, the schema as a whole proceeds according to the graduated stages characteristic of the Idealists – even Schopenhauer, who bitterly unmasks Hegel's metaphysical conceit in presenting the graduated series of beings and forces in nature not as an attempt at the adequate embodiment of the Idea, but as the "adequate objectivity of the will" (*World as Will* 2:371). Hegel's narrative, moreover, seems organized by a kind of dialectical embryology described by Coleridge in *his* account of the graduated stages of nature, where he writes that "the Vita uterina" of higher forms is found in the lower, which "present problems that first find their solution in a superior order." "Parts are seen, the ... full purpose" of which is "realized higher up in the scale," so that the higher finds its "history" in the lower (*Shorter Works* 2:1194). Thus, the Classical claims to be the solution of the Symbolic, while the Romantic finds its history in the Symbolic.

Yet, as I have argued before, the *Aesthetics* has the form of a skewed dialectic in which the synthesis comes in the middle ("Toward" 53–4). For Classical art resolves "the double defect" of the Symbolic, and achieves "the completed Ideal." It is what "true art is in its essential nature" (*Aesthetics* 1:76, 427). The Symbolic artist had been bound by the materiality of a thought that was not yet Concept, and by the limitations of material forms that prevented him from grasping the Idea, whereas the Classical artist, we are told, is a "clear-headed man." Yet Hegel finds this very adequacy inadequate. For one thing, the Classical artist receives his content "already determined for imagination" from "national faith and myth," and his clarity comes from the fact that he now needs to work only on the "external artistic appearance" (1:438–9). By contrast, there is something more vital in Symbolic art, which "tosses about in a thousand forms," as part of the labour of the negative in which consciousness is still "producing its content and making it clear to itself" (1:438). Hegel thus finds himself drawn back to the dissonance and opposition of the Symbolic. He returns to it, admittedly, in the higher form of the Romantic, which is Christian and spiritual rather than pagan and uncouth. But one cannot but suspect that the Romantic is an alibi for revisiting the profoundly generative matrix of the Symbolic. Or, if nothing else, it brings back the "problem" of the Symbolic, making the Classical, which resolved Symbolic deficiency, a problem of its own. But then the Romantic too proves inadequate, and is abandoned in the sequence of forms for philosophy. And philosophy, too, fails to find its adequate form in Hegel's *History of Philosophy*, which paradoxically ends with Schelling, of whom Hegel's major complaints are that he thinks philosophy as art, and that he keeps beginning "again from the beginning" because his philosophy is "in process of evolution" (3:515, 542).

Hegel does not use the terms of *Naturphilosophie* in the *Aesthetics*, but in the *Philosophy of Nature* he does put the constructive or artistic instinct in contiguity with the biological processes of excretion and the reproduction of the species (406–9). This conjunction allows us to think art as part of "life" rather than "mind": for Coleridge, "Mind" can be "logically defined" as a "Subject possessing its Object in itself," whereas "life," in his bio-philosophical rendition, is "a Subject" that "produce[s] an Object" in order "to *find* itself" (*Shorter Works* 2:1426–7). The displaced and occluded connections between *Naturphilosophie* and aesthetics are the subject of another paper. But let me suggest three areas in which these connections can be pursued. First, as a graduated series of forms arranged in stages to articulate their differences, the history of art is not a narrative of creative evolution, since it cannot keep the form of a dialectical spiral; it is not a narrative of increasing complexity and integration, as in Herbert Spencer's theory, adapted from the Coleridgeans. Rather, it is, albeit in disavowed form, a dynamic evolution mobilized by an excitability, a restlessness of the negative that produces a series of dis-integrations. The sequence of forms can thus be thought in terms of Schelling's sense in the *First Outline* that a product occurs through the momentary fixing of a proportion of forces, in the form of an imbalance that seeks for balance. Thus, the "restless fermentation" of Symbolic art (*Aesthetics* 1:438) evinces a hyper-irritability, which calls for the balance achieved by the Classical. But then, since excitability is what prevents life from being "exhausted in its product" (Schelling, *First Outline* 160), the Classical gives way to the Romantic, which involves a disproportion of sensibility; and so on.

Second, Schelling, as we recall, had raised the question of how the individual being can survive against universal nature. And Schopenhauer demystifies the graduated stages of nature in Idealist physiogony as an objectification of the will in which "a higher Idea," he says, subdues the "lower ones through *overwhelming assimilation*," even as these lower ideas struggle to survive (*World as Will* 1:129, 134–5, 149, 153–4). Schelling also recognizes that he cannot deduce a sequence of graduated stages without the survival of precisely those products that have been surpassed. If these products are needed as an included exclusion to affirm the superiority of the higher, then they must somehow be fit to survive – an aporia Schelling does not resolve. But if we concede that Idealist history is a kind of will-to-power, the history of art in Hegel is also a kind of memory. For unlike Schelling, who confines the inhibition of the absolute organism to an

Appendix, Hegel tarries with the negative, and specifically with the Symbolic as a form of inhibition. In the *Phenomenology,* he speaks of this tarrying: "the *length* of the path has to be endured," he writes, "because, for one thing, each moment is necessary; and further, each moment has to be *lingered* over, because each is itself a complete individual shape" (17). And at the end, Hegel gives the name *history* to the "preservation" of the shapes that Schelling's "Nature" assimilates, as he describes how history "presents a slow-moving succession of Spirits, a gallery of images, each of which, endowed with all the riches of Spirit, moves thus slowly just because the self has to penetrate and digest this entire wealth of its substance" (492–3). Given the enormous length of the *Aesthetics*, the levels in the graduated stages do, in fact, become spheres in their own right. Indeed, in the history of aesthetics after Hegel, a discipline that he took in a different direction from Kant and Baumgarten, thinkers such as Wilhelm Worringer and Alois Riegl adapted the phenomenological method to develop Egyptian art, a level in the *Aesthetics*, as a sphere in its own right (Rajan, "Towards" 61). Schelling recognizes the possibility of this epigenesis when he describes Nature, which is seen as assimilating difference, as also "organiz[ing] to infinity," so that a "determinate sphere of formation" (Symbolic art for instance), far from stagnating or simply being left behind, will again form "other spheres" within itself (*First Outline* 44).

And finally, this history that is Hegel's major contribution to aesthetics gives a place to art forms that he must judge "defective" in terms of his own claim that "the highest" art unites "Idea and presentation" (*Aesthetics* 1:79, 74). Yet Hegel also refuses to call this art "unsuccessful," since "the specific shape which every content of the Idea gives to itself in the particular forms of art is always adequate to that content" (1:300). More specifically, through the Symbolic, Hegel makes a space for the principle of inhibition in art, not as delimitation, but as involution, distortion, disfiguration. In other work, I have explored how the category of Symbolic art can be used to think forms such as the Gothic that begin to emerge as legitimate forms of art in the Romantic period, even though they violate – in quite different ways from the Kantian sublime – the canons of aesthetics as the art of thinking beautifully and completely (Rajan, "Work of the Negative"). *History,* for Hegel, is what slows down the narratives of sublation he constructs for Nature and Spirit, which assimilate the difference figured in the Symbolic.

But history is a complex notion. In one sense, the granting of autonomy to the individual existence develops from an earlier *natural history* that constructs categories for different forms without narrativizing them. For natural history as the collection of specimens and a systematizing of the older form of the cabinet of curiosities cultivates a curiosity about other forms, as Green recognizes in making it part of his trivium in *Mental Dynamics* (19). In other words, the principles of natural history remain as a resistance within physiogony's desire to channel the gallery of forms into an evolution, and are a point at which the philosophy of nature inhibits what philosophy wants to do with art history. But, in a sense, it is precisely this evolutionary history that Idealism projects in physiogony that is also the basis for thinking deviation in terms of its potentiality, and inhibition in terms of problems of mediation. For Hegel's viewing of the lower from the perspective of the "higher" is not simply a way of dismissing the "lower." Indeed, the Romantic is not higher, since it is less perfect than the Classical. Instead, Hegel's schema is a structure for reading forms through something beyond them, something that they have not yet achieved, which both makes them less adequate and, in the case of the Symbolic and Romantic, orients them to the future.

Thus, Hegel sees forms of consciousness as developing, separating them into cultural stages as a heuristic tool for seeing art as historically generated. Recognizing these forms as still in process, he makes them sites for a labour of the negative in which the Idea is still trying to know itself – a labour from which he cannot free the "higher." So what is this "Idea"? For the purposes of logic Hegel defines it as "Reason identical to itself." But in his phenomenologies, the Idea is nothing but the drive to be the Idea: the drive that links the residual, dominant, and emergent in a process in which balance or synthesis is no more than the equally limited antithesis of inadequacy. In Schelling, too, the Idea is exposed to its de-generation as an adequate concept and a concept of adequacy. In the early *Bruno,* it is conceived neo-Platonically, and protected within "archetypal" as opposed to "productive" nature (125–6). But in the *Freedom* essay, Schelling relocates it in the ground, as a result of which it is no longer a fixed "model" or "type," but is in "ceaseless change" and production (*Bruno* 125, 134). He talks of an "Idea hidden in the divided ground" and of a "blind will" that "has not yet been raised to ... unity with the light" (*Freedom* 31–2). Using the terms Schopenhauer takes over, *Wille* and *Vorstellung*, but more idealistically, Schelling conceives the Idea as the will's "inner, reflexive representation": the "first stirring" in which God "is realized, although only in himself" (30). Or, to adapt Habermas in his essay "Ernst Bloch: A Marxist Schelling," the Idea is "something not yet made good [that] pushes its essence forward" (71). Approaching art's inability adequately to embody the Idea, Hegel thus recovers inhibition and "defectiveness" as potentiality, through the process that Novalis calls "romanticizing," which finds a higher potency in the lower (Krell 46–7). The fuller theorizing of this intuition was, in Hegel's time, still part of an unfinished evolution that allows us to say of the very discipline of aesthetics what Coleridge says of natural forms: namely, that in Hegel, "parts" or possibilities emerge, the full purpose of which, if it is not "realized higher up in the scale," is thought through later in the process.

NOTES

The author acknowledges the support of the Canada Research Chairs Program and the Centre for Advanced Studies at Ludwig-Maximilians-Universität Munich in preparing this article.

1 This is Robert Richards's criticism of Richard Owen's use of the concepts of homology and archetype in *The Romantic Conception of Life* 516–17.
2 Most of Hegel's "texts," including the *Aesthetics*, were not published in his lifetime, and are based on his lectures, edited by his students. While a bullet-form version of *The Philosophy of Nature* was part of the published *Encyclopaedia*, the longer text we have, published by K.F. Michelet (1842), is an eclectic text including *Zusätze* from the early Jena lectures to the late Berlin ones (1805–30). On the text, see J.N. Findlay's Foreword to the text (v-ix). While there is a similar problem of publication with much of Schelling's work, *On University Studies* and *The First Outline*, though lectures, were published by Schelling. *The Philosophy of Art* was not.
3 As Bernard Bosanquet notes, the generation of Hegel and Schelling "was pregnant with the theory of evolution," not in the sense that they "anticipated Darwin" (a myopically contemporary use of the term "evolution"), but as this theory existed in various different forms in the work of "Buffon, Goethe, Erasmus Darwin, Treviranus, and Lamarck" (196–7).
4 An example of Schelling's use of evolution to mean preformation occurs in the *First Outline*: "the metamorphosis of insects does not occur by virtue of the *mere* evolution of already preformed parts" (38n; cf. also 47n).
5 According to Baumgarten's *Aesthetica*, "Aesthetics (theory of the liberal arts, inferior cognition, art of beautiful thinking, art of reasoning by analogy) is the science of sensitive cognition [*Aesthetica (theorialiberalumartium, gnoseologia inferior, arspulchrecogitandi, arsanalogirationis) estscienciacognitionissensitivae*]" (qtd. in Wenzel 6).
6 When the German text has been cited, references to Schelling's *Ausgewählte Werke* are given by page and volume number after the semicolon.
7 On this point, see *First Outline* 110, 122, 126–7 and "On the Nature" 212–13.
8 Hegel describes the whole *Encyclopaedia* as a "circle of circles" (*Encyclopaedia* #6, 51).
9 Green is also frustrated by the term "natural *history*," describing it in his 1827 Hunterian lectures as a "misnomer, an erratum in the nomenclature of Science" ("Introduction" 312).
10 The distinction is made by Sloan, who notes that descent was the more common method of exposition in early nineteenth-century comparative anatomy in Britain or on the Continent, except for Lamarck and Green, who drew on Lamarck (34–5).
11 What Sloan calls the "Schelling revision" gives the Idea(s) a "realist" rather than "regulative" status (33).

12 Schelling writes: "I do not yet want to evoke the general principle that no individual preformation, but only *dynamic* preformation exists in organic nature, and that organic formation [*Bildung*] is not evolution, but the epigenesis of individual parts." And further on, having talked about the metamorphosis of insects, he adds that it "does not occur by virtue of the mere evolution of already preformed parts, but through actual epigenesis and total transformation" (38–9n; 46–7n).

13 Schelling uses the word "evolution" throughout the *First Outline*, but initially in a general way (11, 16, 18, 21n). He uses the words "preformation" and "epigenesist" only twice, in footnotes. When "evolution" is used *after* these footnotes, it is with the greater specificity given by his discussion of the graduated series of stages in nature, and sometimes in conjunction with "involution" (77, 187, 188).

14 Rupke distinguishes three theories of the origin of species: transmutation, or the gradual change of one species into another due to environmental pressures; autogeny, or spontaneous generation from primordial germs; and heterogeny, "by many of its advocates combined with a limited degree of autogeny, whereby lower species were thought to have originated spontaneously but higher ones by major mutations of embryonal and other germs" (147).

15 Richards locates the explicit introduction of recapitulation theory slightly later than the *First Outline*, in the work of Schelling's student Friedrich Tiedemann (*Zoologie* 1808–14), and in Karl von Baer's *Entwickelungsgeschichte der Thiere* (1828). He traces the beginnings of the theory to Karl Friedrich Kielmeyer's "Ueber der Verhältnisse der organischen Kräfteuntereinander in der Reihe der verschiedenen Organisation" (1793) (Richards, *Meaning* 19–20, 42–8). In *The Romantic Conception of Life* (244–6), Richards makes a bolder claim for Kielmeyer.

WORKS CITED

Bataille, Georges. *The Accursed Share: An Essay on General Economy, Vol. I: Consumption.* 1967. Trans. Robert Hurley. New York: Zone Books, 1991. Print.

Berman, Antoine. *The Experience of the Foreign: Culture and Translation in Romantic Germany.* Trans. S. Heyvaert. Albany, NY: SUNY P, 1992. Print.

Bosanquet, Bernard. "The History of Philosophy." *Germany in the Nineteenth Century.* Second Series. Ed. C.H. Herford. London: Manchester UP, 1915. 185–215. Print.

Bourdieu, Pierre, and Loic J. Wacquant. *An Invitation to Reflexive Sociology.* Chicago: U of Chicago P, 1992. Print.

Brown, John. *Elements of Medicine.* 1780. Trans. John Brown. London: Joseph Johnson, 1788. Print.

Bubner, Rüdiger. "Schelling: Introduction." *German Idealist Philosophy.* Ed. Rüdiger Bubner. Harmondsworth: Penguin, 1997. 161–7. Print.

Chaouli, Michel. *The Laboratory of Poetry: Chemistry and Poetics in the Work of Friedrich Schlegel.* Baltimore: Johns Hopkins UP, 2002. Print.

Coleridge, Samuel Taylor. *Biographia Literaria; or Biographical Sketches of my Literary Life and Opinions.* Ed. James Engell and W. Jackson Bate. Princeton: Princeton UP, 1983. Print.

Coleridge, Samuel Taylor. *Collected Letters.* Ed. E.L. Griggs. 6 vols. Oxford: Clarendon P, 1956–71. Print.

Coleridge, Samuel Taylor. *Notebooks.* Ed. Kathleen Coburn and Anthony J. Harding. 5 vols. Princeton: Princeton UP, 1957–. Print.

Coleridge, Samuel Taylor. *Shorter Works and Fragments.* Ed. H.J. and J.R. de J. Jackson. 2 Vols. Princeton: Princeton UP, 1995. Print.

Coleridge, Samuel Taylor. *Theory of Life. Shorter Works and Fragments.* 1:481–557. Print.

Deleuze, Gilles. *Francis Bacon: The Logic of Sensation.* Trans. Daniel W. Smith. London and New York: Continuum, 2003. Print.

Derrida, Jacques. "The Age of Hegel." *Who's Afraid of Philosophy?: Right to Philosophy I.* Trans. Jan Plug. Palo Alto: Stanford UP, 2002. 117–57. Print.

Derrida, Jacques. *Points ... Interviews 1974–1994.* Ed. Elisabeth Weber. Trans. Peggy Kamuf et al. Palo Alto: Stanford UP, 1995. Print.

Desmond, Adrian. *The Politics of Evolution: Morphology, Medicine and Reform in Radical London.* Chicago: U of Chicago P, 1986. Print.

Ferris, David S. "Post-modern Interdisciplinarity: Kant, Diderot, and the Encyclopaedic Project." *Modern Language Notes* 118.5 (2003): 1251–77. Print.

Findlay, J.N. "Foreword." Hegel. *Philosophy of Nature.* Trans. A.V. Miller. Oxford: Clarendon P, 1970. v–xxv. Print.

Förster, Eckhart, "Introduction," *Opus Postumum* by Immanuel Kant. Trans. and ed. Eckart Förster and Michael Rosen. Cambridge: Cambridge UP, 1993. xv–lv. Print.

Foucault, Michel. *The Order of Things: An Archaeology of the Human Sciences.* New York: Vintage, 1973. Print.

Green, Joseph Henry. "Introduction to the Natural History of the Birds" (1827). In Richard Owen, *The Hunterian Lectures in Comparative Anatomy May and June 1837.* Ed. Philip Sloan. Chicago: U of Chicago P, 1992. 307–21. Print.

Green, Joseph Henry. *Mental Dynamics or Groundwork of a Professional Education, The Hunterian Oration Before the Royal College of Surgeons of England, 15th Feb 1847.* London: William Pickering, 1847. Print.

Green, Joseph Henry. *Spiritual Philosophy: founded on the teaching of the late Samuel Taylor Coleridge.* Ed. John Simon. London: Macmillan, 1865. Print.

Green, Joseph Henry. *Vital Dynamics: The Hunterian Oration Before the Royal College of Surgeons in London, 17th February 1840.* London: William Pickering, 1840. Print.

Habermas, Jürgen. "Ernst Bloch: A Marxist Schelling." *Philosophical-Political Profiles.* Trans. Frederick G. Lawrence. Cambridge, Mass.: MIT P, 1985. 63–80. Print.

Hegel, G.W.F. *Aesthetics: Lectures on Fine Art.* Trans. T.M. Knox. 2 vols. Oxford: Clarendon P, 1975. Print.

Hegel, G.W.F. *Encyclopaedia of the Philosophical Sciences in Outline.* 1816. Trans. Stephen A. Taubeneck. *Encyclopaedia of the Philosophical Sciences in Outline and Critical Writings.* Ed. Ernst Behler. New York: Continuum, 1990. 45–263. Print.

Hegel, G.W.F. *Lectures on the History of Philosophy.* Vol. 3: *Medieval and Modern Philosophy.* Trans. R.F. Brown and J.M. Stewart. Berkeley: U of California P, 1990. Print.

Hegel, G.W.F. *Phenomenology of Spirit.* 1807. Trans. A.V. Miller. Oxford: Oxford UP, 1977. Print.

Hegel, G.W.F. *Philosophy of Nature: Part Two of the Encyclopaedia of the Philosophical Sciences.* Trans. A.V. Miller. Oxford: Clarendon P, 1970. Print.

Herder, Johann Gottfried. *Outlines of a Philosophy of the History of Man.* Trans. from the German as *Ideen zur Philosophie der Geschichte der Menschheit* by T. Churchill. London: Joseph Johnson, 1800. Print.

Höffding, H. "The Influence of the Conception of Evolution on Modern Philosophy." Ed. Ernst Haeckel, J. Arthur Thomson et al. *Evolution in Modern Thought.* Boston: Indypublish, 2008. 197–222. Print.

Kant, Immanuel. *Critique of the Power of Judgment.* 1790. Trans. Paul Guyer and Eric Matthews. Cambridge: Cambridge UP, 2000. Print.

Kant, Immanuel. *Critique of Pure Reason.* 1781. Trans. and ed. Paul Guyer and Allen Wood. Cambridge: Cambridge UP, 1998.

Kant, Immanuel. *Metaphysical Foundations of Natural Science.* 1786. Trans. and ed. Michael Friedman. Cambridge: Cambridge UP, 2004. Print.

Kant, Immanuel. "On the Use of Teleological Principles in Philosophy." 1788. Trans. Günter Zöller. *Anthropology, History, and Education.* Ed. Günter Zöllerandand Robert B. Louden. Cambridge: Cambridge UP, 2007. 192–218. Print.

Kant, Immanuel. *Opus Postumum.* Trans. and ed. Eckart Förster and Michael Rosen. Cambridge: Cambridge UP, 1993. Print.

Kant, Immanuel. "Review of J.G. Herder's Ideas for the Philosophy of the History of Humanity, Parts 1 and 2." 1785. Trans. Allen Wood. *Anthropology, History, and Education.* 121–42. Print.

Krell, David Farrell. *Contagion: Sexuality, Disease and Death in German Idealism and Romanticism.* Bloomington: Indiana UP, 1998. Print.

Latour, Bruno. *The Politics of Nature: How to Bring the Sciences Into Democracy.* Trans. Catherine Porter. Cambridge, Mass.: Harvard UP, 2004. Print.

Lovejoy, Arthur O. *The Great Chain of Being: The Study of the History of an Idea.* Cambridge, Mass.: Harvard UP, 1942. Print.

Marquard, Odo. "Several Connections Between Aesthetics and Therapeutics in Nineteenth-Century Philosophy." *The New Schelling*. Ed. Judith Norman and Alistair Welchman. New York: Continuum, 2004. 13–29. Print.

Novalis (Friedrich von Hardenberg). *Notes for a Romantic Encyclopaedia: Das Allgemeine Brouillon*. Ed. and trans. David W. Wood. Albany, NY: SUNY P, 2007. Print.

Rajan, Tilottama. "The Abyss of the Past; Psychoanalysis in Schelling's *Ages of the World* (1815)." *Romantic Psyche and Psychoanalysis*. Ed. Joel Faflak. *Romantic Circles Praxis* (December 2008): http://www.rc.umd.edu/praxis/psychoanalysis/index.html. Web. 17 May 2012.

Rajan, Tilottama. "Difficult Freedom: Hegel's Symbolic Art and Schelling's Historiography in *Ages of the World* (1815)." *Inventions of the Imagination: Romanticism and Beyond.* Ed. Richard Gray, Nicholas Halmi et al. Seattle: U of Washington P, 2011. 121–40. Print.

Rajan, Tilottama. "(In)Digestible Material: Illness and Dialectic in Hegel's *The Philosophy of Nature*." *Cultures of Taste/ Theories of Appetite: Eating Romanticism*. Ed. Timothy Morton. London: Palgrave, 2004. 217–36. Print.

Rajan, Tilottama. "Toward A Cultural Idealism: Negativity and Freedom in Hegel and Kant." *Idealism without Absolutes: Philosophy and Romantic Culture*. Ed. Tilottama Rajan and Arkady Plotnitsky. Albany, NY: SUNY P, 2004. 51–72. Print.

Rajan, Tilottama. "The Work of the Negative: Symbolic, Gothic and Romantic in Shelley and Hegel." *Studies in Romanticism* 52.1 (2013): 3–32. Print.

Richards, Robert J. *The Meaning of Evolution: The Morphological Construction and Ideological Reconstruction of Darwin's Theory*. Chicago: U of Chicago P, 1992. Print.

Richards, Robert J. *The Romantic Conception of Life: Science and Philosophy in the Age of Goethe*. Chicago: U of Chicago P, 2002. Print.

Robinet, Jean-Baptiste. *Considérations philosophiques de la gradation naturelle de formes de l'être; Ou les essais de la nature qui apprend à faire l'homme*. Amsterdam: E. van Harrevelt, 1761. Print.

Robinet, Jean-Baptiste. *Considérations philosophiques de la gradation naturelle de formes de l'être; Ou les essais de la nature qui apprend à faire l'homme*. Amsterdam: E. van Harrevelt, 1768. Print.

Rudwick, Martin. *Bursting the Limits of Time: The Reconstruction of Geohistory in the Age of Revolution*. Chicago: U of Chicago P, 2006. Print.

Rupke, Nicolaas. *Richard Owen: Biology Without Darwin, a Revised Edition*. Chicago: U of Chicago P, 2009. Print.

Schelling, F.W.J. *The Ages of the World.* 1815. Trans. Jason M. Wirth. Albany, NY: SUNY P, 2000. Print.

Schelling, F.W.J. *Ausgewählte Werke*. 10 vols. Darmstadt: Wissenschaftliche Buchgesellschaft, 1966–8. Print.

Schelling, F.W.J. *Bruno, or on the Natural and the Divine Principle of Things.* 1802. Trans. Michael Vater. Albany, NY: SUNY P, 1984. Print.

Schelling, F.W.J. *First Outline of a System of the Philosophy of Nature.* 1799. Trans. Keith R. Peterson. Albany, NY: SUNY P, 2004. Print.

Schelling, F.W.J. *Ideas for a Philosophy of Nature, as Introduction to this Science (1797); Second Edition 1803.* Trans. Errol E. Harris and Peter Heath. Cambridge: Cambridge UP, 1988. Print.

Schelling, F.W.J. "Introduction to the Outline of a System of the Philosophy of Nature, Or, *On the Concept of Speculative Physics and the Internal Organization of a System of this Science.* 1799. *First Outline of a System of the Philosophy of Nature.* Trans. Keith Peterson. 193–232. Print.

Schelling, F.W.J. *Philosophical Investigations Into the Essence of Human Freedom.* 1809. Trans. Jeff Love and Johannes Schmidt. Albany, NY: SUNY P, 2006. Print.

Schelling, F.W.J. *The Philosophy of Art.* Trans. Douglas Stott. Minneapolis: U of Minnesota P, 1989. Print.

Schelling, F.W.J. *System of Transcendental Idealism.* Trans. Peter Heath. Charlottesville: UP of Virginia, 1978. Print.

Schelling, F.W.J. "On the Nature of Philosophy as Science." Trans. Marcus Bullock. *German Idealist Philosophy.* Ed. Rüdiger Bubner. Harmondsworth: Penguin, 1997. 210–43. Print.

Schelling, F.W.J. *On University Studies.* 1803. Trans. E.S. Morgan. Athens, OH: U of Ohio P, 1966. Print.

Schopenhauer, Arthur. *On The Will in Nature.* 1836/1854. Trans. E.F.J. Payne. New York: Berg, 1922. Print.

Schopenhauer, Arthur. *The World as Will and Representation.* 1818/1844/1859. Trans. E.F.J. Payne. 2 vols. New York: Dover, 1959. Print.

Shelley, Percy Bysshe. *A Defence of Poetry. Shelley's Poetry and Prose.* Ed. Donald Reiman and Sharon Powers. New York: Norton, 1977. 480–508. Print.

Shelley, Percy Bysshe. "The Triumph of Life." *Shelley's Poetry and Prose.* 455–70. Print.

Sloan, Philip. "Introductory Essay: On the Edge of Evolution." In Richard Owen. *The Hunterian Lectures in Comparative Anatomy, May and June 1837.* Ed. Philip Sloan. Chicago: U of Chicago P, 1992. 1–73. Print.

Vater, Michael G. "Introduction." F.W.J. Schelling. *Bruno.* Ed. and trans. Michael G. Vater. 3–107. Print.

Wenzel, Christian Helmut. *An Introduction to Kant's Aesthetics: Core Concepts and Problems.* Malden, Mass.: Blackwell Publishers, 2005. Print.

Wirth, Jason. "Mass Extinction: Schelling and Natural History." Unpublished paper, courtesy of the author. Print.

Chapter Eleven

Degeneration: Inversions of Teleology

JOAN STEIGERWALD

Degeneration entered the discourses of natural history before Darwin made variations a norm and while ideal types still populated a chain of being as the divine plan of nature. Notions of degeneration existed prior to the eighteenth century, but focused upon deviance from a true genealogical lineage, with a decline from a noble birth or pure form carrying social and moral as well as biological connotations. In the nineteenth century, discourses of cultural and personal decline were tied to theories of evolutionary or developmental degeneration. Indeed, describing deviance, disease, decay, and death as degenerations has a long history.[1] In the latter-eighteenth century, the term "degeneration" became common in natural history discourses, but to mark natural variations produced through nutrition and generation under changed material circumstances. In the context of an expanding project for natural history during the course of the eighteenth century, through the growth of natural economies, collections, and imperial exchanges, questions were posed of the extent of possible degenerations of plants and animals. Degeneration continued to mean a deviation from an ideal type and lineage, but it was no longer confined to negative notions of decline. The term marked the effects of the material world on organic forms, but also the capacities of living forms to respond variously to alterations in their physical living conditions. This new attention to deviations in genealogy raised questions regarding the processes producing such variations; the monumental *Natural History* (1749–89) of Georges-Louis Leclerc, Comte de Buffon, tied these questions to the propagation of species and physical descent. The capacities for generation, degeneration, and regeneration were found to be especially evident in the lowest forms of life. In the German context, Gottfried Reinhold Treviranus's *Biology* (1802–22) introduced a new science of life by demarcating the boundary of living with lifeless nature; degeneration and regeneration became central to his exploration of this boundary zone through the simplest and first forms of life. By the turn of the nineteenth century, then, the history of life became coupled to the descent of life, contingent variations and their material conditions became regarded as necessary for the lively diversity of organic forms, and the boundary between living and lifeless nature became subject to interrogation.

If in the histories of nature of the latter eighteenth century degeneration did not simply mark a negative decline, it did invert the prior teleological narratives of natural history by turning to the material conditions of generation and degeneration as opposed to their end products. How was the apparent purposiveness of life, the *telos* of individual living organisms and of the history of life, to be grasped in the face of degeneration? How was the organization of life, its self-formation and constancy of form, to be understood? The bald appeal to a hypothetical interior mould or vital power to explain the generation and constancy of living organisms in the face of the continual stimulus of the external world, as proposed by Buffon and Treviranus respectively, pointed to a problem rather than resolved it. Even the careful experiments demonstrating epigenesis by Caspar Friedrich Wolff and Johann Friedrich Blumenbach in fact only traced instrumentally the gradual appearance of form concretely in organic matter, without providing a clear conceptualization of how such a self-organization could occur that included the variable and the contingent. The notion of degeneration in natural history was also complicated by its relations with human history, as it became associated with artificial alterations produced through human cultivation. To what extent was variable generation a natural process and to what extent was it a product of human intervention? Questions of teleology had to be reconceived in the face of degeneration, and the organic reconceived through its entanglement with both the inorganic and the human.

Reading German natural history in the years around 1800 against the grain of most idealist expositions, this paper argues that not only was natural history entangled with human concerns, but also the recognition of the coupling of generation to degeneration brought to the fore a concern with the material and contingent conditions of life. Immanuel Kant was the first to suggest a teleological logic for generation in the face of degeneration in a critical and systematic way. He made explicit the tensions between the mechanical and purposive, the necessary and the variable, in our natural histories of living forms. In Kant's analysis, the circularity of our conception of organized and self-organized beings as natural purposes – in which all the parts and the whole are reciprocally causes and effects of one another – reflects the circularity of our judgment – as it moves between the experience of these beings as natural products and their conceptualization as purposive. For Lorenz Oken and Friedrich Wilhelm Joseph Schelling, it is the circularity of the processes of life — as they circle back into themselves; outward activities inverting into inward activities — that marked the boundary of living nature. They suggested these processes at the boundary of life can be further conceptualized through the physical and chemical processes they involve, while the outward activity and inward constraint producing simple natural products lead to the development of more complex organizations through further iterations of these dynamics. Indeed, Schelling argued that each natural product could only be conceived through its particular boundary conditions with higher and lower processes, each marking a relative stage within the endless becoming of nature. While Kant regarded the tensions specific to our conceptions of living organisms as limiting their validity, Schelling contended that such tensions are fundamental. For Schelling, the concept of the living being as the band or boundary of the opposed tendencies of (de) generation reflects the dialectic of our judgments.

In natural histories at the turn of the nineteenth century, then, the generation of living form could only be grasped through its degeneration and a descent to the boundary of organic with inorganic nature; the teleological logic of organic formation included the inversion of teleology as fundamental. Moreover, natural history had a complex relationship with the history of human cultivation and intervention. In German philosophies of nature, conceiving the history of life involved investigating not only its boundary conditions with the physical world, but also the dialectics of judgment in conceiving those boundaries.

I. (De)generation

The project of natural history in the eighteenth century extended beyond concerns with the definition and classification of species, and questions of genealogy. It raised questions regarding the physical, social, and moral order of nature; regarding natural, cultural, and sacred history; and regarding the place of human beings in nature. It was tied to projects concerned with the improvement and prosperity of the human estate, with the management of the natural economy and the progress of the state, and accordingly had broad social, commercial, and political implications. It was also tied to the building of state museums and botanical gardens, to the development of collections and networks of exchange of information and specimens, and thus to imperial enterprises. Such interests led to the study of the effects of climates and the potentials of cultivation on varieties of plants and animals. Collections and exchanges of specimens provided laboratories for natural history, with collections allowing the comparisons of species from different parts of the world, and exchanges of living specimens allowing trials on acclimatization.[2] Evidence of the extent of degeneration through experiments with the transplantation and cultivation of plants and animals also acted as evidence for variable conditions of reproduction, making epigenesis a new problematic. (De)generation marks this movement in both directions, inverting teleology and involving it in the material and contingent.

But these experiments and inquiries in natural history were complicated by their relationships with human history. Human skills in cultivation and domestication enabled a countering of the effects of nature upon living organisms, as wild nature was tamed and plant and animal kinds altered. If experiments in cultivation and domestication could be justified through arguments regarding human needs and improvements, these transformations produced degenerations of the natural order. Concerns were expressed not only over human reason creating artificial systems of classification, but also over human hands artificially deforming natural forms to suit human interests. If these new projects of natural history opened a space for exploring the processes of generation, the potential for progressive change, and the extent of degeneration, they also complicated judgments of natural forms by implicating human beings in the production of those forms and their degenerations.

Buffon played a central role in these developments, through his position as director of the royal gardens in Paris from 1739 and the influence of his widely circulated *Natural History*, published in thirty-six volumes (1749–89). Buffon proposed a foundation for natural history in physiological theory and natural philosophy, by introducing his study of natural history with an account of generation as well as a history and theory of the earth. He defined species by physical descent rather than logical types, opening up a space for thinking about the possibilities of degeneration, and investigating actual instances. Arguing in 1753 that it is "the constant succession and renovation of ... individuals, which constitutes the species" (IV:165), he posited a history of generation and degeneration in which changed climate, soil, and diet produce varieties. If initially rejecting a transformation from one species into another, arguing that species descend from the original progenitors of a first creation and a relative constancy of form secured by interior moulds, Buffon increasingly broadened his notion of propagation to include wider degrees of material relationships and variation in the face of new specimens from colonies and exploratory voyages. He thus offered an image of historical cycles of generation, degeneration and death, with all living beings composed of primitive, indestructible organic molecules, like the fragments of polyps, which are integrated into interior moulds during generation, and into which they disintegrated on demise. Buffon claimed evidence for the existence of organic molecules through a series of experiments with John Needham in 1748, in which they found infusions of decaying organic matter, even after boiling and sealing, were soon teeming with microscopic moving bodies. Buffon also conducted experiments on the cross breeding of species, even if these only rendered reproductive fertility an ambiguous sign of species identity and left unexplained how interior moulds could account for the phenomena of hybrids.[3] Many practicing naturalists, in fact, criticized Buffon's definition of species and speculations upon generation and degeneration as lacking application to the concrete work of identifying and ordering species. But his work helped transform the theory and practice of natural history by encouraging attention to the effects of nutrition, climate, and breeding; to studies of hybridization, monstrosities, and varieties; and thus to studies of degeneration or of what might be termed the "irrational backside of natural order" (Larson 61).

Treviranus' *Biology* made evident the extent of studies in degeneration through natural effects and artificial cultivation in the half century since Buffon introduced his *Natural History*. In the first three volumes, appearing from 1802 to 1805, Treviranus offered a compendium of recent literature in natural history, physiology, geography, and geology, but framed as a history of physical life that explored the boundaries between living and lifeless nature. The first volume presented a natural history based upon physiology, proffering a comparative physiology of different plants and animals that placed generation at the forefront, following Buffon (I:155–74). The second volume focused upon how the physical and geographical characteristics of the earth affected the distribution and varieties of living beings. Treviranus introduced new physical and chemical studies of the effects of atmospheric gases; of climate; of temperature, water, and light; and of atmospheric pressure and electricity on living organisms, to account for the degeneration of the different kinds of living organisms in these different regions. The effects of the external world on the degeneration of living nature recounted in volume two were given a historical dimension in volume three. Drawing upon contemporary theories of geological change, and sequences of rock formations and fossilized forms of life associated with these, he presented a picture of an earth in endless transformation. He concluded that "through these transformations also living nature must be changing," some forms becoming extinct and new forms emerging (III:8). Once living beings appeared, they began to transform the world, changing its airs, waters and earths, and producing new materials that would then combine into new living forms. Treviranus included human beings in this history of life, their appearance producing further transformations of the physical earth and of life on earth: "Each kind, like each individual, has certain periods of growth, flourishing and death, but that its death is not dissolution, as with the individual, but degeneration" (III:225–6).

Despite this portrayal of the history of physical life as under the continual influence of the material world, in positing the principles for a science of biology, Treviranus was concerned with demarcating living from nonliving nature. In the Introduction to *Biology*, he claimed the distinctive character of life to be its similarity or uniformity (*Gleichförmigkeit*) of appearances under the condition of dissimilar or contingent influences (*ungleichförmige oder zufällige Einwirkungen*) of the external world (I:23, 38). He contended that the continual stimulus of external effects is as necessary to the activities of life as is their capacity to resist them. But he posited a vital power (*Lebenskraft*) to police the boundary between the living body and the rest of nature, to prevent the living organism from succumbing to the vortex of natural activity, and to make the organization and purposive relations of the parts of the living organism more excellent than that of lifeless bodies. Treviranus also posited an organic or a viable matter – a "matter capable of life [lebensfähige Materie]" – inseparably bound with the *Lebenskraft* through which all living beings possess life, enlisting Needham's and Buffon's infusion experiments from 1748 and arguing for a confirmation of their results through his own trials. Formless or lacking organization, viable matter takes specific determinate forms through the interplay between vital powers and external influences; the changing effects of external stimuli are filtered through the vital power specific to each organic form, which resists certain external stimuli and allows others (I:37–8, 51–2, II:267–95, 319–52).

A striking tension is evident between Treviranus's introductory account of the demarcation of life from lifeless nature and the exploration of the degeneration of living forms under varied physical conditions in the subsequent volumes. The notion of a uniform organization of living forms became dissolved into an image of continual dis-assembly and re-assembly of organic beings, as death and extinction became regarded as necessary to processes of new generation. Pursuing life to its margins, Treviranus argued that the lowest forms of life – what he termed infusoria and *Thierpflanzen, Zoophyta* and *Phytozoa* – are capable of being continually formed from viable matter under the appropriate external conditions (II:3–30, III:39–40). The notion of a special vital power thus also became troubled, as the formation of simple organisms, the origin of living being from the nonliving, became regarded as occurring at all times and in all places. Indeed, Treviranus' conception of a biology inhabited a troubled epistemic space, as in positing the possibility of a science of life he simultaneously effaced its clear conception. In exploring the border zones of life, the space of demarcation expanded and consumed any clear boundary. Instead, life became reimagined as a movement in two directions, towards both the living and the nonliving. The notion of a *telos* of life was inverted, through an image of ongoing transformation of the forms of life under changing material conditions. Treviranus offered a bold narrative of (de)generation; descending to the lowest living forms and the material conditions of life, he dramatized continual cycles of the spontaneous formation, destruction and new formation of living beings (Steigerwald, "Treviranus' *Biology*"). But with his emphasis upon contingent and continual transformation, how organization might emerge and be sustained became unclear. He provided neither concrete details of the processes of the generation and maintenance of living form nor an explanation of how these processes occurred. Rather his claims for the extent of degeneration posed with renewed urgency questions regarding generation, and his claims for the contingent history of life posed questions regarding its apparent teleology.

Yet demonstrations of degeneration were cited as the main grounds for the epigenetic formation of organisms in the latter-eighteenth century. The variation of plants and animals under changed conditions; the effects of transplantation and cultivation as well as cross-breeding and hybridity; and the natural occurrence of monstrosities – all these modes of degenerations provided evidence of deviant yet regular formations through contingent disruptions of development. Such evidence challenged established arguments for pre-existent germs determining organic forms and species, and acted as a stimulus for the turn to epigenetic accounts of generation by figures such as Wolff and Blumenbach. Indeed, Blumenbach modeled his work on natural history and generation on Buffon, introducing his treatment of the varied kinds of plants and animals with an extended discussion of how generative processes under changed material conditions produced degenerations. Both Wolff and Blumenbach were able to test their theories of generation and degeneration in the laboratories of important natural history collections. Wolff, as professor of anatomy and physiology at the St Petersburg Academy of Science from 1766, was able to use its extensive collection of monstrosities for his research, even dissecting some specimens. Blumenbach, as professor of medicine and curator of the natural history museum at the University of Göttingen from 1776, benefitted from the university's British connections, which resulted in gifts of artefacts from the voyages of Captain Cook and the naturalist Joseph Banks; he obtained an impressive collection of skulls and natural historical materials for his comparative studies of natural kinds. Both Wolff and Blumenbach found in these collections concrete evidence of degeneration that exhibited the capacity for variability characteristic of epigenetic formation.[4] But in claiming the capacity of organic matter to self-organize, supporters of epigenesis needed to demonstrate not only the variable but also the regular formation of living organizations. Critics insisted on the necessity of some pre-existing structures to account for that regularity, and effectively contested the apparent gradual generation of form as simply the solidification and growth of those pre-existing structures. Indeed, displaying and conceptualizing the material processes of generation proved a challenge.

The technique for demonstrating epigenetic formation that Wolff employed in his 1759 *Theory of Generation* was to trace the effective action of generative processes through his experimental instruments. Referring to a slice of plant stem under the microscope, he recounted how, with the help of a needle, he could drive small drops of the visible fluids to carve out vessels and vesicles, and alter their form (§§1–24). To explain this process, Wolff attributed the distribution of organic fluids he enacted through his instruments to a *vis essentialis*, an essential power, as the sufficient cause of generation. At first, Wolff contended that the *vis essentialis* is a power analogous to other natural powers, if distinct from merely mechanical forces. But in the face of criticisms of how a mere power could effect organization, he gave more emphasis to the qualities of the organic substances in which the *vis essentialis* acted, attributing differences in structure to differences in the quality of the substances in which the structures developed. Indeed, throughout his publications he was reluctant to characterize the *vis essentialis* beyond the effective actions he was able to exhibit in his experiments: "[I]t is enough that we know it is there, and that we know it according to its effects" (*Theorie* 160). Wolff demonstrated these effects by taking on the role of the *vis essentialis* himself and enacting it through his instruments.[5]

In his 1780 *On the Formative Drive,* Blumenbach also intervened experimentally in the generative process. He conducted experiments with polyps, comparing the regeneration of mutilated or fragmented polyps to the healing of wounds and bones of patients in his medical care. Such experimental interventions showed not only epigenetic development of form, but also the capacities of living organisms to deviate from characteristic development in response to altered or pathological conditions. To account for the gradual appearance of organization, Blumenbach appealed to a formative drive or *Bildungstrieb.* He gave the *Bildungstrieb* more formative power than Wolff's power of distribution, arguing it could take different directions and generate different organic forms under different conditions. But he also emphasized its dependency upon the organic, if relatively unorganized, material found in the seminal fluids of living organisms. In the end, however, Blumenbach's *Bildungstrieb*, like Wolff's *vis essentialis*, seems little more than the re-description of the self-organization of organic matter that he displayed through his experimental instruments.[6]

Wolff and Blumenbach were able to demonstrate epigenetic formation in ways that were persuasive for many of their contemporaries, although some protested that the human hands in these epigenetic experiments confused their results. Moreover, Wolff's and Blumenbach's attempts at a theory of generation contained ambiguities that they themselves recognized. Their naming of a power or drive was but a naming of an activity made evident experimentally. In tracing instrumentally generative, regenerative, and degenerative processes, enacting these transformations with their tools, they posed rather than resolved the problem of how to grasp the complex phenomena of self-organization.

Kant's application of critical philosophy to the questions posed by studies of degeneration and generation brought them into theoretical focus. In reflecting upon the gradual and variable formation of organized beings, he also reflected upon the form of our judgments of this self-formation and more generally upon the self-formation of reason. But it was in his capacity as a natural philosopher that Kant initially contributed to debates in natural history. In writings from the 1750s, he offered theories of the history and alterations of the physical earth, and puzzled over how to understand the formation of the earth's creatures within such histories. In a 1775 essay on human races, he proposed that such histories of nature include the history of plants, animals, and even human beings, and their degenerations, and followed the tradition initiated by Buffon in attributing both the capacity for and constraints upon the modification of species to reproduction. To account for the generation and variation of living organisms, Kant postulated germs and latent predispositions in original species, with particular climates inducing the unfolding of particular predispositions (II:429).[7] If Kant's archaeological musings speculated upon nature's most ancient revolutions and the descent of all living forms from a common original mother, he cautioned that we only know modes of reproduction in which the product is homogenous in its organization with that which propagates it. By his 1790 *Critique of the Power of Judgment*, he had found in Blumenbach the appropriate method for the study of the production of organized beings; Kant held that in starting from an original organization, from the seminal matter found in organized beings, Blumenbach could attribute a large share of the process of generation to natural mechanisms (V:418–20, 424). But Kant recognized that in calling the generative capacity of the seminal matter in organized beings a formative drive, Blumenbach did not explain this process of self-organization; rather than providing a determinant concept of generation, Blumenbach offered only a regulative concept to guide our study of living beings, a concept of their self-formative generative capacities.

Kant's critical writings, however, have an epistemic focus. His *Critique of the Power of Judgment,* in particular, provided a new articulation of the problems we face in trying to comprehend the organized and self-organizing character of living organisms. Kant argued that the apparent purposiveness of living organizations requires that we conceive of their final form as, in some sense, the cause of the forming and combining of parts. Since such organization appears contingent with regard to the mechanisms of nature, we account for its possibility by analogy with the purposive reasoning and designs of human beings. Yet living organisms are generated by natural processes, rather than produced on the basis of an extrinsic idea. Accordingly, their generation cannot be grasped solely through their final form, but also requires a concept of the processes of their self-formation. We need to regard their complex organization as being generated gradually through the forming and combining of parts, and thus to conceive them as causes and effects of themselves (V:369–72). Kant concluded that recognizing organisms as natural products, yet judging them to be purposive on analogy with human reason, regarding them as at once organized and self-organizing, we can only comprehend them through a concept of a natural purpose.

Several ambiguities and circularities complicate Kant's critical examination of our judgments of organisms as natural purposes. Lacking a domain in his critical philosophy, properly a part of neither theoretical philosophy nor practical philosophy, the concept of natural purpose is amphibious, enlisting both concepts of nature and concepts of reason in its formation. Moreover, empirical investigations and teleological judgments of living beings seem co-constitutive, with reflections upon their capacities generative of the forms of judgments needed to recognize those capacities. Kant acknowledged the circularity of our teleological judgments. He contended that we arrive at the concept of natural purpose only through a reflective activity of judgment, as it moves between an encounter with these unique natural products and their possible conceptualization. When judgment reflects, it must give itself a principle to guide its reflective activity, as an instrument of judgment, with a purely subjective principle acting as the basis for the possible concepts of its empirical objects. In the case of organized beings, the principle that judgment gives itself to guide its reflective activity is that "an organized product of nature is that in which everything is a purpose and reciprocally also a means" (V:376). This principle of purposiveness is a product of the reflective activity of judgment; yet this principle arrived at as the end of such a reflective judgment is reciprocally also the means by which judgment reflects. The concept of natural purpose is the end product of the reflecting activity of judgment — the concept of a being that is at once cause and effect of itself, with each part existing not only as the end of all the other parts and the whole, but also reciprocally being the means producing the other parts and the whole. The reciprocal form of reflective activity and its principle provides, then, not only the means for judging organisms but also the form of the concept produced by such judgments. But the concept of natural purpose is "solely a concept of the reflecting power of judgment for its own ends" (XX:236), and does not provide a determinate concept of organized and self-organizing beings. Indeed, the amphibious nature of the concept of natural purpose is a product of the reflective activity of the judgment producing it (Steigerwald, "Natural Purposes"). Through this circular reasoning, Kant made explicit the difficulties of making sense of the apparently teleological character of organic generation as a natural process, the approach he found so promising in the work of Blumenbach.

Kant's critical examination of our judgments of the formative capacities of living organisms is related to the broader ambition of the *Critique of the Power of Judgment* to examine our judgments of the possible unity of the empirical laws of nature. Reflective judgments discern the unity in diversity and synthesize empirical particulars into a concept or law. Such teleological judging enables us to anticipate what we do not yet know, and to project a systematically unified whole onto the diverse, contingent, and empirically given. It is future orientated, and aims at an indeterminate end (Zuckert 1–86). Teleological judgments form part of the grander project of Kant's critical philosophy, the cultivation of human reason. Kant presented the historical development of human reason organically, like his predecessors drawing analogies between human history and natural history. In a series of essays in the 1780s and 1790s, Kant argued that fostering participation in critique within individuals, and within culture more generally, offers the prospect of progressive enlightenment. Reason, he contended, had the potential to generate or cultivate itself; human beings can learn to philosophize, to exercise their talent to reason in accordance with universal principles, but it is reason itself that must develop and recognize its principles. Reason cannot establish a science unless it has an idea to base it upon, but reason recognizes its idea only when it has become actual; human reason must become what it must already be in order to become reason (Shell 178–81; Clark, "Kant's Aliens" 209–10). Kant's concept of the self-organization of organisms could be regarded as derived from his concept of the self-organization of reason. The analogy could also be regarded as travelling in the inverse direction, however; it could be argued that Kant began to think of reason organically because of his interests in natural history. Indeed, Kant's natural philosophical investigations into generation and degeneration and his epistemic reflections on teleological judgments and the self-formation of reason appear co-constitutive, each developing through its other.[8] Human reason, as at once cause and end of itself, then, seems as difficult to comprehend as the living organism.

In the late-eighteenth century, the (de)generative capacities of living organisms was thus entangled with human interventions into natural processes through experiments in cultivation and generation, as well as with human reflections upon the processes of their self-formation. Yet exactly such entanglements of human history with natural history made evident the contingent and material and complex character of these processes. The capacity of organized beings to organize themselves variably and yet regularly in response to changing physical conditions was precisely what made the grasp of (de)generation so difficult. Kant, in making explicit the circularity and tension in our reflective judgments, argued they marked the limitations of our teleological thinking. Oken and Schelling instead emphasized the productive insights of this circularity and tension.

II. Oken: Epigenesis with Catagenesis

Oken often serves as the exemplar of the worst excesses of Romantic *Naturphilosophie* and idealistic natural history. He is notorious for his systems of analogy, relating all living forms to one another and to all parts of nature, and relating all to the human form as the highest being that unites all in itself. But his 1805 work *Generation* exacts a different reading, with its focus upon the matter of life and generation. Indeed, Oken emphasized, and reiterated this emphasis at key junctures of the text, that in relating all living organisms to the human being, he in fact related the human being to the lowest forms of life, leading back "the generation of human beings to the birth of worms" and finding "the nature of both one" (I:108).[9] He framed his study of generation in terms similar to those of Treviranus, arguing for simple primordial living beings, or infusoria, "in which the chaos of creation daily renews itself, and disappears," and concentrating on the lowest forms of life in which the primary functions of life can be seen most clearly (II:1–2). Like Treviranus, Oken drew upon Needham's experiments, but also used the infusoria he found demonstrated in such experiments to modify more recent accounts of generation, such as Blumenbach's. Oken offered a theory of generation, a natural-philosophical rather than empirical account of the formation of life; insofar as he did offer details of the stages of development, he cited the descriptions of others. He sought to offer means to comprehend the processes of epigenesis rather than simply to defend and describe it. His analogical reasoning – his tracing analogies between processes in the highest and lowest forms of life; and between processes in living and non-living entities – if at times producing excesses, was introduced as an instrument for comprehending generation. It is this theory of generation that provides the basis for his history of nature and systems of natural history, and in this theory, he argued it is infusoria, as primordial living elements, that engender all potential generation.

Infusoria are the starting point and end point of Oken's *Generation*. He posited them as the *Urstoff*, the primordial matter of life, arising with creation as generally and indestructibly as earth, air, and water, the matter from which all living organisms are formed and into which they all decay. Oken invoked the infusion experiments conducted by Needham in the mid-eighteenth century, but also more recent confirmations by Treviranus and others, as demonstrations that infusoria arise, not from eggs or inorganic matter, but from putrefying organic matter: "Fermentation is not a chemical, but an organic process, only with inverted direction [umgekehrter Richtung] – a true development [Entwicklung], degeneration [Entzeungung], catagenesis [Katagenesis]" (I:19). As all flesh disintegrates into infusoria or primordial animalcules (*Urthierchen*), so all flesh is a synthesis of infusoria, and even in its smallest elements each animal is animalistic (*thierisch*). Oken also invoked experiments with polyps, simple organisms similar to infusoria, to demonstate both the indestructibility of infusoria and the disappearance of their individuality in the synthesis of new or higher organisms. Experiments with polyps show how each fragment of the polyp can live on to propagate new polyps, many being thus formed out of one, and one out of many. They also show how the individuality of each is lost when two are joined together; if one polyp is placed within another, it transforms into the other entirely and all functions of the two become one (I:11–12, 30–1). All emergence, all growth, all flourishing of organisms, Oken concluded, occurs through the synthesis of the infusoria spread throughout all of nature, their mass continually rejuvenated through the death and destruction of previous living organisms. He at first appears to be reverting to the account of generation introduced by Buffon a half-century earlier. But he criticized Buffon for turning organic molecules into the elements of an otherwise mechanistic account of generation. He instead praised Blumenbach's account of the formative drive of seminal matter, but populated the seminal matter with infusoria (I:97–107). Oken represented infusoria as the *Urstoff* of all forms of life, as well as the domain of the simplest forms of organic beings next to plants and animals. Generation of complex organisms, then, becomes not only a synthesis of infusoria, but also a development from infusoria to higher forms, transformations from an infusorian stage through a plant stage to an animal stage. As the following discussion shows, Oken traced these transformations variously – through sexual differences and their role in reproduction; through physiological functions and their comparison across species; through physical and chemical processes accompanying such functions; and through a mathesis of progressive development. His iterative cycling through these different means of conceiving the processes of generation, and exploring the analogies between them, was an attempt to provide instruments for comprehending generation beyond simply describing the appearances of gradual formation. But in each instant, he brought generation back to its starting point in infusoria, inverting the apparent teleology of the construction of the highest organisms by deconstructing them into their material elements, and pointing to the "complete identity of the way of emerging of human beings with the lowest polyps" (I:108).

Sexual difference is the major means through which Oken tracked the cycles of generation, following it through infusorian, plant, and animal domains. His account can plainly be read as a rehearsal of the normative sexual hierarchies of his society, but he used these sexual conventions to invert the logic of conventional theories of generation. Following tradition, Oken gave the male principle primary significance, as the predominant and active element in sexual reproduction, and the female principle the secondary role, as the passive and vegetative element. He thus inscribed in nature conventional sexual hierarchies, in the process naturalizing those hierarchies through purported biological necessity.[10] Also in keeping with his contemporaries, Oken held that sexual difference first emerges with plants. Infusoria are sexless, enclosed within themselves as wholes, and thus propagating themselves from themselves through mere division. In plants, a duality emerges between male and female parts as it strives for a supplement outside itself. It is a duality bound in one individual, however; only in animals do the sexes rupture into two distinct individuals. But in specifying the contributions of the male and female principles of reproduction, Oken reversed conventional hierarchies of matter and form. The male principle becomes the matter of generation, making matter the active or animating principle of life. It is the infusoria, the pollen and semen, that are the primary determinants of organic life. The female principle becomes the power producing the type, making the "merely forming," then, the receptive or animated principle. Form is what is passive or plant-like in reproduction, the vesicle in which the embryo takes shape, if also providing nutrition (I:121, 102–6). If the female principle is the form binding the infusoria, it is always dependent upon the material male principle and never appears without it. In Oken's theory, the male principle remains predominant over the female, as in conventional sexual hierarchies, but by making infusoria the male principle, *contra* convention, the matter of generation takes precedence over form.

Oken traced the processes of generation from their basis in infusoria through the formation of plant-like structures to animal forms. In doing so, he foregrounded the physiological functions found in fully formed organisms, and the formation of their rudimentary structures in generation. He also related these functions and formations to physical and chemical processes and their materials. His analogies based upon the tripartite structure of infusoria, plants, and animals are often perplexing, and can be read as forced and indulgent. Thus, the infusoria, or simple polyps, as active principles, are associated with the circulatory system and the muscular and skeletal structures of movement; plants, as receptive principles, are associated with the lymphatic system and the liver; and animals, as the union of infusoria and plants, are associated with the digestive and nervous systems. And so on. Yet these analogies can be read as providing means for comprehending generative processes. Oken maintained that in the simplest forms of life, physiological functions can be more readily understood than in the more complex organizations of higher animals. His emphasis on primary physiological functions enabled him to draw analogies between the generative development of different organisms, and to draw analogies in the nature of these functions across species, from the lowest to the highest. He thus suggested not a comparative anatomy, but a comparative physiology. His emphasis on organic functions also helped him conceive how more complex organization developed from and through simpler ones, both in individual generation and the history of life. Oken then traced the physical or chemical processes that accompanied generative processes in the movement from infusoria through plants to animals and their correspondent physiological systems. Infusoria or polyps display simple forms of cohesion characteristic of earths and metals. In plant-like structures, including those of an embryo rooted in an egg or uterus, the processes of oxidation and deoxidation, and the exchange of airs and nutrients, are displayed. In animal-like structures, able to move freely, chemical processes of digestion through water and salts come to the fore. For the empirical details of these processes, Oken relied on the work of others, but the logic was his own. The cycles of analogies across tripartite structures, traced iteratively, may seem to inscribe increasingly strained systems onto the empirical phenomena described by others. Yet there is a rhythm and development to these iterations and analogies, as Oken used them as instruments in his attempt to comprehend how more complex organizations might develop from the simple elements of life – generatively, historically, physiologically, physicochemically, materially. He also tried to conceive degeneration, development in an inverse direction, by tracking the physicochemical conditions of functions and their generation, and by following the disintegration or decomposition of these processes into their material elements. Infusoria, as the primordial elements of life, remained the animating principles of all these organic transformations.

Oken also attempted to formalize the logic of organic transformation, of generation and degeneration, mathematically. This mathesis is perhaps the move in his theory of generation and natural philosophy most difficult to follow, yet it is in part a response to the insistence, by Kant among others, that in any doctrine of nature there can be only as much science proper as there is mathematics therein.[11] In Oken's mathesis, the tripartite structure of analogies is again reiterated. Infusoria propagate by dividing themselves through their own power, in an endless repetition of one as one; this numerical repetition is figured through the line, with the movement of propagation a linear locomotion (*Fortbewegung*). In plants, the male is bound inwardly with the female genitals in one individual, each driven to completion through the other; this inner opposition, a heterogeneity within homogeneity, is figured through the circle. A circle is a line insofar as it is only possible through its diameter, as a plant is possible only through a polyp, but it is also a circle through its periphery. It is the tension between line and periphery, between male and female, that forms the circle of the plant and its endless cycles of reproduction, the movement of its generation. The animal, as the union of polyp and plant, of line and circle, is figured as the ellipse, the most beautiful and harmonious form (I:109–24). These mathematical figures, as the basic models for polyps, plants, and animals, are the roots of the higher mathematics of more complex forms that Oken developed in other works in convoluted speculations.

In *Generation*, if the form of this mathesis remains crude, it nevertheless suggests a logic of progressive development from infusoria through plants to animals. This development is figured mathematically as a movement circling back into itself to produce an opposition and active tension between the line and circle in analogy to the opposition between the male and female.

This mathesis reveals the idealistic influences upon Oken's theory of generation, complementing his emphases on the matter of generation and degeneration. The philosophy of nature, he argued, is the science of the eternal transformation of this absolute into the world, a transformation he traced in his 1809 *Textbook of the Philosophy of Nature*. Thinking the formation of the world mathematically, he conceived the basis of living forms, indeed all natural forms, through their origin in an absolute figured as zero, 0, nothing (*Nichts*), as the undifferentiated basis of all. Real becoming emerges from the absolute or zero, its first form that of duality, the opposition of positive and negative, which makes possible the series of numbers, and the individuality and diversity of definite forms. In a manner analogous to the development of living forms, a tension between line and periphery generates the world as a sphere, as the successive repetition of numbers that produces motion and time becomes restricted and produces stasis and spatial form. Thinking the formation of the world logically, Oken also traced the origin of our concepts of the world. God appears in his account as a figure for the formation of our cognition of the world, with the becoming of the world as the self-revelation of God mirroring the emergence of self-reflective consciousness in human thought. The pure activity of God restricts itself, dividing itself into the ideal and real, reflecting the division of thought into subject and object, and giving rise to the world of becoming and our concepts of determinate objects. But Oken insisted that the first origins of this process of the emergence of the world from the absolute or God, or the emergence of self-reflective consciousness of the world, cannot be thought, except negatively. 0 or nothing represents this unthinkable origin. 0 itself has no predicates; it is not ideal or real, positive or negative, many or one, formed or unformed, being or nonbeing. In arguing that all is "actually created out of nothing" (II:28), Oken signaled an origin impossible to conceive. Oken's philosophy of nature can only begin with a world already differentiated and becoming.

From such abstract reflections upon the problem of first origins, Oken proceeded to speculations upon the historical unfolding of the universe, the solar system, the earth, and even the first origin of life. In *Generation*, Oken emphasized that "no organism emerges out of the inorganic, so each organism that emerges must emerge out of the organic itself" (I:18). But in his *Textbook of the Philosophy of Nature*, he suggested how, as a singular event in the historical formation of the earth, life was first formed. Under specific conditions, with the right admixture of waters, airs and earths, of carbon and salts, through particular chemical, oxidizing, and cohesive processes, at the sea shore, with light shining upon the water, a primordial mucous (*Urschleim*) was first produced and took the form of infusoria. Once it was produced, organic life became cycles of generation and destruction through this primary organic matter, and death became not annihilation but only change, with one individual emerging out of another endlessly: "The human being too is a child of the warm and shallow parts of the seashore;" "not created, but developed [*entwickelt*]" (II:§856, 950). It was this development of living form, the history of life from the simplest polyp to human beings, that Oken attempted to conceive.

Oken thus represented the formation of the world mathematically and conceptually as well as physically, and the emergence of living beings and even human beings as both an ideal and a real process. In both the *Textbook for the Philosophy of Nature* and *Generation,* a tension exists between the mathesis and the material processes of this formation. The analogies between lines, circles and spheres and the matter and form of living beings and the world remain undeveloped. But Oken was attempting to conceive the natural history of the world and its (de)generation through logical concepts as well as through its material basis as organic and physical processes.

III. Schelling: Boundary Concepts and the Dialectic of Judgment

The tension in Oken's philosophy of nature between the idealistic conception and material processes of the becoming of the natural world and the generation of life is fundamental to Schelling's work. In his various philosophical writings, Schelling shifted between the discourses of natural philosophy and transcendental idealism, making explicit what remained implicit in Kant: how the production of knowledge reflects the productivity of nature, even as the productivity of nature is constructed through reflection upon the productivity of knowledge. Schelling recognized the speculative and artificial aspects to even the experimental sciences, in the ways that experiments provoke particular kinds of responses from natural processes through the particular kinds of questions and instruments employed in investigations. He also recognized that no philosophical science can stand outside the particularity of its positioning; the perspective of each is necessarily limited by its position within the world and as a product of the world, even as it attempts to conceptualize the world. Ever in the midst of things, the finite subject produces a succession of partial concepts of the natural world, each nevertheless enlivened by the dialectic of judgment constituting their production. Schelling's history of nature, like that of his contemporaries, was both an ascent and a descent, in that his account of the progressive development of organization was also an account of its material and contingent conditions. In trying to comprehend both the individual products and the whole of this historical development, he did not rest with any determinate representation of this tension as primary. Rather he conceived each natural product as the band or boundary of opposed tendencies in the endless becoming of nature, in which there is no point of origin and no fixed end, only different stages of activity and analysis. Drawing upon Kant, Schelling argued that both individual living organisms and nature as a whole are self-enclosed systems that turn back into themselves in reciprocal relationships of causes and effects. Drawing upon the sciences of his time, he further conceived each living being through a double involution – an involvement with its environment, and an involvement with itself – that effaced any definitive ground of organic life.

Throughout his writings in the philosophy of nature, Schelling enlisted the language of Kant's account of living organisms, stating that "Every organic product carries the ground of its existence in *itself*, for it is cause and effect of itself." But he also extended this representation to the whole of nature, arguing that "nature becomes a circle, which returns into itself, a system enclosed within itself," and thus emerges as "a reciprocal connection of *means* and *ends* (*Ideas* 30–1, 40–1/*Werke* 1,5:94, 106/*SW* 1,2:40, 54).[12] In his 1798 *On the World Soul*, Schelling asserted that in general the world is an organization, and that in the end the organism is the condition of the mechanisms of nature. Such assertions have been read as offering an organicist and a purely idealistic representation of the natural world, in which nature as whole and each natural product is an organized structure modeled upon an idea or archetype (Richards; Warnke). But, as discussed above, even for Kant our conception of organized and self-organizing beings as natural purposes reflects the circularity of our reflective judgments rather than an idea. Schelling did not try to resolve the contradictions of our concepts of living organisms, but rather he incorporated them as inevitable and productive. He developed a philosophy of life that rendered all of nature as alive through the play of opposed processes.

Indeed, arguably the concept of excitability (*Erregbarkeit*) is more central to Schelling's philosophy of life than the concept of organization. One of the problems posed by living organisms is comprehending how organic products are able to preserve their characteristic organization and activities in distinction from and yet in relationship to the world around them. Schelling portrayed the boundary of life in terms much like those Treviranus would use in his *Biology*, enlisting the concept of excitability to think through how such a boundary might be enacted. An individual organism must maintain its own sphere of activity against the activities of the universal organism. Yet it must also prevent itself from falling into a condition of complete stasis if it is to preserve its vitality, and accordingly needs continual stimulus from its surroundings. Schelling argued that self-preservation requires that each organism assimilate or organize all for itself in order not to be assimilated or organized into nature as a whole. This activity depends upon generating an opposition between inner and outer – for the individual organism to remain receptive to the stimulus of external material influences, but also to engage in an activity resisting them. Receptivity and activity only obtain reality through their reciprocal determination, with the life of each organism pulsating between different proportions of this relationship, and receptivity or activity predominating under the different conditions of life. All living matter distinguishes itself from the external world, however, by its receptivity to impressions being antecedently conditioned by its character as a special sphere of activity, even as that sphere of activity is also conditioned by its receptivity to stimulus. This general capacity of excitability is distinguished more evidently in higher organisms through the organs of sensibility and irritability, which perform more complex interactions between receptivity to stimulus and response (*First Outline*/*Werke* 1,7:117–34, 180–90, 230–45/*SW* 1,3:69–91, 155–72, 220–39; Rajan). Unlike Treviranus, Schelling did not baldly posit a vital power to police the boundary of life, but enlisted excitability as a "boundary concept [Grenzebegriff]" for distinguishing living and nonliving spheres (*Werke* 1,6: 81–2/*SW* 1,2: 386). He argued that the reciprocal accommodation of organic and inorganic nature is explained from the "*common physical origin* of both" (*First Outline*/*Werke* 1,7: 132–4/*SW* 1,3: 92–3; emphasis in original). But the organism's unique capacity to respond to, and yet to distinguish itself from, the external world is dependent upon a reciprocal receptivity and activity within itself. Thus like Kant, Schelling regarded organic life as cause and effect of itself. But he expanded Kant's account to depict an involution in two directions – an involvement of the organic with the inorganic, as well as an involvement of the organic with itself. Living organisms are distinguished by the receptive inward and outward activities they contain within themselves. This double involution marks the boundary of living being, but this boundary remains precarious and porous, its contingency and variability essential to the free activity of life.

Each sphere of organic life has a form of excitability, marking its boundary with the inorganic world, as a dynamic relationship between outward and inward activity. Excitability then passes over into formative activity – on the one hand, as the inward activities of nutrition and the maintenance of form; and, on the other hand, as the outward expansive activity of propagation and the evolution of higher forms of life. Schelling framed these different processes through iterating an opposition between negative and positive principles. Like Oken, he was offering a philosophy of nature, not an empirical account of the organic formation. But also like Oken, his formulaic iteration of opposed general principles was tempered by relating these principles to contemporary empirical inquiries. Schelling represented the formation of specific organic forms as processes of the individuation of organic matter. The general chemical qualities of matter acquire more individuated form and composition in organic matter, and the organic matter and nutritive juices gradually refined through the processes of individuation are formed into specific organs and specific kinds of organisms. This chemical process depends upon inorganic materials and chemical reactions, what Schelling termed "the negative conditions of processes of life"; but it is also a vital process dependent upon organic materials and organs, positive activity and stimuli ensuring that individuation results in living beings (*Werke* 1,6:183–5, 196–235/*SW* 1,2:493–5, 507–47).

Schelling also enlisted contemporary accounts of the processes of generation and degeneration, drawing upon Blumenbach and Kant in particular. He appealed to Blumenbach's concept of a formative drive or *Bildungstrieb* as the expression of a capacity of organic matter to produce itself through continual processes of nutrition, growth, and reproduction. He also appealed to Kant's concept of germs or predispositions of generative matter, arguing that particular predispositions informing particular directions of the formative drive become fixed under the influence of external causes and passed on through propagation. The formative drive and dispositions of germ matter working together with excitability incorporate materials from the external environment to generate and sustain organic bodies in ongoing activities of formation and individuation. This individuation of organic matter and form is then passed on in reproduction. Through the dynamic interplay between negative and positive processes – between the chemical and vital processes of individuation, and between germs or predispositions and the formative drive – Schelling sought to account for both the regular formation and propagation of organic beings and their variations. In enlisting the *Bildungstrieb*, Schelling insisted that it should not be understood as an immaterial power or first cause of the organization of living forms; since it can only be effective in organic matter, it cannot be the original cause of the organization of organic matter. Indeed, he made clear that he posited neither a special vital power nor chemical process as constituting life. What distinguishes organic matter and its capacities for self-formation are its greater degree of freedom than the lawfulness of inorganic matter. The *Bildungstrieb* is "only an expression [Ausdruck] of that union of freedom and lawfulness in all natural formations, but not a ground of explanation of this union itself." The formative drive is thus a "synthetic concept [synthetischer Begriff]," or a boundary concept like excitability, marking the difference of organic matter from inorganic matter and their relationship (*Werke* 1,6:215–19,/*SW* 1,2:526–30; *First Outline*/*Werke* 1,7:101–12 /*SW* 1,3:42–62).

For Schelling, life is the interplay between these primary organic activities or functions – excitability, irritability and sensibility, and formative activity – each organism existing in an interaction with the larger organism of nature as well as inverting into itself to demarcate its specific sphere of organic functions. These functions are the conditions of life – completely physical and regular in operation, yet also free. Accordingly, excitability also means the capacity for deviation from regular activity, or disease, and generation also the capacity for degeneration, for variation and decay (*First Outline/Werke* 1,7:117–34, 180–90, 230–45/*SW* 1,3:69–91, 155–72, 220–39; Krell, 100–14). In the history of life – in both the life of individual organisms and the historical relationships of different organisms – Schelling argued different relative proportions of these organic functions can be found in a graduated series of stages. His writings upon the philosophy of nature presented this history largely as a descent to the boundaries of life, with its focus upon the reciprocal determination of organic and inorganic nature. But he also offered suggestions for conceiving the ascent to higher forms of life through the changed proportions of the primary organic functions of generation, irritability, and sensibility. He represented a gradation of forms of life, in which reproductive activity predominates in lowest forms of life, and then irritability and finally sensibility predominate in higher and more individuated forms (*First Outline/Werke* 1,7:210–30/*SW* 1,3:195–220).[13]

Schelling rejected the idea that the different kinds of organisms actually develop from one another in a chain of becoming from a single original product or towards a common end. Such a development would only be thinkable if nature had an absolute archetype (*Urbild*) for all its members before its eyes, so that each organism could be represented as different approximations to this absolute through a comparative anatomy. But, he insisted, such an absolute product nowhere exists; there is no absolute origin or end to which all individual organisms might be compared. Rather than comparative anatomy, Schelling advocated a comparative physiology that analysed and reciprocally compared living organisms through the boundary concepts and functions correspondent to their stage of development or activity (*First Outline/Werke* 1,7:112–13/*SW* 1,3:62–5). In cycling through the tripartite functions in different kinds of organisms, and in the generation and life of individual organisms, and suggesting correspondent physical and chemical processes associated with each, Schelling might be read as offering a variation upon Oken's tripartite analogies as a means to conceive organic development. But Schelling's logic of development introduced a more complex set of concepts in its attention to boundaries of life and the involution of life with itself at and through these boundaries.

The world soul (*Weltseele*) Schelling introduced in his 1798 text, *On the World Soul*, is thus two-faced or duplicitous, facing in contrary directions, doubled within itself, at once world and soul, grounded in neither matter nor spirit. For Schelling the conflicted boundary work of the emerging science of life became an important stimulus for his philosophy of nature. Indeed, his philosophy of nature might be described as a philosophy of life, in which the duplicitous nature of organic being became understood as exemplary of natural processes. His philosophy of life was not an organicism that portrayed all natural products as organized wholes. Rather Schelling argued that each natural product can be conceived as the band or border between opposed tendencies. He depicted these opposed principles in general terms as the interplay between the productivity of nature and its constraint. But he also explored boundary concepts in concrete contexts, arguing that different methods of inquiry and different modes of conceptualization are needed to comprehend the different states of natural phenomena. Organic life constitutes one such boundary. Excitability and the formative drive mark the double involution of organic life as a movement both toward and away from inorganic nature and its inversion into itself. What is striking in Schelling's philosophy of nature is that he did not take any natural power as foundational, but rather explored the dynamic interplay of opposed tendencies in a series of natural processes and powers – from organic life through chemical processes to the powers of inorganic matter. He also explored the relationships between these processes in different configurations in different texts. He thus constructed the world from the different boundary conditions delimited at particular sites and at particular moments in the history of the world. Boundary concepts are meant to account for both what is necessary and what is variable or spontaneous in nature. They also highlight how our understandings of nature are always relative and site specific. If we project an idea of nature, we have no perspective upon its form as a whole. We can only work from within nature to conceive states of relative stability. Any such boundary work is always open to disassembly through alternative analysis or investigation. The natural history of the world soul Schelling's works depicted is a history recounted from within nature and is subject to the particularity of its positioning.

The double essence of the world soul reflects the double essence of the human soul. The human soul is "the *Mitwissenschaft*, the co-science," of the world soul, subject to the necessities of nature, and yet also possessing a capacity for freedom of action and thought (*Ages*/*SW* 1,8:200). As a finite being in the world, the human soul is constrained and produced by nature, yet it is also able to contemplate itself and its natural history. Schelling's philosophy of nature constructs that natural history by examining the epistemic conditions of our concepts of natural processes. He turned to the philosophical apparatus of transcendental idealism to examine the genesis of our cognition of the natural world, tracing the emergence of conscious thought from immersion in sensation to the activities of reflection and judgment. He constructed the concept of matter as a dynamic opposition of positive and negative, repulsive and attractive, powers from the oscillations of the activity of thought and its constraint in productive intuition. He constructed the concept of organic life as a double involution from the activity of consciousness reflecting upon its own self-formation. Philosophical reflection analyses our cognitive activity into its different products – sense and intuition, intuition and concepts, thought and action. But it then poses the problem of their relationship. Reflection lifts cognitive activity out of the sphere of givenness and blind habits of thought, so that that we might think critically and freely. Judgment (*Urtheil*) separates and compares what is unconsciously united, intuitions and concepts, so that they might be related consciously and reflectively. But a duplicity and boundary is accordingly generated that must be traversed with the aid of a band (*Band*) or mediating link (*Mittelglied*) (*System*/ *Werke* 1,9/ *SW* 1,3). Each concept is in turn an enfolded judgment (*entfaltete Urtheil*), a holding together of opposed tendencies, reflecting the dialectic of antagonism and juncture of its construction (*Ages*/*SW* 1,8:214; Heidegger). Schelling thus sought to demonstrate how the boundary concepts of his philosophy of nature reflect the dialectics of the mind traced in transcendental idealism. He concluded that "No objective existence is possible without a mind to know it" (*Ideas* 177/*Werke* 1,5:213/*SW* 1,2:222), but in that mind the whole of reality emerges through the dialectics of consciousness in its unending productions and reproductions. The limitations of our understanding of nature are thus due to the limitations of our ways of thinking.

Schelling was critical of the tendencies of the modern age to repress and to deny the conflicts and contradictions of our conceptions, and to unite all in a coherent system. What it does not recognize is that unity itself is founded in opposition, and that "the construction of this contradiction is the highest task of science." The main weakness of all modern philosophy lies in the lack of "intermediate concepts [mittleren Begriffe]": "But the intermediate concepts are precisely the most important concepts, indeed the only concepts that actually explain in the whole of science" (*Ages*/ *SW* 1,8:321, 286). Schelling did project an absolute idea of the world, as the realization of an absolute reason that knows all. Coming of philosophical age in the context of German idealism, he always had his eye upon the infinite. But he acknowledged that each of our philosophical systems remains a particular system. Indeed, his repeated oscillation between the discourses of the philosophy of nature and transcendental idealism was due to the recognition that each only offered a partial perspective on our world. Philosophical science "is the development of a living, actual being that presents itself in it" (199). As finite beings, not only do we stand in nature and recount the history of nature from our particular position, but we think from the perspective of human consciousness, subject to the dialectics and contradictions of its cognitive acts. Not only the world soul, but also the human soul, is duplicitous and doubled.

In later works, Schelling sought a new language for philosophy, a new figure for the form of philosophy, in the symbolic language of mythology. His 1809 *Philosophical Investigations into the Essence of Human Freedom* and 1811–15 drafts of *The Ages of the World*, for example, offered a myth of the creation of the world and the self-revelation of God. In a manner similar to Oken, Schelling portrayed a gradual emergence of knowledge with consciousness, analogous to the gradual unfolding of nature; but he now portrayed this emergence as mythical and historical. Also similar to Oken, in these myths the first origin of the world is inconceivable, or in Schelling's terms "unprethinkable [unvordenklich]," an eternal beginning that has "*happened* since all eternity (and as still always happening)" (*Ages*/*SW* 1,8:219, 225; emphasis in original). Schelling's God is always already a living God, his existence grounded in a bond with nature. In these works Schelling valued the symbolic language of mythology in contrast to the dialectic form of reflective thought. But it is important not to read into Schelling's account modern definitions of the symbol as an object or figure that represents or stands for something else. As he argued in his lectures upon the philosophy of art in 1802–4, mythological figures are symbolic not because they "signify ideas," but because they "*are* without reference to anything else" and are "significant for themselves." In a symbol, "meaning is simultaneously being itself, passed over into the object and one with it" (*Philosophy of Art,* 49, 75/*SW* 1,5:411, 447; emphasis in original). What was the meaning that Schelling expressed through the symbolic language of mythology? Some have contended that Schelling's project of a new mythology was motivated by the search for significance missing in the modern world. The unity of part and whole in the symbolic figures of mythology allows meaning to shine through that cannot be expressed discursively. Individual experiences obtain their significance through the continuity of life, as life is given meaning and coherence as a whole (Whistler; Halmi, 141–61). But Schelling held that symbols are "significant for themselves" and "without reference to anything else," and thus they remain no more than themselves. Schelling took as his model Greek mythology, fascinated by how, through imaginative fantasy, it offered a world populated with a diversity of living gods – gods blessed and beautiful, yet limited and marked by all too human frailties. These gods are for art what ideas are for philosophy, images of the divine intuited in actuality. But rather than providing a coherent meaning to life beyond the conflicts of modern life, Greek mythology entangled the gods in the profane world of humanity, with all its strife and turmoil. The symbolic language of mythology did not offer a significance to life beyond its contradictions and confusions, but drew attention to its tragic consequences. In Schelling's creation myths, God is always already divided in himself, at once freedom and necessity. Schelling depicted a living God inflected with a dialectical tension to account for the life of his creatures, but then mapped back onto God the divisions obtaining in his creatures (Clark, "Necessary Heritage" 86). Like the organism of nature, "in the circle out of which everything becomes, it is no contradiction that that through which the one is generated is itself in turn generated by it" (*Freedom*/*SW* 1,7:358). Schelling's mythical creation story reveals the same tensions and circular logic as the history of organic life in his philosophy of nature. The symbolic language of mythology, then, does not resolve the contradictions of reflection, but presents them in yet another form.

Schelling's symbolic language of mythology portrays a living God in the midst of the world, divided in itself, productive of all the forms of creation, but always already mired in the messy diversity of material being. His creation myths thus have the same inverted and circular logic as the natural histories of his time. Schelling introduced boundary concepts to depict the (de)generations of this natural history, its movement in two directions, and the tensions and dialectics of our judgments. Yet, he held that if we stand in nature and can only recount the history of nature from the particularity of our position, we can nevertheless productively conceive the boundary conditions of specific natural products at specific sites in the becoming of nature.

Schelling's philosophy of life placed human beings within nature; reflecting upon nature and yet a product of nature, human beings can only offer a partial perspective upon our world. Kant also emphasized the limits of our grasp of nature as a whole and of living organisms in particular. He highlighted how the circularity of our conception of organized and self-organized beings as natural purposes reflects the circularity of our judgment. Schelling, however, found the tensions and reflective character of our concepts of nature and organisms as productive. He argued that exploring the boundary conditions of natural products through intermediate concepts is precisely what a philosophy of life requires. Schelling's philosophy of life spoke to broader concerns emerging in natural history in the years around 1800 – the coupling of the development of life to the descent of life, the boundary of organic with inorganic nature, and the relationship of natural history to human history. As works from Buffon through to Treviranus traced the extent of degeneration in the history of life and the involvement of the living beings with their material environment, accounting for the regular and yet varied formation of living forms without appeal to speculative postulates of interior moulds or vital powers became a problem. Even Wolff's and Blumenbach's careful experiments upon epigenetic formation simply enacted it with their tools rather than explicated it. In his account of generation, Oken enlisted the concept of self-organization as circles of causes and effects resulting from such reflections. Depicting an inversion and a circling of material processes into themselves, he drew analogies between the higher and lower functions of life, and between organic and inorganic processes, arguing that it is the iterative occurrence of these different processes that is generative of complex forms. Oken thus highlighted the material processes of generation, and the similarities of the generation of human beings to the lowest polyps. But in trying to render the logic of these material processes, Oken appealed to a mathesis and extended analogies that often confused rather than clarified. Schelling's use of boundary concepts suggests a means of making sense of the world from within the world, while accepting the tensions and limitations enfolded within those concepts. If Schelling's philosophy of life conceived all natural products through opposed processes, he was able to conceive living organisms in particular through their double involution – their involvement with themselves to demarcate distinct spheres of activity, and their continued involvement with the inorganic world. Drawing upon the work of his contemporaries, he enlisted excitability and the formative drive not as constitutive powers of living organisms, but as concepts to think through the boundary conditions of specific living forms in specific contexts. Schelling accepted that the tensions and limitations of our boundary concepts are a product of the dialectic of our judgments. He thus did not seek a way out of the circular and inverted logic of our natural histories, but a way into it as a general condition of our life in the world.

NOTES

1 See Finucci and Brownlee; Chamberlain and Gillman; Hurley.
2 See Osborne; Koerner; Spary; Müller-Wille.
3 See Spary 99–154; Sloan "Idea of Racial Degeneracy," "Buffon, German Biology"; Eddy; Needham; Roe.
4 See Hagner; Heesen and Spary; Gascoigne; Little and Ruthenberg.
5 See Wolff, *Von der eigenthümlichen Kraft* 50n, 66–7n; Steigerwald, "Intruments" 86-92; Rodolph 78; Detlefsen.
6 Compare Blumenbach, *Anthropological Treatises* 69–71, *Bildungstrieb*, and *Beyträge* 24–5. See Steigerwald, "Instruments" 92–8.
7 Volume and page numbers for Kant's works are from the *Akademie* edition, included in the editions cited.
8 Compare Mensch, who traces the influence of organic metaphors in the development of Kant's understanding of theoretical reason, but does not consider the circularities in his account of reflective judgment.
9 Page numbers are from the original edition of *Die Zeugung*, cited in this edition.
10 See Laqueur; Schiebinger; Reill. But compare MacLeod; Steigerwald "Figuring Nature."
11 See Kant, *Metaphysical Foundations* IV:470; Proβ.
12 Emphasis in original. Page numbers for Schelling's works are from the *Sämmtliche Werke* (cited as *SW*) when included in the editions cited. When an edition does not reference *SW*, its pagination is given separately.
13 Schelling here drew upon Kielmeyer's widely cited 1793 lecture *Über die Verhältnisse der organischen Kräfte*. See Richards 294–306.

WORKS CITED

Blumenbach, Johann Friedrich. *The Anthropological Treatises of Johann Friedrich Blumenbach.* Trans. Thomas Bendysche. London: Langman, Green, Longman, Roberts & Green, 1865. Print.

Blumenbach, Johann Friedrich. *Beyträge zur Naturgeschichte.* Göttingen: Dieterich, 1790. Print.

Blumenbach, Johann Friedrich. *Über den Bildungstrieb und das Zeugungsgeschäfte.* 1st ed. Stuttgart: Gustav Fischer Verlag, 1971. Print.

Buffon, Georges-Louis Leclerc, Comte de. *Natural History.* Trans. William Smellie. London: T. Cadell & W. Davies, 1812. Print.

Chamberlin, J. Edward, and Sander L. Gilman, eds. *Degeneration: The Dark Side of Progress.* New York: Columbia UP, 1985. Print.

Clark, David L. "Kant's Aliens: The *Anthropology* and its Others." *The New Centennial Review* 1:2 (2001): 201–89. Print.

Clark, David L. "'The Necessary Heritage of Darkness': Tropics of Negativity in Schelling, Derrida, and de Man." *Intersections: Nineteenth-Century Philosophy and Contemporary Theory.* Ed. Tilottama Rajan and David L. Clark. Albany, NY: SUNY P, 1995. 201–89. Print.

Detlefsen, Karen. "Explanation and Demonstration in the Haller-Wolff Debate." *The Problem of Generation in Early Modern Philosophy: From Descartes to Kant,* Ed. Justin Smith. Cambridge: Cambridge Studies in Philosophy and Biology, 2006. 235–61. Print.

Eddy, J.H.R. "Buffon, Organic Alterations, and Man." *Studies in History of Biology* 7 (1984): 1–45. Print.

Finucci, Valeria, and Kevin Brownlee, eds. *Generation and Degeneration: Tropes of Reproduction in Literature and History from Antiquity to Early Modern Europe.* Durham, NC: Duke UP, 2001. Print.

Gascoigne, John. "Blumenbach, Banks, and the Beginnings of Anthropology at Göttingen." *Göttingen and the Development of the Natural Sciences.* Ed. Nicolaas Rupke, 86–98. Göttingen: Wallstein Verlag, 2002. Print.

Grant, Ian Hamilton. *Philosophies of Nature after Schelling.* New York: Continuum, 2008. Print.

Hagner, Michael. "Enlightened Monsters." *The Sciences of Enlightened Europe.* Ed. William Clark, Jan Golinski and Simon Schaffer. Cambridge: Cambridge UP, 1999. 175–217. Print.

Halmi, Nicholas. *The Genealogy of the Romantic Symbol.* Oxford: Oxford UP, 2007. Print.

Heesen, Anika te, and E.C. Spary, eds., *Sammeln als Wissen.* Göttingen: Wallstein, 2001. Print.

Heidegger, Martin. *Schelling's Treatise on the Essence of Human Freedom.* Trans. Joan Stambaugh. Athens, OH: Ohio UP, 1985. Print.

Hurley, Kelly. *The Gothic Body: Sexuality, Materialism, and Degeneration at the fin de siècle.* New York: Cambridge UP, 1996. Print.

Kant, Immanuel. *Critique of the Power of Judgment.* Trans. Paul Guyer and Eric Matthews. Cambridge: Cambridge UP, 2000. Print.

Kant, Immanuel. *Kants gesammelte Schriften.* Berlin: Königlichen Preuschen Akademie der Wissenschaften, 1902–83. Print.

Kant, Immanuel. *Metaphysical Foundations of Natural Science.* Trans. Michael Friedman. Cambridge: Cambridge UP, 2004. Print.

Kant, Immanuel. "Of the Different Human Races." Trans. Jon Mark Mikkelsen. *The Idea of Race.* Ed. Robert Bernasconi and Timothy L. Lott. Indianapolis: Hackett, 2000. 8–22. Print.

Kielmeyer, Carl Friedrich. *Über die Verhältnisse der organischen Kräfte.* Ed. K.T. Kanz. Marburg an der Lahn: Basilisekn-Presse, 1993. Print.

Koerner, Lisbet. *Linnaeus: Nature and Nation.* Cambridge, Mass.: Harvard UP, 1999. Print.

Krell, David Farrell. *Contagion: Sexuality, Disease, and Death in German Idealism and Romanticism.* Bloomington: Indiana UP, 1998. Print.

Laqueur, Thomas. *Making Sex: Body and Gender from the Greeks to Freud.* Cambridge, Mass.: Harvard UP, 1990. Print.

Larson, James L. *Interpreting Nature: The Science of Living Form from Linnaeus to Kant.* Baltimore: Johns Hopkins UP, 1994. Print.

Lenoir, Timothy. *The Strategy of Life: Teleology and Mechanics in Nineteenth-Century German Biology.* Chicago: U of Chicago P, 1982. Print.

Little, Stephen, and Peter Ruthenberg, eds. *Life in the Pacific of the 1700s: The Cook/ Forster Collection of the Georg August University of Göttingen*, 3 vols. Honolulu: Honolulu Academy of Arts, 2006. Print.

MacLeod, Catriona. *Embodying Ambiguity: Androgyny and Aesthetics from Winckelmann to Keller.* Detroit: Wayne State UP, 1998. Print.

Müller-Wille, Staffan. "Figures of Inheritance, 1650–1850." *Heredity Produced: At the Crossroads of Biology, Politics and Culture, 1500–1870.* Ed. Staffan Müller-Willeand and Hans-Jörg Rheinberger. Cambridge, Mass.: MIT P, 2007. 177–204. Print.

Needham, John Tuberville. "A Summary of Some Late Observations upon the Generation, Composition, and Decomposition of Animal and Vegetable Substances." *Philosophical Transactions* 45 (1748): 615–66. Print.

Oken, Lorenz. *Gesammelte Werke.* Ed. Thomas Bach, Olaf Breidbach and Dietrich von Engelhardt. Weimar: Hermann Böhlaus, 2007. Print.

Osborne, Michael A. *Nature, the Exotic, and the Science of French Colonialism.* Indianapolis: Indiana UP, 1994. Print.

Proβ, Wolfgang. "Lorenz Oken: Naturforschung Zwischen Naturphilosophie und Naturwissenschaft." *Die Deutsche Literarische Romantik und die Wissenschaften*. Ed. Nicholas Saul. München: Iudicium, 1991. 44–71. Print.

Rajan, Tilottama. "Excitability: The (Dis)Organization of Knowledge from Schelling's *First Outline* (1799) to *Ages of the World* (1815)." *European Romantic Review* 21:3 (2010): 309–25. Print.

Reill, Peter Hanns. "The Scientific Construction of Gender and Generation in the German Late Enlightenment and in German Romantic *Naturphilosophie*." *Gender, Race and Reproduction in Philosophy and the Early Life Sciences*. Ed. Susanne Lettow. Albany, NY: SUNY P, 2014. 65–82. Print.

Richards, Robert. J. *The Romantic Conception of Life: Science in the Age of Goethe*. Chicago: U of Chicago P, 2002. Print.

Roe, Shirley A. "John Tuberville Needham and the Generation of Living Organisms." *Isis* 74 (1983): 159–84. Print.

Rudolph, Gerhard, "La méthode hallérienne en physiologie." *Dix-Huitième Siècle* 23 (1991): 75–84.Print.

Schelling, Friedrich Wilhelm Joseph. *The Ages of the World*. Trans. Jason M. Wirth. Albany, NY: SUNY P, 2000. Print.

Schelling, Friedrich Wilhelm Joseph. *First Outline of a System of the Philosophy of Nature*. Trans. Keith R. Peterson. Albany, NY: SUNY P, 2004. Print.

Schelling, Friedrich Wilhelm Joseph. *Ideas for a Philosophy of Nature*. Trans. Errol E. Harris and Peter Heath. Cambridge: Cambridge UP, 1988. Print.

Schelling, Friedrich Wilhelm Joseph. "Introduction to the Outline of a System of the Philosophy of Nature." *First Outline of a System of the Philosophy of Nature*. Trans. Keith R. Peterson. Albany, NY: SUNY P, 2004. 193–232. Print.

Schelling, Friedrich Wilhelm Joseph. *Philosophical Investigations into the Essence of Human Freedom*. Trans. Jeff Love and Johannes Schmidt. Albany, NY: SUNY P, 2006. Print.

Schelling, Friedrich Wilhelm Joseph. *The Philosophy of Art*. Trans. Douglas W. Stott. Minneapolis: U of Minnesota P, 1989. Print.

Schelling, Friedrich Wilhelm Joseph. *Sämmtliche Werke*. Ed. Karl Friedrich August Schelling. Stuttgart: Cotta, 1856–61. Print.

Schelling, Friedrich Wilhelm Joseph. *Werke. Historisch-kritische Ausgabe*. Ed. Jörg Jantzen et al. Stuttgart: Frommann-Holzboog, 1976– . Print.

Schiebinger, Londa. "The Private Lives of Plants: Sexual Politics in Carl Linnaeus and Erasmus Darwin." *Science and Sensibility: Gender and Scientific Inquiry 1780–1945*. Ed. Marina Benjamin. Oxford: Blackwell, 1991. 121–43. Print.

Shell, Susan Meld. *The Embodiment of Reason: Kant on Spirit, Generation, and Community*. Chicago: U of Chicago P, 1996. Print.

Sloan, Phillip. "Buffon, German Biology, and the Historical Interpretation of Biological Species." *The British Journal for the History of Science* 12 (1979): 109–53. Print.

Sloan, Phillip. "The Idea of Racial Degeneracy in Buffon's *Histoire Naturelle.*" *Racism in the Eighteenth Century*. Ed. Harold E. Pagliaro. *Studies in Eighteenth-Century Culture.* Vol. 3. Cleveland: Press of Case Western Reserve University, 1973. 293–321. Print.

Spary, E.C. *Utopia's Garden: French Natural History from Old Regime to Revolution.* Chicago: U of Chicago P, 2000. Print.

Steigerwald, Joan. "Figuring Nature, Figuring the (Fe)male: The Frontispiece to Humboldt's *Ideas Towards a Geography of Plants.*" *Figuring it Out: Science, Gender and Visual Culture.* Ed. A. Shteir and B. Lightman. Lebanon, NH: UP of New England, 2006. 54–82. Print.

Steigerwald, Joan. "Instruments of Judgment: Inscribing Organic Processes in Late Eighteenth-Century Germany." *Studies in History and Philosophy of Biological and Biomedical Sciences* 33 (2002): 79–131. Print.

Steigerwald, Joan. "Natural Purposes and the Reflecting Power of Judgment: The Problem of the Organism in Kant's Critical Philosophy." *Romanticism and Modernity.* Ed. Thomas Pfau and Robert Mitchell. New York: Routledge, 2011. 29–46. Print.

Steigerwald, Joan. "Treviranus' *Biology*: Generation, Degeneration and the Boundaries of Life." *Reproduction, Race, and Gender in Philosophy and the Early Life Sciences.* Ed. S. Lettow. Albany, NY: SUNY P, 2014. 105–27. Print.

Treviranus, Gottfried Reinhold. *Biologie; oder, Philosophie der lebenden Natur für Naturforscher und Aerzte.* Göttingen: Johann Fridrich Röwer, 1802–22. Print.

Warnke, Camilla. "Schellings Idee und Theorie des Organismus und der Paradigmawechsel der Biologie um Wende zum 19. Jahrhundert." *Jahrbuch für Geschichte und Theorie der Biologie* 5 (1998): 187–234. Print.

Whistler, Daniel. *Schelling's Theory of Symbolic Language: Forming the System of Identity.* Oxford: Oxford UP, 2013. Print.

Wolff, Caspar Friedrich. *Theoria generationis.* Trans. Paul Samassa. Ostwald's Klassiker, 85. Leipzig: Engelmann, 1896. Print.

Wolff, Caspar Friedrich. *Theorie von der Generation in zwo Abhandlungen erklärt und bewiesen.* Berlin: Birnstiel, 1764. Print.

Wolff, Caspar Friedrich. *Von der eigenthümlichen und wesentlichen Kraft der vegetabilischen sowohl als auch der animalischen Substanz.* St Petersburg: Kayserliche Academie der Wissenschaft, 1789. Print.

Zuckert, Rachel. *Kant on Beauty and Biology: An Interpretation of the Critique of Judgment.* Cambridge: Cambridge UP, 2007. Print.

Contributors

Gillian Beer is King Edward VII Professor of English Literature at University of Cambridge, Dame Commander of the Order of the British Empire, and past President of Clare Hall, University of Cambridge. Her publications include *Alice in Space: The Sideways Victorian World of Lewis Carroll* (2016), *Darwin's Plots* (1983; 3rd ed. 2009), *Open Fields* (1997), *Virginia Woolf: The Common Ground* (1996), *Forging the Missing Link* (1992), *Arguing with the Past* (1989), *George Eliot* (1986), and *Meredith: A Change of Masks* (1970). Beer also holds an Honorary Doctor of Letters from Oxford University and from Harvard University, and has served on and chaired the Adjudication Panel for the Man Booker Prize.

Alan Bewell is Professor at the University of Toronto. He is author of *Natures in Translation: Romanticism and Colonial Natural History* (2017), *Romanticism and Colonial Disease* (1999), and *Wordsworth and the Enlightenment* (1989); coauthor of *Educating the Imagination: Northrop Frye, Past, Present, Future*; and editor of *Medicine and the West Indian Slave Trade* (1999). He is currently at work on another book, *Romanticism and Mobility*. He is the recipient of a Guggenheim Fellowship and several SSHRC grants as well as the Northrop Frye Award at the University of Toronto for excellence in teaching and research, and is a Fellow of the Royal Society of Canada. He serves on the editorial boards of *Studies in Romanticism*, *Nineteenth-Century Literature*, and the Palgrave series *The Enlightenment, Romanticism, and the Cultures of Print*.

Joel Faflak is Professor of English and Theory at Western University, where he is Director of the School for Advanced Studies in the Arts and the Humanities. He is author of *Romantic Psychoanalysis* (2008), coauthor of *Revelation and Knowledge* (2011), editor of De Quincey's *Confessions of an English Opium-Eater* (2009), and editor or coeditor of eight volumes, including *The Romanticism Handbook* (2011),

The Handbook to Romanticism Studies (2012), *The Public Intellectual and the Culture of Hope* (2013), and *Romanticism and the Emotions* (2014). He has been the recipient of several SSHRC grants, and won the 2002 Polanyi Prize for Literature.

Noah Heringman is Associate Professor of English at the University of Missouri, where he teaches on poetry, aesthetic theory, and the cultural history of science. He has published *Sciences of Antiquity: Romantic Antiquarianism, Natural History, and Knowledge Work* (2013), *Romantic Rocks, Aesthetic Geology* (2004), and the edited collection *Romantic Science: The Literary Forms of Natural History* (2003), as well as co-editing "Romantic Antiquarianism" (with Crystal B. Lake) for *Romantic Circles Praxis.* His articles have appeared in *Studies in English Literature 1500–1900, Studies in Romanticism, Earth Sciences History,* and *The Huntington Library Quarterly,* and in *A Concise Companion to the Romantic Age, A Companion to Werner Herzog,* and *The Routledge Companion to Literature and Science.*

Theresa M. Kelley is Marjorie and Lorin Tiefenthaler Professor of English at the University of Wisconsin – Madison. She has received awards and fellowships from the National Endowment for the Humanities, the Simon D. Guggenheim Foundation, Henry E. Huntington Library, and the Institute for Research in the Humanities at the University of Wisconsin-Madison. She is author of *Clandestine Marriage: Botany and Romantic Culture* (2012), *Reinventing Allegory* (1998), and *Wordsworth's Revisionary Aesthetics* (1988); the editor of the issue "Romantic Frictions" for the *Romantic Circles Praxis* series; and co-editor of *Romantic Women Writers: Voices and Countervoices* (1995) and *The Matter of the Archive* (forthcoming). She is at present writing *Reading for the Future* and *Color Matters.*

Maureen N. McLane, Professor of English at New York University, is the author of *Balladeering, Minstrelsy, and the Making of British Romantic Poetry* (2008), *Romanticism and the Human Sciences* (2000), and co-editor of *The Cambridge Companion to British Romantic Poetry* (2008). She has written extensively on romantic mediality, comparative poetics, the human sciences, and contemporary poetries in English. She has also published five books of poetry, most recently *Mz N: the serial: a poem-in-episodes* (2016) and *Some Say* (2017). Her book *My Poets* (2012), an experimental hybrid of memoir and criticism, was a Finalist for the National Book Critics Circle Award in Autobiography. Ongoing projects include work on rhyme, "compositionist" poetics, "choratopes," and lyric discontents. She continues to work on (and with) poets dead and alive.

Andrew Piper is Associate Professor in the Department of Languages, Literatures, and Cultures and an associate member of the Department of Art History and Communication Studies at McGill University. He is also co-founder of FQRSC-funded research group *Interacting with Print: Cultural Practices of Intermediality, 1700–1900.* Piper has published *Book Was There: Reading in Electronic Times* (2012) and *Dreaming in Books* (2009). His articles have appeared in *European Romantic Review*, *RaVoN*, *Goethe Yearbook*, *The New Everyday*, and *PMLA*, and in collections such as *Women in German Yearbook*, *Bookish Histories*, *Spatial Turns*, and *Book Illustration in the Long Eighteenth Century.*

Tilottama Rajan is Distinguished University Professor, Canada Research Chair (Tier 1), and Director of the Centre for the Study of Theory and Criticism at Western University. She is a Fellow of the Royal Society of Canada, and held the H.I. Romnes Professorship at the University of Wisconsin, Madison. She has published *Romantic Narrative* (2011), *Deconstruction and the Remainders of Phenomenology* (2002), *The Supplement of Reading* (1990), and *Dark Interpreter* (1980); edited Mary Shelley's *Valperga* (1998); and co-edited *Idealism without Absolutes* (2004), *After Poststructuralism* (2002), *Romanticism and the Possibilities of Genre* (1998), and *Intersections: Nineteenth-Century Philosophy and Contemporary Theory* (1995).

Robert J. Richards is Morris Fishbein Professor of the History of Science and Medicine, University of Chicago, where he is also Professor in the Departments of Philosophy, History, and Psychology, Director of the Fishbein Center for the History of Science and Medicine, and Professor on the Committee on Conceptual and Historical Studies of Science. He is author of *The Tragic Sense of Life* (2009), *The Romantic Conception of Life* (2002), *The Meaning of Evolution* (1992), and *Darwin and the Emergence of Evolutionary Theories of Mind and Behaviour* (1987), and is co-editor of *The Cambridge Companion to* The Origin of the Species (2009) and *Darwinian Heresies* (2004). He has also been Mellon Professor at Grinnell College (2007).

Matthew Rowlinson is Professor of English at Western University. He is the author of *Real Money and Romanticism* (2010) and *Tennyson's Fixations* (1994), and editor of Tennyson's *In Memoriam* (2014). He has published essays in *The Oxford Encyclopedia of British Literature, The Blackwell Companion to Victorian Poetry, Boundary 2, differences, PMLA, Romantic Circles, Victorian Poetry, Victorian Studies*, and *Yale Journal of Criticism.*

Joan Steigerwald is Associate Professor in the Department of Humanities, and the Graduate Programs in Humanities, Science and Technology Studies, and Social and Political Thought, at York University. She has just completed a book entitled

Experimenting at the Boundaries of Life: Organic Vitality in Germany around 1800. She has edited two special issues for *Studies in the History and Philosophy of the Biological and Biomedical Sciences*, "Entanglements of Instruments and Media in Exploring Organic Worlds" in 2016, and "Kantian Teleology and the Biological Sciences" in 2006. She has published numerous articles on Goethe, Humboldt, Kant, Schelling and the German life sciences. Her new project is *Object Lessons of a Romantic Natural History.*

Gábor Áron Zemplén is Associate Professor at ELTE Business School (Budapest) and Senior Researcher at the MTA BTK Lendület Morals and Science Research Group. He lectured at the Budapest University of Technology and Economics, Hungary (BME, 2002–2017, Philosophy and the History of Science), received postdoctoral fellowships from the Max Planck Institute for the History of Science in Berlin and the Collegium Budapest, and held teaching positions in England, Bavaria (where he was also researcher at the Deutsches Museum, Munich), and Switzerland (writing a textbook on theories of light, colour, and vision, University of Bern). As an intellectual historian, he traced the influence of Goethe's methodology on Otto Neurath's philosophy of science, wrote contextualist case-studies on the history of modificationist colour-theories (sixteenth to nineteenth centuries), and a monograph on Newton.

Index

www.ingramcontent.com/pod-product-compliance
Lightning Source LLC
LaVergne TN
LVHW090151080826
844660LV00013B/754/J